Methods in Enzymology

Volume 221
MEMBRANE FUSION TECHNIQUES
Part B

METHODS IN ENZYMOLOGY

EDITORS-IN-CHIEF

John N. Abelson   Melvin I. Simon

DIVISION OF BIOLOGY
CALIFORNIA INSTITUTE OF TECHNOLOGY
PASADENA, CALIFORNIA

FOUNDING EDITORS

Sidney P. Colowick and Nathan O. Kaplan

TABLE I
COMMERCIALLY AVAILABLE CYTOPLASMIC FLUOROPHORES[a]

Fluorophore[b]	Molecular weight[c]	Fluorescence of ester derivative[d]	pH sensitivity[e]	ϵ_{max}[f]	$(\lambda_{ex})_{max}$	$(\lambda_{em})_{max}$
CFDA	460	−	+	72,000	490	515
CNFDA	476	−	−	49,000	598	660
CDCFDA	445	−	+	93,000	504	529
BCECF-AM	520	−	+	77,000	508	531
CEDA	692	−	+	100,000	519	542
Calcein-AM	623	−	−	83,000	496	520
CCB-AM	365	+	−	10,000	321	448

[a] Fluorophores can be loaded into cells as acetoxymethyl ester or diacetate derivatives. From Haugland.[19] (See Haugland[19] for further details and other derivatives.) Other, higher molecular weight compounds, such as indo 1-AM, fura 2-AM, or SNARF 1-AM, presently used as pH or ion indicators, can also be used as probes of transcellular movement under appropriate conditions.

[b] CFDA, Carboxyfluorescein diacetate; CNFDA, carboxynaphthofluorescein diacetate; CDCFDA, carboxydichlorofluorescein diacetate; BCECF-AM, biscarboxyethylcarboxyfluorescein, acetoxymethyl ester; CEDA, carboxyeosin diacetate; CCB-AM, carboxycalcein blue-AM.

[c] Molecular weight of uncleaved compound. The cleaved compound will have a lower molecular weight.

[d] Nonfluorescent (−); fluorescent (+).

[e] pH sensitivity of the cleaved compound: pH sensitive (+); pH insensitive (−).

[f] Extinction coefficient, and the maximum excitation and emission wavelengths, refer to the cleaved compound.

exchange proteins[18] can be used to catalyze incorporation of those lipid fluorophores that do not spontaneously insert into cell membranes.

We have made use of the RBC membrane anion transporter to incorporate a particular aqueous fluorophore (NBD-taurine) into the RBC[12]. However, a number of soluble probes can be loaded into cells as the charge-neutral, membrane-soluble form[19] (Table I). These diffuse across the plasma membrane into the cell cytoplasm, where non-specific, cytoplasmic esterases cleave off the ester groups. The cleavage converts the probe into a negatively charged molecule, to which the plasma membrane is less permeable. The efflux rate varies with the probe, depending on size and net negative charge. Multicharged molecules such as biscarboxyethyl-carboxyfluorescein (BCECF) and calcein have $t_{1/2}$ values (for efflux from cells) greater than 2 hr at 37°. Most of the derivatives are nonfluorescent

[18] K. W. Wirtz and T. W. Gadella, Jr., *Experientia* **46**, 592 (1990).
[19] R. P. Haugland, "Molecular Probes: Handbook of Fluorescent Probes and Research Chemicals." Molecular Probes, Eugene, OR, 1992.

until the ester groups are cleaved off; however, there are several (notably the ion indicator dyes) that also fluoresce in their inactive form. Also, many of the dyes partition into noncytoplasmic compartments. Special care must be taken to ensure that the dye is in a form and location to report the activity being studied. However, because of the possibility of making (near) simultaneous measurements on both membrane and cytoplasmic dyes during fusion, many of these ester derivatives may be used to study effects of charge or molecular weight. A partial list of such dyes are included in Table I.

Transfer of fluorescently labeled macromolecules into cells can be achieved by a variety of methods,[20] for example, permeabilization by brief hypotonic treatment, lytic agents, high-voltage electric fields, or microinjection. The advantage of using macromolecules is that they remain in cells for long periods of time,[4] and the size of initial fusion junctions can be determined by monitoring the rate of transfer as a function of size of the macromolecule. A key problem in cell fusion experiments is the possibility of false-positive artifacts arising from transfer of dye without fusion taking place. Despite their highly hydrophobic nature, many lipid-soluble compounds rapidly exchange from one bilayer to another, presumably either through equilibrium with a small pool of water soluble molecules, microlenses, or via contact. Soluble dyes may leak out of the cell cytoplasm and then be taken up by unlabeled neighbors. In viral fusion there is often a built-in control for dye transfer in the absence of fusion. For instance, influenzavirus hemagglutinin is expressed on the cell surface in the form of an uncleaved precursor HA_0, which promotes binding to sialoglycoprotein and lipid receptors but does not induce fusion or fluorescence dequenching. Mild treatment of the cells with trypsin cleaves HA_0 into a form that induces pH-dependent membrane fusion. Incubating fibroblasts expressing HA_0 complexed to labeled RBCs at 37° and low pH for long periods of time does not result in fluorescence dequenching if HA_0 is not cleaved. Neither will fluorescence dequenching occur at 37° and neutral pH with the cleaved precursor.[11,12]

In the following sections we will describe some protocols for labeling of RBCs with aqueous and lipid fluorophores.

Labeling of Erythrocytes with Octadecylrhodamine (R18) or DiI

For a discussion of the properties of R18, see other chapters in this volume[2,3] The fluorescent probe is inserted into the red cell membrane by injecting 3 μl of a 14 mM R18 solution (in ethanol, with vigorous vortex-

[20] T. Uchida, *Exp. Cell Res.* **178,** 1 (1988).

ing) into 2 ml phosphate-buffered saline (PBS) containing 1% (v/v) RBCs in a 15-ml polystyrene conical tube. The suspension is incubated for 15 min at room temperature. Then 13 ml of Dulbecco's modified Eagle's medium supplemented with 10% (v/v) fetal calf serum (DMEM/S) is added to the tube and the cells are incubated for 20 min at room temperature to absorb unbound or loosely bound probe. The cells are washed six times by centrifugation (300 g, 10 min at 4°) and resuspension in 15 ml PBS, each time using a clean tube. This procedure results in incorporation of approximately 6 pmol of R18 per 10^6 cells, which is about 4% of total RBC lipid. The fluorescence intensity of intact R18–RBCs is 5 to 30% that of labeled RBCs treated with 0.1% (v/v) Triton X-100. Erythrocyte ghosts made from R18–RBCs by hypotonic lysis show the same percentage dequenching in Triton as the intact RBCs, indicating that the quenching is not due to hemoglobin. The R18-labeled RBCs show no change in fluorescence when incubated at 37° for 30 min either at pH 7 or 5, or at 4° at pH 7 for 1 week.

The same labeling procedure is used for DiI. Excitation and emission maxima in RBC membranes are, respectively, 550 and 570 nm for DiI, and 565 and 590 nm for R18. Both types of labeled RBCs are highly fluorescent when viewed through a rhodamine epifluorescence filter set.

Entrapment of NBD-taurine in Erythrocytes

Ten milliliters of a 1% hematocrit in PBS is pelleted at 300 g for 10 min at 4°. The pellet is suspended in 1 ml of 10 mM NBD-taurine in PBS and incubated at 37° for 30 min. The cell suspension is cooled on ice for 5 min. To block efflux of NBD-taurine after entrapment, the erythrocytes are treated with 4,4-diisothiocyanodihydrostilbene 2,2-disulfonic acid (DIDS), which is known to inhibit anion transport irreversibly. The incubation mixture is washed twice in 40 ml cold N-2-hydroxyethylpiperazine-N'-2-ethanesulfonic acid (HEPES)–sulfate buffer and finally suspended in 40 ml cold 0.05 mM DIDS in HEPES–sulfate. The resulting suspension is incubated on ice for 10 min, followed by incubation at 37° for 20 min. At the end of incubation, cold DMEM/S is added to a final volume of 50 ml. The RBCs are spun down and washed four times with 50 ml cold PBS. The cells are finally suspended in 10 ml cold PBS and can be used for up to a week if stored at 4°.

The NBD-taurine-labeled erythrocytes are highly fluorescent. Addition of 0.1% (v/v) Triton X-100 results in a three- to sixfold increase in fluorescence. Anti-NBD-taurine antibodies, which quench the NBD-taurine fluorescence on binding,[21] are used to assess leakage from erythrocytes. DIDS-

---

[21] O. Eidelman and Z. I. Cabantchik, this series, Vol. 172, p. 122.

treated NBD-taurine-labeled RBCs show little leakage of the soluble fluorophore when stored at 4° for several days.

Loading with Acetate or Acetoxymethyl Ester Derivatives

About 15 μl of a 2.5 mM solution of the soluble acetoxymethyl ester or diacetate dye in DMSO is added to 2 ml of a 1 to 2% suspension of erythrocytes and incubated for 20 min at 37°. The labeled cells are then washed three to four times by centrifugation in 50 ml cold PBS to remove unincorporated label, then stored on ice. The cells are highly fluorescent and show two- to fivefold dequenching of soluble probe after detergent lysis. They must be used within 0.5 to 1 hr of the end of the initial incubation.

Incorporation of Fluorescent Macromolecules into Erythrocytes

A variety of methods have been developed for loading macromolecules into RBCs.[20] A detailed protocol based on Rechsteiner's preswell technique has been described.[22] The RBCs are lysed in a hypotonic medium containing the macromolecule, incubated on ice for 2 min, and resealed under isotonic conditions. About 60% of the hemoglobin is released, and the trapping efficiency depends on the molecular weight of the molecule.[22] If hemoglobin needs to be removed completely, RBC ghosts can be prepared in the presence of the fluorescent macromolecule. A protocol used in our laboratory is given in Clague *et al.*[23] It should be noted that there are differences in the kinetics of viral fusion between intact RBCs and ghosts, or between lipid-symmetric and -asymmetric ghosts.[24]

Double Labeling with Lipid and Aqueous Fluorophore

Double labeling with R18 (or DiI) and NBD-taurine is carried out using the same procedure as for entrapment of the aqueous fluorophore, except that 15 μl of a 1-mg/ml solution of R18 in ethanol is added to the incubation mixture. Double labeling R18- or DiI-labeled RBCs with acetoxymethyl ester derivatives can be carried out as above, with the substitution of 5 to 20 μl/ml of 2.5 mM ester derivative in DMSO per milliliter of 1 to 2% RBCs. Alternately, if a series of experiments is to be carried out with identical membrane probe loading, labeling can be performed in two steps: first the cells are labeled with the lipid-soluble probe; then the labeled cells are incubated with acetate or acetoxymethyl ester derivatives as described.

[22] H. Ellens, S. J. Doxsey, J. S. Glenn, and J. M. White, *Methods Cell Biol.* **31**, 155 (1989).
[23] M. J. Clague, C. Schoch, L. Zech, and R. Blumenthal, *Biochemistry* **29**, 1303 (1990).
[24] M. J. Clague, C. Schoch, and R. Blumenthal, *J. Virol.* **65**, 2402 (1991).

The first step is best performed on freshly collected RBCs. Addition of the soluble probe can be made up to 4 days after the membrane label, if cells are stored in PBS at 0 to 4°.

For experiments using simultaneous excitation of both fluorophores, the NBD-taurine fluorescence will have significant overlap with the R18 or DiI spectra and will distort images of the membrane-bound fluorophore. Preliminary experiments (S. J. Morris, unpublished data, 1990) show that a series of BODIPY analogs,[19] which have much narrower emission spectra, can be taken up by band III and sealed inside the RBCs by DIDS. They have higher dequenching ratios than NBD-taurine and should be good substitutes for the latter compound.

Binding Erythrocytes to Hemagglutinin-Expressing Cells

We have examined fusion of labeled RBCs to cells expressing hemagglutinin (HA) protein. The first experiments used a line (GP4F) of bovine papillomavirus-transformed NIH 3T3 cells that constitutively expresses HA from the influenzavirus strain A/Japan/305/57 at high surface density. However, the same experiments have been performed with CV1 cells (a monkey kidney line) infected with influenzavirus from different strains, or with Simian virus 40 (SV40) containing cDNA of HA and mutants of HA.[24,25] Details about the GP4F line and culture conditions can be found in Ellens *et al.*[22]

Preparation of Erythrocyte–GP4F Complexes for Spectrofluorimetry

The fibroblasts are grown to 70 to 80% confluence in Falcon (Becton-Dickinson, Oxnard, CA), Costar (Cambridge, MA), or Corning (Corning, NY) T75 flasks, washed twice with 10 ml DMEM, and then treated with 5 ml of 5 μg/ml trypsin, 0.22 mg/ml neuraminidase in PBS for 10 min at room temperature. Some care is required not to lift the cells from the growth substrate. Neuraminidase treatment enhances binding and fusion, but is not required. This enzyme solution is removed and the cells are washed once with 10 ml DMEM/S or 1 mg/ml soybean trypsin inhibitor in DMEM, followed by two washes with PBS. A suspension of 5 ml labeled RBCs (0.1% hematocrit) in PBS is added to 2×10^7 GP4F cells in the flask and incubated at room temperature for about 10 min, with occasional gentle agitation, to form RBC–GP4F complexes. Unbound RBCs are removed by three washes with DMEM, gently running the solution over the cells and aspirating. We find it convenient to follow the progress of the

[25] A. Puri, F. Booy, R. W. Doms, J. M. White, and R. Blumenthal, *J. Virol.* **64**, 3824 (1990).

binding by placing the flask(s) on the stage of an inverted microscope and observing the cells. After gentle agitation, unbound RBCs will float away from the fibroblasts. Binding can be terminated when the desired amount of decoration is achieved. For more RBCs per fibroblast, more RBCs can be added as needed.

The decorated cells are lifted from the flask by incubation for 10 min at 37° in 1.5 ml of 0.5 mg/ml trypsin and 0.2 mg/ml EDTA in PBS; DMEM/S (8.5 ml) is then added to the flask, and the suspension is triturated by pipetting up and down three times in a 10-ml pipette, to break up any large clumps of cells. The suspension is transferred to a 15-ml conical centrifuge tube and washed once in Ca–Mg PBS or DMEM. The decorated cells are resuspended in 0.5 ml PBS or DMEM and placed on ice until use. For HA_0 controls, the cells can be lifted with chymotrypsin in the presence of 1 mg soybean trypsin inhibitor per milliliter in DMEM.

Preparation of Erythrocyte–GP4F Complexes for Video Microscopy or Patch Clamping

Cells are decorated and lifted as above. The cell complexes are suspended in DMEM/S, plated onto glass coverslips at various concentrations, and placed in an incubator at 37°. Polylysine pretreatment of the coverslip (50 to 500 μg polylysine per milliliter H_2O for 2 min, followed by rinsing with distilled water) helps initial adhesion. As in our other methods, R18 does not transfer nonspecifically from the RBCs, and NBD-taurine-labeled RBCs retain their soluble dye with a half-life of more than 12 hr under these conditions. Thus the reattached fibroblasts will display fusion activity, once fusion is induced, for many hours following reattachment.

Alternately cells can be grown on 10×10 mm squares of #00 coverslips [Corning (Corning, NY) or Carolina Biological Supply (Burlington, NC)] in 12-well plates, trypsin activated, and decorated with RBCs as for spectrofluorometry, except that after washing away the excess RBCs the wells are filled with DMEM/S and the cells are returned to the incubator until used (2 hr maximum). These cells are excellent for video microscopy. However, they do not form good seals for patch clamping, presumably because of partial damage on treatment with trypsin.

Recording Fluorescence Changes on Fusion

Two types of measurements are performed on viral envelope-mediated cell fusion: Spectrofluorometric measurements of populations of cells and video microscopic measurements of single cells. Spectrofluorometric measurements may be used to screen many different sets of experimental conditions rapidly, and relatively little material is required. However, the examination takes place under nonphysiological conditions for the cells. Because the cells must be in suspension, contributions from cell matrix and cytoskeletal structures cannot be assessed. Results represent the average of a large number of events. However, microfluorometry on single cells can be performed by attaching photomultiplier tubes with the appropriate filters to the microscope.[15]

Imaging experiments, on the other hand, provide detailed observations of single cells. Physiological growth conditions are preserved. Spatial redistribution of dye becomes a measurable experimental parameter. However, the experiments require more expensive equipment and take longer to set up and perform.

Spectrofluorometry of Hemagglutinin-Induced Erythrocyte–Cell Fusion

The spectrofluorimetric assay, which has previously been used to study the kinetics of fusion of enveloped viruses with cells,[2,3] can be applied to cell–RBC fusion.[11,12] For details about spectrofluorimetric measurements (e.g., equipment, modes of triggering of the fusion reaction, analysis of the data) see Volume 220 [21] in this series.[3] The RBC membrane is labeled with R18 under conditions in which the fluorophore is self-quenched,[11] and the cells are loaded with the aqueous marker NBD-taurine under conditions in which its fluorescence is quenched by hemoglobin inside the RBC.[12] On fusion with cells, both dyes are diluted and the ensuing fluorescence increase is measured spectrofluorimetrically, using both criteria for fusion, that is, lipid mixing and cytoplasmic continuity. Figure 1 shows the kinetics of pH-induced cell fusion of RBCs with the HA-expressing fibroblasts, using both dyes.

After lowering the pH, the fluorescence of both markers increases after a given time lag. Figure 1 shows the temperature dependence of fluorescence dequenching. The maximal extents of fluorescence dequenching of R18 are usually considerably less than those of NBD-taurine. The incomplete dequenching of R18 may be due to the many RBCs bound per GP4F cell: because the ratio of the surface areas of the two cells is about 6, fusion of more than one RBC per GP4F cell will result in insufficient dilution of R18 for maximal dequenching. The quenching of NBD-taurine is relieved by escape from the hemoglobin environment, and therefore movement into the fibroblast will result in full dequenching. However, at each temperature measured the lag in the onset of fusion measured with NBD-taurine is the same as that measured with R18. The continuous monitoring of fluorescence changes shows that at 37° fusion is rapid; the maximal extent is reached within minutes. Because the two fluorescent events show the

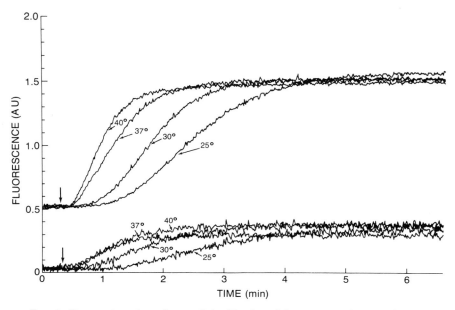

Fɪɢ. 1. Temperature dependence of the kinetics of fluorescence changes of R18 and NBD-taurine on fusion of HA-expressing cells with RBCs. The RBCs were double labeled with NBD-taurine and R18. RBC–GP4F cell complexes were formed, washed, and suspended as described in the text. Fifty microliters of the R18–RBC–GP4F cell complex was injected into a cuvette containing 2 ml PBS, pH 7.4, prewarmed to different temperatures (marked) in a circulating water bath. About 1 min later the pH in the medium was lowered to pH 5.0 (large arrows). Upper curves and lower curves are NBD-taurine and R18 fluorescence changes, respectively. (From Sarkar et al.[12])

same lag time as a function of temperature, the data indicate that the lag in onset is not due to effects of dye diffusion, but rather is due to initial events following pH activation of HA but preceding lipid continuity and cytoplasmic mixing. Moreover, the correspondence of the kinetics of the aqueous and lipid probes indicate that the cytoplasmic connections form as rapidly as the outer bilayers mix and that there is no long-lived "partial-fusion" intermediate. Leakage of soluble dye during fusion could be accounted for by performing the experiment in the presence and absence of an agent (e.g., antibody) capable of quenching the fluorescence of the released material.[12]

Fluorescence Light Microscopy

Initial observations of cell fusion can be made on suspended fibroblast–RBC complexes by first treating the complexes with low pH, then examining and photographing the samples as "whole mounts" in an upright fluorescence microscope.[12] The amount of light needed for such observations rapidly photobleaches the samples. The working times for the dyes can be extended better than 10-fold by the inclusion of *n*-propyl gallate (NPG) in the examination buffers.[26] Control experiments established that at 1 mM, NPG does not change the kinetics of fusion as judged by fluorometry, photomicroscopy, or video microscopy.

Fluorescence light microscopy of whole-mounted cells provides a rapid screen for fusion and is easy to perform. Three microliters of decorated cells suspended in DMEM containing 1 mM NPG is placed within a 10-mm raised circle printed on a microscope slide (Roboz Scientific, Washington, DC), covered with an 18 × 18 mm #1 coverslip, and examined with a Plan-Neofluor × 100/1.3 oil immersion objective on a Zeiss (Goettingen, Germany) Axioplan microscope, either by phase contrast or epifluorescence. Cells are photographed using Kodak (Rochester, NY) Ectachrome 160 or Tri X pan film. The best reproduction of fluorescence produced by these dyes is made on Cibachrome prints. NBD-taurine or fluorescein fluorescence is visualized using the Zeiss fluorescein filter set (Zeiss 450–490 excitation filter, ft 510 dichroic mirror, and lp 520 emission filter). R18 or DiI fluorescence is observed with the rhodamine fluorescence set (BP 546 excitation filter, ft 580 dichroic mirror, and lp 590 emission filter). Spectral overlap is not a problem, except for an occasional reddish outline on the double-labeled cells when viewing the NBD-taurine fluorescence. This can be greatly reduced by using a narrower excitation filter. See Sarkar *et al.*[12] for color micrographs of HA-induced cell fusion.

Fluorescence Video Microscopy

Due to photodynamic damage, kinetics of fusion of single GP4F – RBC complexes are not observable by standard fluorescence microscopy at high light levels. We have employed microchannel plate intensifiers coupled to video cameras to measure fluorescence at low light levels in a Zeiss inverted fluorescence microscope.[12] Several reviews of fluorescence video microscopy have been published.[27] All types of decorated cell preparation described above can be used. Cell complexes attached to #0 coverslip fragments are placed in an appropriate cell buffer in the environment chamber on an inverted microscope. Alternately, a 10- to 20-μl sample of decorated, lifted cells are allowed to settle onto a polylysine-coated coverslip.

[26] H. Giloh and J. W. Sedat, *Science* **217,** 1252 (1982).
[27] T. M. Jovin and D. J. Arndt-Jovin, *Annu. Rev. Biophys. Biophys. Chem.* **18,** 271 (1989).

All solutions used for video microscopy contain 1 mM n-propyl gallate to reduce photooxidation and photobleaching. Images formed on the microchannel intensifier plate, coupled with either a CCD or Nuvicon camera, are acquired and recorded at high resolution on ¾″ tape. The pH is lowered by adding 100 to 500 μl of appropriate citrate-buffered balanced salt solution over a 5- to 10-sec period. Figure 2 shows redistribution of fluorescent dyes in fusing cells.

Either of the fluorophores could be followed after RBC-decorated GP4F cells are exposed to pH 5. Redistribution of both NBD-taurine and R18 after triggering fusion takes 25 to 60 sec to complete. In contrast to

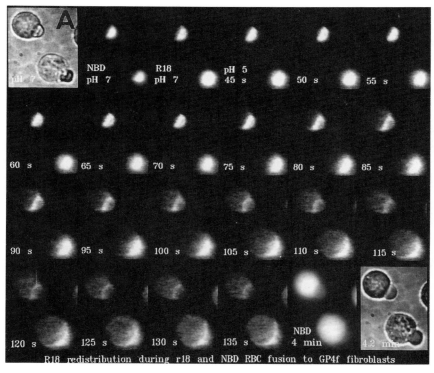

FIG. 2. Processed images of fusion of HA-expressing cells with RBCs, as detected by low light-level fluorescence video microscopy. GP4F cells were treated with trypsin–neuraminidase, decorated with double-labeled RBCs, and bound to polylysine-coated coverslips in 1.0 ml of PBS (pH 7.4) in the environmental chamber of an inverted microscope, as described in text. When the temperature reached 37°, the pH was changed to 5.0 over 5 to 10 sec by the addition of PBS containing citric acid. Fusion events began 30 to 45 sec later, as would be expected from the lag period at this temperature (see Fig. 1). Within the lag period,

R18, the NBD-taurine appears to be redistributed homogeneously throughout the GP4F cell immediately after crossing the RBC–GP4F barrier, although its total fluorescence in the GP4F cell increases slowly (Fig. 2A). If the cytoplasmic continuity junction formed after fusion is 100 nm (the smallest structure seen by electron microscopy[4]), then NBD-taurine redistribution would occur within 0.1 sec. However, the half-time for redistribution of the dye is about 25 sec (Fig. 2A). Therefore we conclude that movement of fluorophores between effector and target is restricted during the initial events in fusion, consistent with the opening of small junctional pore(s).[12]

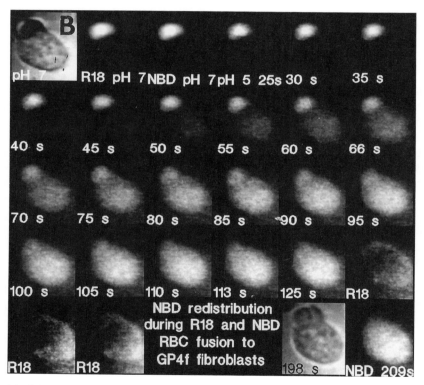

bright-field, NBD-taurine, and rhodamine fluorescence were recorded and one of the two fluorophores was followed for the next 5 to 10 min. The time in seconds after the pH change is noted at the bottom of the frames. (A) Bright-field, NBD-taurine, and rhodamine fluorescence redistribution. (B) A similar experiment in which NBD-taurine redistribution is followed. An average of eight frames (267 msec) was used for the bright-field images and an average of 32 frames (1.06 sec) was used for the fluorescent images. This entire plate was contrast enhanced with the same gray scale. (From Sarkar et al.[12].)

Figure 2B shows movement of R18 from RBCs to GP4F cells in about 50 sec after lowering the pH, as indicated by a brightened fluorescence in the interfacial region between the membranes. The diffusion coefficient of R18 incorporated into human RBC plasma membranes is $3-4 \times 10^{-9}$ cm^2/sec at 22°, as measured by fluorescence photobleaching recovery measurements.[28] The relatively low diffusion coefficient (as compared to aqueous diffusion of NBD-taurine, $D = 5 \times 10^{-6}$ cm^2/sec) gives rise to an observable wave of R18 redistribution after crossing the RBC–GP4F cell barrier (see Fig. 2B). The calculation for redistribution of a lipid marker that must diffuse through a pore with a given circumference is highly complex,[29] but a simple calculation with only the orifice as rate-limiting yields a redistribution time of 40 sec through a single 4.5-nm junction or a number of junctions with the same cross-sectional area.

Data Analysis

When analyzing data from video micrographs, care must be taken to normalize for inhomogeneities in both spatial intensity of the incident light as well as spatial sensitivities of the detector system.[30] In addition, each step in data collection and transfer to appropriate digital analysis systems involves the setting of arbitrary gain and offsets, which are rarely reproducible. It is essential to include some standard set of images for controls. More care is needed in this situation than in "ratio" methods (see the next section below), because taking a ratio in effect normalizes for these geometric and electronic factors. For the single-camera system, we have developed software to floating-point divide the test image by the control image of a small concentration of dye in solution (NBD-taurine), and scale the resultant image into a 16-bit buffer for further analysis. In this way, the average or total intensity of any area of an image can be plotted as a function of time, under different conditions of pH and temperature. There are a number of commercial hardware and software configurations available for continuous data capture. It is beyond the scope of this chapter to provide a comprehensive review, which will probably be quickly out of date in this fast-moving field. System requirements will vary, depending on the experimental requirements and the budget of the investigator.

Multiimage Systems

A video microscope has been developed[31,32] that can simultaneously excite two or more vital dyes placed in living cells and sort the emission

[28] B. Aroeti and Y. I. Henis, *Exp. Cell Res.* **170,** 322 (1987).
[29] R. J. Rubin and Y. Chen, *Biophys. J.* **58,** 1157 (1990).
[30] S. Inoue, "Video Microscopy." Plenum, New York, 1986.

I.

II.

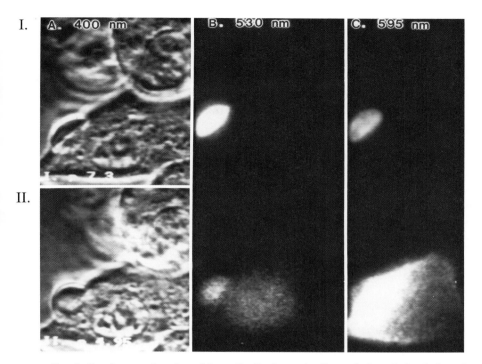

FIG. 3. Simultaneous imaging of dye redistribution and cell morphology during fusion of an HA-expressing cell to an erythrocyte. The GP4F cells were grown on #00 glass coverslips and decorated with RBCs, and labeled with DiI and calcein as described in text. Three images were formed simultaneously on three intensified CCD cameras. (A) Bright-field phase contrast at 400 nm by transillumination; (B) calcein fluorescence at 530 nm; (C) DiI fluorescence at 595 nm. The fluorophores were excited by an Omega optical dual epifluorescence filter and dichroic mirror.[31] Row I: The cell complex before the pH change. Both fluorophores are confined to the RBC, which appears biconcave when examined at several focal planes. Row II: The same cell 2 min after low pH-triggered fusion. Both dyes have redistributed and the RBC has swollen into a round cell. Each image was originally 170×480 pixels. They have been reduced to 170×240 pixels to be presented on the same 512×480 video screen. Four video frames from the beginning and the end of the experiment were averaged for the images in rows I and II, respectively. The same 512×240 area of interest was cut from each image and placed one on top of the other, using the software video editor. Lettering was added and the RGB display screen photographed as described by Inoue.[30]

fluorescence into each of four intensified cameras. This allows the rapid, simultaneous observation of both membrane and cytoplasmic markers at video camera rates. A phase or differential interference contrast image can

[31] S. J. Morris, *BioTechniques* **8,** 296 (1990).
[32] S. J. Morris, *in* "Optical Microscopy: Emerging Methods and Applications" (B. Herman and J. J. Lemasters, eds.). Academic Press, San Diego, CA, 1993.

also be formed at a different wavelength. Figure 3 shows the beginning and the end of a fusion experiment between HA-expressing cells (GP4F) and erythrocytes. It shows three types of simultaneously acquired images (i.e., phase contrast, redistribution of a membrane dye, and of a cytoplasmic marker) before and after fusion. Analysis of the continuous video recording indicates that redistribution of the cytoplasmic dye, calcein, is retarded compared to the membrane dye.[31] Furthermore swelling of the RBC begins after the dyes have started to move (S. J. Morris, unpublished observations, 1991). Thus the multiparameter imaging approach will yield answers that cannot be determined easily in separate experiments run in parallel.

Section II

Conformational Changes of Proteins during Membrane Fusion

[5] Protein Conformational Changes in Virus–Cell Fusion

By ROBERT W. DOMS

After binding to the cell surface, enveloped animal viruses must undergo a membrane fusion event to deliver their genome into the host cell cytoplasm. Viruses that fuse at neutral pH, such as the paramyxoviruses and the human immunodeficiency viruses (HIV), can do so directly with the plasma membrane of the host cell. A large number of virus families, however, express their fusion activity only at acid pH. These viruses, which include the orthomyxoviruses, rhabdoviruses, flaviviruses, and alphaviruses, can infect cells only after internalization and delivery to endosomes, where the acidic milieu allows fusion to occur.[1–4]

In addition to representing a critical step in the infectious entry pathway, virus–cell fusion offers what is probably the best model system with which to study protein-mediated membrane fusion. Detailing the molecular events underlying virus–cell fusion may provide insights into the ubiquitous intracellular fusion events involved in membrane transport and may suggest new antiviral strategies as well. Although the precise molecular details of virus–cell fusion remain obscure, it is clear that fusion is catalyzed by specific viral membrane proteins, many of which have been identified and characterized in great detail.[1–4] The first step of infection, binding to a cell surface receptor, may be mediated by the viral fusion protein itself or by a second viral membrane protein, as in the case of the paramyxoviruses. After binding, the viral fusion protein must undergo a conformational change that in turn leads to fusion between the viral envelope and the cellular membrane. For acid-dependent viruses, the low pH of the endosome lumen serves as the trigger that leads to the conformational change. The trigger for viruses that fuse at neutral pH is unknown, but may involve receptor binding. The purpose of this chapter is to review the techniques most commonly used to study the conformational changes that occur during virus–cell fusion. Chapters elsewhere in this volume describe techniques that can be used to characterize the properties of the fusion reaction itself.

[1] D. Hoekstra and J. W. Kok, *Biosci. Rep.* **9,** 273 (1989).
[2] J. M. White, *Annu. Rev. Physiol.* **52,** 675 (1990).
[3] M. Marsh and A. Helenius, *Adv. Virus Res.* **36,** 1071 (1989).
[4] T. Stegmann, R. W. Doms, and A. Helenius, *Annu. Rev. Biophys. Chem.* **18,** 187 (1989).

Properties and Examples of Virus Fusion Proteins

Virus – cell fusion is generally mediated by a single viral envelope protein, although several studies suggest that large DNA viruses may encode several fusion factors.[5,6] Despite considerable differences in primary structure, virus fusion proteins share a number of important similarities. All viral fusion proteins identified thus far are integral membrane proteins, the vast majority are glycosylated, many are fatty acylated, and all are oligomeric (trimers or tetramers).[1-4] Many, although not all, viral fusion proteins are synthesized as fusion-inactive precursors that become fusion competent only after a posttranslational proteolytic cleavage. The cleavage generates two subunits: the C-terminal subunit, which spans the viral membrane, and the N-terminal subunit, which remains bound to it by covalent and/or noncovalent interactions. For the paramyxoviruses, orthomyxoviruses, and some retroviruses, the newly formed amino-terminal sequence on the membrane-spanning subunit is highly conserved between different virus strains and is strongly hydrophobic. This region plays a critical role in the fusion reaction and has often been referred to as the fusion peptide.

By far the best characterized viral fusion protein is the influenzavirus hemagglutinin (HA).[7] Hemagglutinin is synthesized as a fusion-inactive precursor termed HA_0, which forms trimers shortly after synthesis. A posttranslational proteolytic cleavage generates a pair of disulfide-linked subunits, HA_1 and HA_2. HA_1 resides entirely outside the viral membrane and contains the major antigenic epitopes of the molecule as well as the receptor-binding site for sialic acid. HA_2 spans the viral membrane and is notable for a highly conserved N-terminal hydrophobic sequence that is involved in the fusion reaction. Nearly the entire ectodomain of HA can be released in water-soluble form by bromelain digestion. The resulting molecule, termed BHA, has been crystallized and its three-dimensional structure determined to high resolution.[8] When incubated at acid pH, BHA undergoes an irreversible conformational change that is similar to that undergone by the intact protein.[9,10] Thus, BHA has proved extremely useful as a model for studying the fusion-inducing conformational change.

The other viral proteins that are referred to below are also noncovalently associated trimers that catalyze fusion at acid pH. The Semliki

[5] M. Butcher, K. Raviprakash, and H. P. Ghosh, *J. Biol. Chem.* **265,** 5862 (1990).
[6] R. W. Doms, R. Blumenthal, and B. Moss, *J. Virol.* **65,** 4884 (1990).
[7] D. C. Wiley and J. J. Skehel, *Annu. Rev. Biochem.* **56,** 365 (1987).
[8] I. A. Wilson, J. J. Skehel, and D. C. Wiley, *Nature (London)* **289,** 366 (1981).
[9] J. J. Skehel, P. M. Bayley, E. B. Brown, S. R. Martin, M. D. Waterfield, J. M. White, I. A. Wilson, and D. C. Wiley, *Proc. Natl. Acad. Sci. USA* **79,** 968 (1982)
[10] R. W. Doms, A. Helenius, and J. White, *J. Biol. Chem.* **260,** 29731 (1985).

Forest virus (SFV) spike glycoprotein contains three copies each of the E_1, E_2, and E_3 subunits. Digestion with proteinase K releases the ectodomain of the spike protein in water-soluble form. Like HA, the SFV spike protein undergoes a posttranslational proteolytic cleavage and undergoes an irreversible conformational change at acid pH. However, unlike HA, it does not contain a N-terminal hydrophobic domain. The vesicular stomatitis virus (VSV) G protein, unlike HA, does not undergo a posttranslational proteolytic cleavage and contains no obvious hydrophobic region analogous to the HA_2 fusion peptide. Although G protein catalyzes fusion at acid pH, its conformational change appears to be fully reversible.[11]

Conditions under Which to Study Conformational Change

The conditions under which a viral fusion protein will undergo a relevant conformational change *in vitro* are likely to reflect those under which virus–cell fusion occurs. Thus, at least a rudimentary characterization of viral fusion activity is necessary. The most important experimental parameter is pH. Acid-induced conformational changes are easier to study because acid pH by itself is generally sufficient to trigger the conformational change in either isolated virions or proteins. Thus, by adjusting pH it is possible to trigger the conformational change in a rapid and reproducible manner.

In addition to pH, several other factors may greatly affect virus fusion activity and so may have dramatic effects on the conformational change as well. Because virus–cell fusion displays a strong temperature dependence, acid treatments should initially be performed at 37°. Lower temperatures may prove of benefit, however. Reduced temperatures may slow the kinetics of the conformational change to the point at which discrete steps may be identified.[12] Lower temperatures may also block the conformational change at an intermediate step, a potentially valuable finding in characterizing the sequence of structural rearrangements that ultimately lead to fusion.[13] Other factors to consider include ionic conditions and the presence of receptors or other potential cofactors. An interesting example of this is provided by the SFV fusion protein. Semliki Forest virus fusion is absolutely dependent on the presence of cholesterol in the target membrane.[14] A water-soluble ectodomain fragment of the SFV spike protein reflects this property, undergoing a conformational change at acid pH only when cholesterol is present.[15] This finding underscores the importance of

[11] R. W. Doms, A. Helenius, and W. Balch, *J. Cell Biol.* **105,** 1957 (1987).
[12] J. M. White and I. A. Wilson, *J. Cell Biol.* **105,** 2887 (1987).
[13] F. Boulay, R. W. Doms, I. Wilson, and A. Helenius, *EMBO J.* **6,** 2643 (1987).
[14] J. White and A. Helenius, *Proc. Natl. Acad. Sci. USA* **77,** 3273 (1980).
[15] M. C. Kielian and A. Helenius, *J. Cell Biol.* **101,** 2284 (1985).

correlating the parameters that affect virus fusion activity with the conditions used to study the conformational change.

In addition to defining the conditions under which the conformational change occurs, it is also necessary to consider whether the change is reversible. For most acid-dependent viruses, the acid-induced conformational change is irreversible. Incubation of an acid-dependent virus at low pH in the absence of a target membrane typically leads to inactivation of viral fusion activity. However, vesicular stomatitis virus and certain strains of influenzavirus are not inactivated by pretreatment at acid pH.[16,17] The vesicular stomatitis virus G protein, in fact, apparently undergoes a fully reversible conformational change.[11] Thus, viral proteins that undergo reversible conformational changes will likely have to be studied at acid pH, in contrast to other viral proteins that can be acid treated and then returned to neutral pH for subsequent assays.

Finally, it is necessary to determine in what context the conformational change is to be studied. Isolated virions may be used, either in solution or after binding to the cell membrane. The protein may be examined when expressed on the plasma membrane of virus-infected cells. Alternatively, the fusion protein may be extracted with nonionic detergents and studied in solution. An additional approach has been to use ectodomain fragments of viral fusion proteins generated by proteolytic digestion or by expression of truncated genes that lack the transmembrane and cytoplasmic domain coding sequences.[9,10,18] An advantage of this approach is that it becomes easier to detect changes in the hydrophobicity of the molecule that may ensue following exposure of a fusion peptide sequence, as in the case of influenzavirus HA. Ultimately, the choice of a model system depends in part on the assays used to monitor the conformational change, on the characteristics of the fusion activity of the virus, and on the specific questions being addressed.

Techniques with Which to Study Conformational Changes in Virus–Cell Fusion

A battery of standard biochemical, immunological, genetic, and biophysical techniques has been employed to study conformational changes that occur during virus–cell fusion. Unfortunately, it is not possible to predict which technique will provide a useful assay for any given viral fusion protein. Thus, several may have to be tried.

[16] R. Blumenthal, A. Bali-Puri, A. Walter, D. Covell, and O. Eidelman, *J. Biol. Chem.* **262**, 13614 (1987).
[17] A. Puri, F. P. Booy, R. W. Doms, J. M. White, and R. Blumenthal, *J. Virol.* **64**, 3824 (1990).
[18] B. Crise, A. Ruusala, P. Zagouras, A. Shaw, and J. K. Rose, *J. Virol.* **63**, 5328 (1989).

Morphological Changes

Negative stain and cryoelectron microscopy have been used to study the acid-induced conformational changes in both isolated viral fusion proteins as well as in intact virions. Morphological changes have been observed at two levels: in individual viral spike proteins and in the distribution of spikes within the viral envelope. Morphological studies on isolated influenzavirus HA suggest that significant changes in tertiary structure occur in the stem domain of the molecule, and that the globular HA_1 domains undergo at least partial dissociation from their adjoining subunits.[19] By contrast, examination of intact virions incubated at acid pH has been problematic owing to the high density of viral spike proteins. A way to avoid this is to use virus particles in which a fraction of the spikes have been removed by proteolytic digestion, making observation of individual spikes more practical.[20] A study using this approach has suggested that higher order structures of HA are formed after acid treatment.[20] In addition to revealing structural changes in individual fusion proteins, electron microscopic studies of VSV have shown that at acid pH the viral fusion protein (G protein) reversibly clusters at the ends of the virions.[21] Interestingly, on return to neutral pH the G protein spikes return to their original, more diffuse distribution. Obviously, studying isolated viral proteins by themselves would not reveal this interesting phenomenon, whereas studying intact virions themselves makes it difficult to observe changes in individual viral spike proteins. A combined approach is perhaps the best.

Changes in Antigenic Structure

The structural rearrangements that lead to membrane fusion are likely to result in significant and detectable changes in the antigenic structure of a protein.[22-25] Although some antigenic epitopes may be lost, others are created or become accessible following the conformational change. Identifying these epitopes can provide relatively detailed structural information about which regions of the protein are affected. The first antigenic changes that were correlated with a fusion-inducing conformational change were described by several groups in 1983, when it was noted that a subset of monoclonal antibodies to one of four major antigenic epitopes in influen-

[19] R. W. H. Ruigrok, et al., *EMBO J.* **5,** 41 (1986).

[20] R. W. Doms and A. Helenius, *J. Virol.* **60,** 8339 (1986).

[21] J. C. Brown, W. W. Newcomb and S. Lawrenz-Smith, *Virology* **167,** 625 (1988).

[22] J. W. Yewdell, W. Gerhard, and T. Bächi, *J. Virol.* **48,** 239 (1983).

[23] R. G. Webster, L. E. Brown, and D. C. Jackson, *Virology* **126,** 587 (1983).

[24] R. S. Daniels, A. R. Douglas, J. J. Skehel, and D. C. Wiley, *J. Gen. Virol.* **64,** 1657 (1983).

[25] F. Gonzalez-Scarano, *Virology* **140,** 209 (1985).

zavirus HA was unable to precipitate the protein in its acid conformation.[22-24] The location of the epitope, at the tip of the molecule near
the trimeric interface, suggested that structural rearrangements occurred
from the top of the molecule to its base, where the fusion peptide is located.
These studies underline an important point, namely that preexisting
monoclonal antibodies that recognize the native structure of a fusion
protein should be screened for their ability to recognize the fusion-competent form in an attempt to define epitopes that are lost or rearranged.
Alternatively, antibodies that fail to recognize the native protein can be
screened against the fusion-competent form to determine if their previously sequestered epitopes have become exposed. Monoclonals that recognize the fusion protein in Western blots but not by direct immunoprecipitation are attractive candidates.

A more direct approach to obtain antibodies capable of detecting a
conformational change is to raise antibodies against the fusion-active form.
Once again, the influenzavirus HA provides the best characterized example, although antigenic changes have also been detected in the fusion
proteins of both La Crosse bunyavirus and rubella virus. Certain monoclonals raised against acid-treated HA recognize only the acid conformation
of the molecule, indicating that their epitopes become exposed after acid
treatment.[26] Finally, anti-peptide antibodies have also been used to map
antigenic changes in viral fusion proteins.[12] Thus, antibodies to the amino-
terminal fusion domain of the HA_2 subunit react only with the acid
conformation of HA, indicating that this region becomes exposed following acid treatment.

Changes in Protease Sensitivity

Viral membrane proteins often display considerable resistance to proteolysis. This useful property has been exploited both to examine the
folding and assembly of viral membrane proteins into their mature, pro-
tease-resistant forms as well as to study the conditions under which the
conformational change occurs.[9,10,25-27] In the case of HA, new trypsin
cleavage sites become exposed and the molecule becomes susceptible to
digestion with a wide variety of proteases. By contrast, the E_1 subunit of the
SFV spike protein becomes resistant to trypsin cleavage following acid
treatment.[15] Interestingly, acquisition of trypsin resistance occurs only at
acid pH in the presence of cholesterol. Indeed, the fusion activity of SFV is
absolutely dependent on the presence of cholesterol in the target mem-

[26] C. S. Copeland, R. W. Doms, E. M. Bolzau, R. G. Webster, and A. Helenius, *J. Cell Biol.*
103, 1179 (1986).
[27] S. Katow and A. Sugiura, *J. Gen. Virol.* **69**, 2797 (1988).

brane. That the E_1 subunit undergoes its conformational change, as monitored by changes in its proteolytic properties, only at acid pH and in the presence of cholesterol confirms its relevance for the fusion reaction and provides an important model system.[15] These results also show that some viral fusion proteins may, in addition to acid pH, require an additional cofactor or receptor. Examining the cleavage patterns of the fusion protein under study in both its native and fusion active states with a panel of proteases may yield a simple, quantitative assay for the conformational change.

Disulfide Exposure

Another consequence of the fusion-inducing conformational change may be exposure of disulfide bonds. The influenzavirus HA consists of two disulfide-linked subunits, HA_1 and HA_2. These subunits are held together by noncovalent interactions and by a single disulfide bond located in the trimeric interface region near the base of the molecule. This disulfide is not accessible to reduction with dithiothreitol (DTT) at 37°. However, after treatment with acid the disulfide becomes exposed, indicating that a structural rearrangement must take place at the base of the molecule.[28] A similar approach could be taken with other viral proteins by incubating their fusion-active and -inactive conformations with reducing agents under nondenaturing conditions. Iodoacetamide is then added to quench the reaction, and the protein subjected to sodium dodecyl sulfate-polyacrylamide gel electrophoresis (SDS-PAGE) under nonreducing conditions. In the case of acid-treated HA, incubation with DTT at 37° breaks the interchain disulfide bond, enabling the HA_1 and HA_2 subunits to migrate independently under nonreducing conditions. This fortunate situation is unlikely to occur with most other viral proteins, however. Rather, acid treatment may affect the number of intrachain disulfide bonds that can be reduced under nondenaturing conditions. Generally, reduction of disulfide bonds leads to a slight, but easily detectable, decrease in mobility in SDS-PAGE. Thus, changes in gel mobility may be taken as presumptive evidence that disulfide bonds either become exposed (or sequestered) following the conformational change.

Changes in Amphiphilic Character

Certain viral fusion proteins possess hydrophobic domains involved in the fusion reaction. Generally these domains are sequestered in the fusion-inactive conformation of the protein but become exposed on activation.

---

[28] P. N. Graves, J. L. Schulman, J. F. Young, and P. Palese, *Virology* **126**, 106 (1983).

Exposure of a hydrophobic sequence in the ectodomain of the molecule may be detected by changes in its amphiphilic properties. This property is particularly useful if a water-soluble ectodomain fragment of the viral membrane protein is available. The bromelain-solubilized ectodomain fragment of the influenzavirus HA (BHA) provides an excellent example. At neutral pH, BHA exists as a stable, water-soluble trimer. However, when incubated under conditions that elicit the fusion activity of the intact virus (mildly acid pH), BHA undergoes an irreversible conformational change and the hydrophobic fusion peptide at the amino terminus of HA_2 (normally sequestered in the interior of the trimer) becomes exposed. As a consequence, acid-treated BHA exhibits hydrophobic properties.[9,10]

Changes in hydrophobicity can be monitored in several ways. In the absence of lipids or detergents, BHA aggregates in a concentration-dependent manner. Aggregation can be easily detected by gradient centrifugation[9,20] and can be prevented by detergent, which in turn provides an additional assay. Detergent binding exhibited by acid-treated BHA can be measured directly as described by Simons *et al.*,[29,30] or indirectly using Triton X-114. At temperatures above 20°, Triton X-114 solutions separate into aqueous and detergent phases. As a result, hydrophilic proteins will partition into the aqueous phase whereas membrane proteins will often (but not always) partition into the detergent phase.[31] This simple but elegant technique, described by Bordier,[31] has been used to document the exposure of a hydrophobic domain in the HA_2 subunit of influenzavirus HA.[10]

Finally, exposure of a hydrophobic fusion peptide may enable the viral protein to bind artificial or natural membranes. Such an approach has been taken with BHA, which binds to liposomes under conditions that parallel those of membrane fusion.[9,10] A simple liposome-binding assay has been developed to detect such an interaction in a quantitative fashion.[10] Liposomes can be prepared by any of a number of techniques, using a lipid composition that preferably simulates that of a host cell.[14] A trace amount of the viral protein in a volume of 5 μl is added to 45 μl of liposomes (9 mM lipid) in MNT buffer [20 mM morpholinoethanesulfonic acid (MES), 30 mM Tris, and 100 mM NaCl]. The pH of the reaction mixture is adjusted by adding, using a Hamilton syringe, pretitrated amounts of 0.5 N acetic acid. After acid treatment, the solution can be reneutralized or maintained at acid pH, depending on whether the protein undergoes a

[29] K. Simons, A. Helenius, and H. Garoff, *J. Mol. Biol.* **80**, 119 (1983).
[30] R. W. Doms and A. Helenius, *in* "Molecular Mechanisms of Membrane Fusion" (S. Ohki, T.D. Flanagan, S. W. Hui, and E. Mayhew, eds.), p. 385. Plenum, New York, 1988.
[31] C. Bordier, *J. Biol. Chem.* **256**, 1604 (1981).

reversible or irreversible conformational change. After the incubation, 150 μl of 67% w/v sucrose in MNT buffer at the appropriate pH is added. The resulting mixture is placed in the bottom of a 700-μl ultraclear tube (Beckman, Fullerton, CA) and overlaid with 300 μl of 25% sucrose followed by 200 μl of 10% sucrose. The tube is placed in a Teflon insert (Beckman) and centrifuged in a Beckman SW50.1 or SW55 rotor for 3 hr at 40,000 rpm. After centrifugation, seven 100-μl fractions are taken sequentially from the liquid–air interface. Capillary pipettes with a hand-held pump are ideal for this task. The distribution of protein and lipid can then be determined. Lipid will be recovered in the first two or three fractions, as will any liposome-associated protein. Unbound protein will remain in the bottom two fractions. This technique has also been used to measure binding (or fusion) of intact virions to liposomes.[27]

Variations of the above assay may be used to characterize the nature of the protein–lipid interaction. Salts, chaotropic agents, and alkaline pH may be employed to determine if the protein binds to liposomes superficially or in a manner analogous to integral membrane proteins. The liposome composition can be varied to determine if binding displays lipid dependence. Likewise, the pH, ionic conditions, and temperature dependence of binding can be addressed.

Photolabeling

As mentioned above, acid pH may trigger a conformational change that enables a portion of the fusion protein to interact with a target membrane. Although the liposome-binding assay makes it possible to detect and characterize the properties of such an interaction, it does not identify the specific region(s) of the viral protein that interacts directly with the membrane. The use of a photoaffinity label that partitions into the hydrophobic core of the target bilayer may be used to address this question. This approach has been used to identify the region of BHA that interacts with target membranes.[32,33] The probe used, 3-(trifluoromethyl)-3-(m-[^{125}I]iodophenyl)diazirine (TID), is commercially available. After BHA is bound to liposomes by acid treatment, a small volume of TID in ethanol is added. The TID equilibrates into the liposomal membrane almost instantaneously, after which it is briefly photoactivated. As a result, TID becomes covalently bound to lipids as well as protein segments that interact directly with the membrane. The liposomes are solubilized, and the BHA is immu-

[32] F. Boulay, R. Doms, and A. Helenius, *in* "Positive Strand RNA Viruses" (M.A. Brinton and R. R. Ruickert, eds.), p. 103. Liss, New York, 1987.

[33] C. Harter, P. James, T. Bächi, G. Semenza, and J. Brunner, *J. Biol. Chem.* **264,** 6459 (1989).

noprecipitated and analyzed by SDS-PAGE and autoradiography. A review in this series describes the technique in detail.[34]

Changes in Quarternary Structure

Most and perhaps all viral fusion proteins are oligomeric.[2] Thus, a fusion-inducing conformational change may alter subunit–subunit interactions. In the case of VSV G protein, the acid-induced change results in increased stability of the trimeric structure of the molecule.[11] At neutral pH, G protein exists as a trimer that is stable to detergent solubilization but dissociates into monomers on centrifugation. Incubation and centrifugation of the protein at acid pH enables it to remain trimeric. The pH dependence with which G protein acquires stability to centrifugation closely parallels that of virus fusion activity, implying that the conformational change manifested as increased stability is relevant for fusion.[11] In addition, stability is not irreversibly conferred by acid treatment. Recentrifugation of acid-treated molecules at neutral pH results in their dissociation, implying that the conformational change is reversible. Indeed, studies with intact virions have shown that the fusion activity of VSV, unlike that of influenzavirus and SFV, is not inactivated by acid treatment in the absence of target membranes.[16] pH-dependent changes in oligomeric stability have also been reported for the SFV fusion protein.[35]

Other Techniques

A number of more specialized techniques have also been employed to monitor conformational changes in viral fusion proteins. Circular dichroism and tryptophan fluorescence spectroscopy studies on purified influenzavirus HA suggest that the conformational change entails major changes in tertiary structure but without significant changes in secondary structure.[36] Additional fluorescence studies on intact virions have shown that the rotational mobility of HA is drastically and irreversibly reduced when viruses are incubated at low pH, presumably because of aggregation of HA in the viral membrane.[37]

An interesting genetic approach that has been taken to address the role of subunit cooperativity in the conformational change takes advantage of

---

[34] J. Brunner, this series, Vol. 172, p. 6287.
[35] J. M. Wahlberg, W. A. M. Boere, and H. Garoff, *J. Virol.* **63,** 499 (1989).
[36] S. A. Wharton, R. W. H. Ruigrok, S. R. Martin, J. J. Skehel, P. M. Bayley, W. Weiss, and D. C. Wiley, *J. Biol. Chem.* **263,** 4474 (1988).
[37] P. R. Junankar and R. J. Cherry, *Biochim. Biophys. Acta* **854,** 198 (1986).

the oligomeric nature of the influenzavirus HA and the random assembly of subunits in the endoplasmic reticulum.[38] Coexpression of two HA gene products, each of which causes fusion at different pH, leads to the formation of mixed trimers. The hybrid trimers undergo the conformational change at a pH intermediate to that of either parental HA, implying that the conformational change is a highly cooperative event.[38] Another technique that has been used to address the role of spike protein cooperativity in membrane fusion is radiation inactivation. Using this approach, the functional unit for VSV G protein has been proposed to consist of approximately 15 G protein molecules.[39]

Concluding Remarks

Most of the techniques described above allow one to examine what happens to a relatively small region of the viral protein during its conformational change; that is, the loss of an antigenic epitope, or the exposure of a particular cleavage site or disulfide bond. Obviously, a number of assays will have to be used to obtain a more complete picture of the structural rearrangements that lead to fusion. Detailed kinetic studies, and perhaps the use of low temperatures, may reveal the sequence with which these rearrangements occur.

In addition to describing aspects of the conformational change, it is necessary to ask whether the changes observed in an *in vitro* system are relevant for fusion *in vitro*. The conditions under which a given aspect of the conformational change occurs should reflect those under which fusion takes place. Thus, a conformational change that occurs in the paramyxovirus F protein at pH 9.0, to which the virus is never exposed *in vitro,* is of dubious biological significance.[40] Studies using isolated viral proteins should also be critically evaluated, because the conformational change that occurs during interaction of the intact virus with a target membrane may differ from what happens in the absence of a target membrane. Furthermore, most studies on the "acid" conformations of viral fusion proteins have examined structural changes that occur after acid treatment *and* reneutralization. The resulting structures may of course differ in important respects from the nascent molecules shortly after acid treatment. In fact, one study has suggested that for influenzavirus C HA the onset of trypsin susceptibility following acid treatment may actually be a postfusion

[38] F. Boulay, R. W. Doms, R. G. Webster, and A. Helenius, *J. Cell Biol.* **106,** 629 (1988).
[39] K. Bundo-Morita, S. Gibson, and J. Lenard, *Virology* **163,** 622 (1988).
[40] M.-C. Hsu, A. Scheid, and P. W. Choppin, *Proc. Natl. Acad. Sci. USA* **79,** 5862 (1982).

event.[41] This further emphasizes the need for correlating structural changes and the conditions under which they occur with what happens during fusion itself. It is to be hoped that employing a variety of techniques to describe the conformational changes associated with virus–cell fusion will lead to a more detailed understanding of the molecular mechanisms of the fusion reaction.

[41] F. Formanowski, S. A. Wharton, L. J. Calder, C. Hofbauer, and H. Meier-Ewert, *J. Gen. Virol.* **71**, 1181 (1990).

[6] Monitoring Protein Conformational Changes during Membrane Fusion

By Tetsuro Yoshimura

Proteins are known to participate in membrane fusion in numerous cellular events, such as fertilization, myoblast fusion, virus infection, exocytosis, and intracellular protein transport. Various proteins and peptides can induce fusion of biological and/or artificial membranes, and their fusion mechanisms have been studied extensively.[1,2] These studies have shown that a change in conformation, exposure of the hydrophobic moieties of the proteins, and insertion of hydrophobic segments into membranes contribute to the initiation of membrane fusion. For instance, hydrophobic stretches of amino acids of virus envelope proteins, such as the F protein of Sendai virus, the hemagglutinin (HA) protein of influenzavirus, and the E protein of Semliki Forest virus, are thought to be exposed by the action of proteases and pH-dependent conformational changes of the proteins and to mediate fusion between viral and cellular membranes.[2,3] Hydrophobic regions of clathrin,[4,5] bovine serum albumin (BSA) and its fragments,[6] and diphtheria toxin[7] can be exposed through

[1] K. Hong, N. Düzgüneş, P. R. Meers, and D. Papahadjopoulos, *in* "Cell Fusion" (A. E. Sowers, ed.), p. 269. Plenum, New York, 1987.

[2] T. Stegmann, R. W. Doms, and A. Helenius, *Annu. Rev. Biophys. Biophys. Chem.* **18**, 187 (1989).

[3] J. M. White, *Annu. Rev. Physiol.* **52**, 675 (1990).

[4] T. Yoshimura, S. Maezawa, and K. Hong, *J. Biochem. (Tokyo)* **101**, 1265 (1987).

[5] S. Maezawa, T. Yoshimura, K. Hong, N. Düzgüneş, and D. Papahadjopoulos, *Biochemistry* **28**, 1422 (1989).

[6] L. A. M. Garcia, P. S. Araujo, and H. Chaimovich, *Biochim. Biophys. Acta* **772**, 231 (1984).

[7] F. Defrise-Quertain, V. Cabiaux, M. Vandenbranden, R. Wattaiez, P. Falmagne, and J.-M. Ruysschaert, *Biochemistry* **28**, 3406 (1989).

conformational change before membrane fusion. Membrane fusion induced by melittin[8,9] and a synthetic amphipathic peptide[10-13] are thought to be associated with the formation of amphipathic helices and their hydrophobicity. Hydrophobic interactions of the F protein of Sendai virus,[14] HA protein of influenzavirus,[15] diphtheria toxin,[16,17] α-lactalbumin,[18,19] ovalbumin,[20] and cytochrome c[21] with target membranes appear to be responsible for the induction of membrane fusion.

In this chapter various procedures for detection of protein conformational change and exposure of hydrophobic regions during membrane fusion are described. These procedures can be roughly classified into biophysical, biochemical, and immunological methods. Procedures for the detection of the penetration of hydrophobic segments of proteins into bilayer membranes occurring through conformational changes are also outlined. Protein conformational changes and membrane insertion are usually detected under conditions that trigger membrane fusion.

Biophysical Procedures

Fluorescence

Intrinsic Fluorescence

A conformational change in a protein sometimes results in a change in its intrinsic fluorescence. To obtain information on conformational changes in relation to membrane fusion, the fluorescence spectrum of a fusogenic protein in the wavelength range of 300 to 400 nm is measured

[8] C. G. Morgan, H. Williamson, S. Fuller, and B. Hudson, *Biochim. Biophys. Acta* **732**, 668 (1983).
[9] G. D. Eytan and T. Almary, *FEBS Lett.* **156**, 29 (1983).
[10] R. A. Parente, S. Nir, and F. C. Szoka, Jr., *J. Biol. Chem.* **263**, 4724 (1988).
[11] R. A. Parente, L. Nadasdi, N. K. Subbarao, and F. C. Szoka, Jr., *Biochemistry* **29**, 8713 (1990).
[12] S. Takahashi, *Biochemistry* **29**, 6257 (1990).
[13] T. Yoshimura, Y. Goto, and S. Aimoto, *Biochemistry* **31**, 6119 (1992).
[14] S. L. Novick and D. Hoekstra, *Proc. Natl. Acad. Sci. U.S.A.* **85**, 7433 (1988).
[15] C. Harter, P. James, T. Bächi, G. Semenza, and J. Brunner, *J. Biol. Chem.* **264**, 6459 (1989).
[16] E. Papini, G. Schiavo, M. Tomasi, M. Colombatti, R. Rappuoli, and C. Montecucco, *Eur. J. Biochem.* **169**, 637 (1987).
[17] M. E. Dumont and F. M. Richards, *J. Biol. Chem.* **263**, 2087 (1988).
[18] J. Kim and H. Kim, *Biochemistry* **25**, 7867 (1986).
[19] J. Kim and H. Kim, *Biochim. Biophys. Acta* **983**, 1 (1989).
[20] C.-H. Yun and H. Kim, *J. Biochem (Tokyo)* **105**, 406 (1989).
[21] S. Lee and H. Kim, *Arch. Biochem. Biophys.* **271**, 188 (1989).

with an excitation wavelength between 270 and 300 nm under conditions that induce membrane fusion. This spectrum is compared with that obtained under conditions in which no fusion occurs. The fluorescence intensity at 330 nm of diphtheria toxin excited at 280 nm decreases below pH 5,[22] where the protein induces fusion of neutral[23] and acidic phospholipid vesicles.[7] The intrinsic fluorescence of bromelain-treated influenzavirus hemagglutinin (BHA) is also reduced at the pH of membrane fusion, and the pH dependencies of the fluorescence (at 335 nm) of wild-type and mutant BHAs show profiles that are similar to those of their fusion activities.[24]

Fluorescent Probes

Fluorescent compounds are often used as microenvironmental probes to examine the properties of a compound, or an environment, as well as their changes under different conditions.

1-Anilinonaphthalene 8-Sulfonate. The fluorescent reagent 1-anilinonaphthalene 8-sulfonate (ANS) is sensitive to microenvironmental change: an increase in intensity and a blue shift of its fluorescence are indications of its binding to hydrophobic regions of a protein.

Procedure: The reagent ANS, dissolved in distilled water, is added to a solution of a fusogenic protein under conditions that induce membrane fusion, and its fluorescence spectrum is monitored in the wavelength range of 450 to 550 nm, with an appropriate excitation wavelength between 350 and 400 nm. The concentrations of ANS and the protein should be selected so as to obtain clear data. The increase in intensity and blue shift of the fluorescence of ANS, compared with the spectrum under conditions that do not induce fusion, suggest exposure of hydrophobic domains of the protein through conformational changes.

Clathrin induces fusion of vesicles containing phosphatidylserine below pH 6[5,25] and, when ANS is present, its fluorescence concomitantly shifts to lower wavelengths and increases in intensity.[4,5] When ANS is incubated with diphtheria toxin, its fluorescence maximum shifts from 510 to 470 nm at pH 4.2, but does not change at pH 7.2, and the transitional pH for change in fluorescence intensity of ANS at 470 nm in the presence of the toxin is about pH 5.0, which is the same as that for toxin-induced fusion of phospholipid vesicles.[7] Increase in the fluorescence intensity of the fluorophore, with concomitant appearance of fusogenic activity, has

[22] M. G. Blewitt, L. A. Chung, and E. London, *Biochemistry* **24**, 5458 (1985).
[23] V. Cabiaux, M. Vandenbranden, P. Falmagne, and J.-M. Ruysschaert, *Biochim. Biophys. Acta* **775**, 31 (1984).
[24] M. J. Banda, A. G. Rice, G. L. Griffin, and R. M. Senior, *J. Biol. Chem.* **263**, 4481 (1988).
[25] K. Hong, T. Yoshimura, and D. Papahadjopoulos, *FEBS Lett.* **191**, 17 (1985).

also been observed with tetanus toxin[26] and glyceraldehyde-3-phosphate dehydrogenase.[27]

N-(1-Anilinonaphthyl-4)maleimide. The nonfluorescent reagent *N*-(1-anilinonaphthyl-4)maleimide (ANM) becomes fluorescent when bound covalently to sulfhydryl groups of proteins and, as in the case of ANS, the quantum yield of protein-labeled ANM probe is dependent on the polarity of the surrounding environment.[28] The reagent ANM is better than ANS for a quantitative assay of change in protein conformation, because the amount of labeled probe is fixed under any condition. Moreover, as the excitation spectrum of anilinonaphthyl (AN) moieties in an ANM-labeled protein sample shows considerable overlap with the emission spectrum of tryptophan residues of the protein, resonance energy transfer between the two fluorophores can be used to detect conformational changes.

Procedure: An ANM-labeled protein is prepared essentially as described by Ohyashiki *et al.*[29] The protein is treated in a buffer of neutral pH with a 5- or 10-fold molar excess of ANM dissolved in a trace amount of acetone. After incubation at a constant temperature for 30 or 60 min, the reaction is stopped by the addition of a 5-fold excess of 2-mercaptoethanol, and the reaction mixture is dialyzed against the same buffer at 4° overnight. The number of labeled AN groups is determined with a millimolar extinction coefficient of 10.8 mM^{-1} cm^{-1} at 345 nm. Then, the fluorescence spectra of the preparation are monitored in the wavelength range of 400 to 500 nm, with an appropriate excitation wavelength between 300 and 400 nm, under conditions inducing membrane fusion. To examine resonance energy transfer, fluorescence spectra in the wavelength range of 300 to 500 nm are measured at an appropriate excitation wavelength between 270 and 300 nm.

Increase in the fluorescence intensity of clathrin-bound AN labels and greater transfer of resonance energy from tryptophan to AN residues in ANM-labeled clathrin preparations have been observed in the pH region where membrane fusion is induced.[4,5]

cis-Parinaric Acid. Exposure of hydrophobic regions of proteins can be examined by measuring the effective hydrophobicity of proteins, using another fluorescent probe, *cis*-parinaric acid (*cis*-PnA). The hydrophobicity index of a protein, the slope of a plot of the increase in *cis*-PnA

[26] V. Cabiaux, P. Lorge, M. Vandenbranden, P. Falmagne, and J.-M. Ruysschaert, *Biochem. Biophys. Res. Commun.* **128,** 840 (1985).
[27] A. E. L. Viñals, R. N. Farías, and R. D. Morero, *Biochem. Biophys. Res. Commun.* **143,** 403 (1987).
[28] Y. Kanaoka, M. Machida, M. Machida, and T. Sekine, *Biochim. Biophys. Acta* **317,** 563 (1973).
[29] T. Ohyashiki, T. Sekine, and Y. Kanaoka, *Biochim. Biophys. Acta* **351,** 214 (1974).

fluorescence vs. protein concentration, has been shown to correlate well with its effective hydrophobicity as determined by the hydrophobic partition method.[30]

Procedure: A fusogenic protein at an appropriate concentration is added to the test solution containing $0.5-2 \times 10^{-3}\%$ (w/v) sodium dodecyl sulfate (SDS) under conditions that trigger membrane fusion, and a trace amount of an ethanolic solution of *cis*-PnA ($1.8 \mu M$) is promptly added. After incubation at 25° for 1 to 2 min, the fluorescence intensity at 420 nm is measured with an appropriate excitation wavelength of 320 to 325 nm. Measurements are repeated at different concentrations of protein, the fluorescence increment is plotted as a function of protein concentration, and the hydrophobicity index of the protein is determined from the slope of the plot.

The change in hydrophobicity of clathrin has been examined by this procedure.[4,5] As shown in Fig. 1, the plots are linear, and the slopes of the plots increase steeply below pH 6, where membrane fusion starts to occur.

Circular Dichroism

Protein conformational changes are reflected most directly by circular dichroism (CD) spectra: CD spectra in the far- and near-ultraviolet regions are sensitive to changes in peptide backbone structure and the conformation around aromatic groups. For detection of conformational changes, the CD spectrum of a fusogenic protein is monitored in the wavelength range of 190 to 350 nm, under conditions in which the protein induces membrane fusion, and compared with the spectrum obtained under conditions in which the protein has no fusion activity. However, because proteins are usually in an aggregated state under conditions that trigger membrane fusion, differential scattering and absorbance flattening artifacts may affect their CD spectra. Circular dichroic spectra have been useful in detecting conformational changes of fusogenic peptides.

The far-ultraviolet CD spectrum of the low-pH form of influenzavirus HA, which is the form with fusion activity, has been found to be similar to that of its neutral-pH form, whereas the near-ultraviolet CD spectra of the two forms are different. This observation suggests that the conformational change at low pH is accompanied by a significant change in the environment of aromatic residues.[31,32] A synthetic, amphipathic peptide with the

[30] A. Kato and S. Nakai, *Biochim. Biophys. Acta* **624,** 13 (1980).
[31] J. J. Skehel, P. M. Bayley, E. B. Brown, S. R. Martin, M. D. Waterfield, J. M. White, I. A. Wilson, and D. C. Wiley, *Proc. Natl. Acad. Sci. U.S.A.* **79,** 968 (1982).
[32] S. A. Wharton, R. W. H. Ruigrok, S. R. Martin, J. J. Skehel, P. M. Bayley, W. Weis, and D. C. Wiley, *J. Biol. Chem.* **263,** 4474 (1988).

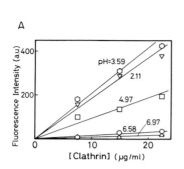
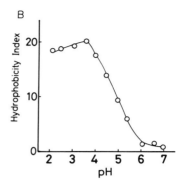

FIG. 1. pH dependence of the effective hydrophobicity of clathrin. An ethanolic solution of *cis*-PnA (1.8 μM) is added to clathrin solutions of different pH values containing 5×10^{-4} % (w/v) SDS. After incubation at 25° for 1 to 2 min, the fluorescence intensity at 420 nm is measured at an excitation wavelength of 320 nm. (A) Fluorescence increments are plotted as a function of clathrin concentration; (B) the slopes of the plots, defined as the hydrophobicity index, are plotted against pH.

repeat unit Glu-Ala-Leu-Ala shows CD spectra typical of a random coil structure at neutral pH, but of a helical structure at low pH, where it induces fusion of small unilamellar phosphatidylcholine vesicles.[10,33]

Difference Spectra

Ultraviolet spectra in the near-ultraviolet regions are also sensitive to conformational changes around aromatic amino acid residues: red and blue spectral shifts are observed on perturbation of tyrosine and/or tryptophan residues by a solvent, and on their exposure to a more polar environment. These respective shifts result in positive and negative absorption bands in the difference spectrum.

Procedure: Difference spectra are measured using pairs of spectrophotometer cells, with each cell having two compartments. One side of one paired cell (A) is filled with a solution of the fusogenic protein under conditions inducing membrane fusion, and the other side of this cell contains a buffer solution in which fusion could not occur. One side of the other paired cell (B) is filled with a solution of the fusogenic protein at the same concentration, but under conditions unsuitable for fusion, and the other side is filled with buffer in which membrane fusion would occur. Then, the difference spectrum, A − B, is monitored in the wavelength range of 250 to 330 nm.

[33] N. K. Subbarao, R. A. Parente, F. C. Szoka, Jr., L. Nadasdi, and K. Pongracz, *Biochemistry* **26,** 2964 (1987).

Difference spectra of bovine serum albumin and its fragment under fusion active and inactive conditions show negative bands at 280 and 288 nm, indicating exposure of buried tyrosine residues through a conformational change.[6]

Triton X-114 Partitioning

A solution of the nonionic detergent Triton X-114 is homogeneous when the temperature of the medium is below its cloud point (about 20°), but separates into an aqueous phase and a detergent phase above this point.[34] Proteins dissolved in Triton X-114 solution at low temperatures are thus partitioned into the aqueous and detergent phases according to their hydrophobicity when the temperature is raised above 20°. When hydrophobic moieties of a protein are exposed through a conformational change, the protein should partition into the detergent phase. This technique has been used to detect the exposure of hydrophobic regions of colicin E3,[35] IgE receptor,[36] clathrin,[4,5] and the spike glycoproteins of Semliki Forest virus[37] and influenzavirus.[38]

Procedure: Before starting experiments, commercial Triton X-114 should be precondensed.[34] Triton X-114 (10 g) containing 8 mg of butylated hydroxytoluene is mixed with 500 ml of 10 mM Tris-HCl (pH 7.4) containing 150 mM NaCl. After dissolution at 0°, the clear solution is incubated at 30° overnight, which results in separation of aqueous and detergent phases. The resulting upper aqueous phase is discarded, the resulting lower detergent phase is mixed with the same volume of the buffer, and condensation is repeated twice under the same conditions. Then a fusogenic protein (50 to 100 μg) is mixed at 0° with buffer solution containing 1% (w/v) precondensed Triton X-114 under conditions inducing membrane fusion, and layered on a cushion of 6% (w/v) sucrose containing 0.06% (w/v) Triton X-114. The sample is incubated at 0° for 5–30 min and then at 30° for 3 to 10 min, and centrifuged at 300 g for 30 min at 30°. After centrifugation, the upper aqueous phase and lower detergent phase are analyzed by sodium dodecyl sulfate-polyacrylamide gel electrophoresis (SDS-PAGE).

As shown in Fig. 2, clathrin is recovered in the aqueous phase at neutral

[34] C. Bordier, *J. Biol. Chem.* **256**, 1604 (1981).
[35] V. Escuyer, P. Boquet, D. Perrin, C. Montecucco, and M. Mock, *J. Biol. Chem.* **261**, 10891 (1986).
[36] G. Alcaraz, J.-P. Kinet, N. Kumar, S. A. Wank, and H. Metzger, *J. Biol. Chem.* **259**, 14922 (1984).
[37] M. Kielian and A. Helenius, *J. Cell Biol.* **101**, 2284 (1985).
[38] R. W. Doms, A. Helenius, and J. White, *J. Biol. Chem.* **260**, 2973 (1985).

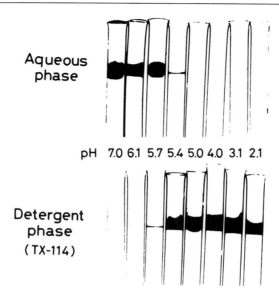

FIG. 2. Effect of pH on clathrin partitioning in Triton X-114 solution. Clathrin (100 μg) is incubated at 0° for 30 min in buffer solutions of various pH values containing 1% (w/v) Triton X-114, and then incubated at 30° for 10 min. The mixtures are centrifuged at 300 g for 3 min at 30°, and the resulting aqueous and detergent phases are analyzed by SDS-PAGE.

pH, in both phases at pH 5–6, and entirely in the detergent phase below pH 5, indicating exposure of hydrophobic domains of the protein below pH 6, where it induces membrane fusion.[4,5] The water-soluble ectodomain fragments of the E_2 protein of Semliki Forest virus[37] and HA protein of influenzavirus[38] are scarcely partitioned into the detergent phase at neutral pH, but are largely recovered in this phase at about pH 5, the pH at which they induce membrane fusion.

Biochemical and Immunological Procedures

Protease Digestion

Proteolytic enzymes, such as trypsin and proteinase K, are often used for obtaining information on conformational changes. The conformational change is monitored by the ability of these proteases to cleave a protein due to exposure of susceptible sites.

Procedure: A solution of the fusogenic protein is incubated at an appropriate temperature for 15 to 30 min under conditions that trigger membrane fusion, adjusted to the optimum conditions for proteolysis, and treated with an appropriate amount of protease (usually at an enzyme-to-substrate ratio of 1 : 10) for 10 to 30 min at 20 or 37°. The reaction is stopped by the addition of a protease inhibitor, and the resulting proteolytic products are analyzed by SDS-PAGE. When the amount of sample is low, radiolabeled proteins such as ^{125}I- and ^{35}S-labeled fusogenic proteins are used, proteolytic products are collected by triacetic acid precipitation, and their radioactivity is counted directly or after separation by SDS-PAGE. For this procedure, the conformational change that occurs under conditions for membrane fusion should be irreversible when changing to the optimum condition for proteolytic digestion, for example, to the optimum pH of the protease. Reversal to the original state and protein aggregation are sometimes prevented by carrying out the incubation and/or protease digestion in the presence of a nonionic detergent (such as Triton X-100 or Brij) or phospholipid liposomes. However, no appreciable difference in results has been observed in the presence or absence of these additions.[17,31,37]

Influenzavirus BHA becomes susceptible to tryptic digestion after incubation below the threshold pH for induction of membrane fusion, either in the presence or absence of Brij 36T.[31] In contrast, the pH dependence of fusion induced by BHA is 0.25 pH units lower than the pH for conversion of the total BHA population to its proteinase K-sensitive form.[38] Although the abilities of some mutants to induce cell–cell fusion are greatly impaired, mutant BHAs show pH-dependent sensitivity to proteinase K.[39] The susceptibilities of the E_1 and E_2 ectodomains of Semliki Forest virus to trypsin digestion change at the pH of fusion, both with and without Triton X-100, and the susceptibility of E_1 glycoprotein depends on the presence of cholesterol, which plays an important role in the fusion reaction.[37] This protease digestion technique is also used to determine the regions of a fusogenic protein inducing membrane fusion.[6,40]

Immunological Assay

Monoclonal Antibody. Monoclonal antibodies can also be used to monitor protein conformational changes. Mice are immunized with a fusogenic protein, and hybridomas are selected for production of antibodies against the protein, under conditions in which it is fusion active and inactive. Then the reactivity of the monoclonal antibodies against the protein is tested under these two conditions by immunoprecipitation, radioimmunoassay, or enzyme immunoassay. This procedure can be used

[39] M.-J. Gething, R. W. Doms, D. York, and J. White, *J. Cell Biol.* **102**, 11 (1986).
[40] S. Maezawa and T. Yoshimura, *Biochemistry* **29**, 1813 (1990).

to detect the conformational change of influenzavirus hemagglutinin at the pH of the membrane fusion.[38,41]

Moreover, by specifying their antigenic sites, monoclonal antibodies can be used as probes to locate the regions of a fusogenic protein that are modified by the conformational change. The regions of the HA protein of influenzavirus that change conformation at pH 5 have been determined by the pattern of reactivity of monoclonal antibodies with defined antigenic sites at neutral and low pH.[42,43]

Anti-Peptide Antibody. Anti-peptide antibodies are also available for the study of some types of protein conformational change. Anti-peptide antisera are raised against synthetic peptides from various regions of a fusogenic protein molecule. They are then incubated with the protein under fusion and nonfusion conditions. The resulting antigen–antibody complexes are quantitated by immunological assays as described above, and the regions of the protein molecule that are exposed or buried as a result of the conformational change are located. This type of analysis is also used to monitor the kinetics and sequence of membrane fusion events, and is most valuable when the three-dimensional structure of the protein is known. Anti-peptide antibodies can be used to probe regions of the HA protein of influenzavirus that change in response to low pH, as well as to detect steps in the conformational change of the protein.[44]

Other Related Procedures

Hydrophobic Labeling

In addition to the conformational change of a fusogenic protein, another important event during membrane fusion is insertion of a segment(s) of a fusogenic protein into the target membrane. The technique of hydrophobic labeling is usually applied to obtain direct evidence of the penetration of a segment(s) into the hydrophobic interior of lipid bilayers. Radiolabeled hydrophobic photoaffinity labels, such as 3-(trifluoromethyl)-3-([^{125}I]iodophenyl)diazirine (TID), 1-palmitoyl-2-(2-azido-4-nitro)benzoyl-sn-glycero-3-phospho-[^{3}H]choline (PC I), 1-myristoyl-2-[12-amino-(4N-3-nitro-1-azidophenyl)]dodecanoyl-sn-glycero-3-phospho-[^{3}H]choline (PC II), and 1-palmitoyl-2-[11-[4-[3–(trifluoromethyl)diazirinyl]phenyl][2 - ^{3}H]undecanoyl] - sn - glycero - 3 - phosphorylcholine

[41] R. S. Daniels, A. R. Douglas, J. J. Skehel, and D. C. Wiley, *J. Gen. Virol.* **64**, 1657 (1983).
[42] R. G. Webster, L. E. Brown, and D. C. Jackson, *Virology* **126**, 587 (1983).
[43] D. C. Jackson and A. Nestorowicz, *Virology* **145**, 72 (1985).
[44] J. M. White and I. A. Wilson, *J. Cell Biol.* **105**, 2887 (1987).

(PTPC/11), and the hydrophobic fluorophore dansyl chloride (DNS-Cl) have been employed as probes. Membrane-penetrating segments as well as labeled residues can also be identified by this procedure.

Procedure. Briefly, the procedure is as follows. Phospholipid liposomes are prepared in the presence of PC I, PC II, or PTPC/11, or TID or DNS-Cl dissolved in a trace amount of ethanol or acetone is added to preformed liposomes. Then a fusogenic protein solution is added to this liposome suspension, under conditions in which membrane fusion can or cannot be induced. After incubation, the samples are irradiated with ultraviolet light, except on labeling with DNS-Cl. The protein is recovered by trichloroacetic acid precipitation, or the liposome fraction is precipitated by ultracentrifugation, and the labeled protein is separated by SDS-PAGE and analyzed. For identification of segments inserted into the membrane, the protein–liposome complex is treated with protease, and the peptide segment(s) remaining buried in the membrane is identified by amino acid or sequence analysis.

Membrane insertion of the F protein of Sendai virus,[14] the HA protein of influenzavirus,[15,45] clathrin,[46] diphtheria toxin,[16,17] α-lactalbumin,[19] ovalbumin,[20] and cytochrome c[21] has been confirmed, using TID or PC I plus PC II for hydrophobic labeling. PTPC/11 has been synthesized,[45] and the segment responsible for the hydrophobic insertion of influenzavirus BHA into membranes has been identified.[15] DNS-Cl has also been used for identification of segments of α-lactalbumin, ovalbumin, and cytochrome c that penetrate into membranes.[18,20,21]

---

[45] C. Harter, T. Bächi, G. Semenza, and J. Brunner, *Biochemistry* **27**, 1856 (1988).
[46] J. Seppen, J. Ramalho-Santos, A. P. de Carvalho, M. ter Beest, J. W. Kok, M. C. P. de Lima, and D. Hoekstra, *Biochim. Biophys. Acta* **1106**, 209 (1992).

[7] Synthetic Peptides as Probes of Function of Viral Envelope Proteins

By Nejat Düzgüneş

Introduction

Lipid-enveloped viruses fuse with cellular membranes to microinject their genome into the cytoplasm of the host cell. In the case of paramyxoviruses, such as Sendai virus, or immunodeficiency viruses, such as human immunodeficiency virus type 1 (HIV-1), the viral membrane undergoes fusion with the plasma membrane at neutral pH.[1-6] Orthomyxoviruses, such as influenzavirus, or togaviruses, including Semliki Forest virus, fuse with the endosome membrane after endocytosis of the virion and mild acidification of the endosome.[2,3,7,8] Membrane fusion is mediated by the viral envelope glycoproteins.[7-10] The N-terminal amino acid sequences of some of these proteins show considerable homology, suggesting that this domain of the proteins is involved in membrane fusion.[7,11,12]

The hydrophobic N terminus of the cleaved hemagglutinin of influenzavirus (HA_2) is exposed only when the pH is lowered.[13,14] The N terminus of the Sendai virus F protein, however, is normally exposed on the protein.[11,15,16] One possible function for this region of the viral envelope proteins is to penetrate target membranes. The destabilization of the target cell and viral membrane lipid bilayers within the area of virus–cell adhesion may be a critical step in membrane fusion.[9,17-19] Thus, the ability of synthetic peptides corresponding to these sequences to penetrate into membranes and destabilize them, and the requirements for particular amino acid sequences and peptide structure for this function, can reveal some of the molecular mechanisms of membrane fusion mediated by viral fusion proteins. Similar mechanisms may also be involved in intracellular

[1] Y. Okada, in "Membrane Fusion in Fertilization, Cellular Transport and Viral Infection" (N. Düzgüneş and F. Bronner, eds.), p. 297. Academic Press, San Diego, 1988.
[2] M. Marsh and A. Helenius, *Adv. Virus Res.* **36,** 107 (1989).
[3] D. Hoekstra and J. W. Kok, *Biosci. Rep.* **9,** 273 (1989).
[4] M. Marsh and A. Dalgleish, *Immunol. Today* **8,** 369 (1987).
[5] F. Sinangil, A. Loyter, and D. J. Volsky, *FEBS Lett.* **239,** 88 (1988).
[6] B. S. Stein and E. G. Engleman, in "Mechanisms and Specificity of HIV Entry into Host Cells" (N. Düzgüneş, ed.), p. 71. Plenum, New York, 1991.
[7] J. White, M. Kielian, and A. Helenius, *Q. Rev. Biophys.* **16,** 151 (1983).
[8] S. Ohnishi, in "Membrane Fusion in Fertilization, Cellular Transport and Viral Infection" (N. Düzgüneş and F. Bronner, eds.), p. 257. Academic Press, San Diego, 1988.
[9] N. Düzgüneş, *Subcell. Biochem.* **11,** 195 (1985).
[10] T. Stegmann, R. W. Doms, and A. Helenius, *Annu. Rev. Biophys. Chem.* **18,** 187 (1989).
[11] M. J. Gething, J. M. White, and M. D. Waterfield, *Proc. Natl. Acad. Sci. U.S.A* **75,** 2737 (1978).
[12] W. R. Gallaher, *Cell (Cambridge, Mass.)* **50,** 327 (1987).
[13] J. J. Skehel, P. M. Bayley, E. B. Brown, S. R. Martin, M. D. Waterfield, J. M. White, I. A. Wilson, and D. C. Wiley, *Proc. Natl. Acad. Sci. U.S.A.* **79,** 968 (1982).
[14] J. M. White and I. A. Wilson, *J. Cell Biol.* **105,** 2887 (1987).
[15] K. Asano, T. Murachi, and A. Asano, *J. Biochem. (Tokyo)* **93,** 733 (1983).
[16] M. C. Hsu, A. Scheid, and P. W. Choppin, *J. Biol. Chem.* **256,** 3557 (1981).
[17] D. Hoekstra and J. Wilschut, in "Water Transport in Biological Membranes" (G. Benga, ed.), p. 143. CRC Press, Boca Raton, FL, 1989.
[18] L. V. Chernomordik, G. B. Melikyan, and Y. A. Chizmadzhev, *Biochim. Biophys. Acta* **906,** 309 (1987).
[19] N. Düzgüneş and S. Shavnin, *J. Membr. Biol.* **128,** 71 (1992).

membrane fusion mediated by certain cytoplasmic or membrane proteins. The ability of such synthetic peptides to induce membrane fusion, in addition to inserting into membranes, not only provides information on the possible mode of action of these segments of viral envelope proteins, but also on the minimal molecular requirements for protein-mediated membrane fusion.

This chapter presents an outline of the methods used to study the interaction of viral peptides with biological and phospholipid model membranes.

Hemolysis

The lysis of erythrocytes (hemolysis) has been used as an indirect indicator of the fusion of several viruses with biological membranes.[20,21] Hemolytic activity can be used to monitor the destabilization of erythrocyte membranes by viral peptides, as a result of the insertion of the peptides into the membrane.[22]

Red blood cells are obtained by layering whole blood, obtained from a blood bank (preferably tested for hepatitis B and HIV-1, and found to be negative), on Ficoll-Paque (Pharmacia, Piscataway, NJ) or Histopaque (Sigma, St. Louis, MO), and centrifuging at low speed [2000 rpm in a Sorvall (Norwalk, CT) RT6000 cell centrifuge for 30 min at 20°]. Although repeated centrifugation and removal of buffy coats is also a possibility, the Ficoll-Paque centrifugation enables a one-step purification of the red blood cells. The red blood cells, which pellet to the bottom, are washed three times with phosphate-buffered saline (PBS), and resuspended in the same buffer at a concentration of 5×10^8 cells/ml. The peptides are added to this suspension at different concentrations and at different pH values, adjusted by introducing aliquots of pretitrated concentrated acetate buffer (e.g., $2\ M$, adjusted to pH 1.5). Alternatively, the cells can be resuspended in buffers of different pH. In this case, however, the cells should not be kept for extended periods of time at extremes of pH.

After the incubation period, the cells are centrifuged in an Eppendorf centrifuge at 4° for 2 min, and the hemoglobin concentration in the supernatant is determined by the absorbance at 545 nm. For calibration, total hemolysis is achieved either by addition of detergent [Triton X-100 at a final concentration of 1%, by adding 50 μl of a 20% (w/w) stock solution into a 1-ml cell suspension], or by resuspension of an identical quantity of

[20] T. Bächi, G. Eichenberger, and H. P. Hauri, *Virology* **85**, 518 (1978).
[21] M. C. Hsu, A. Scheid, and P. W. Choppin, *Proc. Natl. Acad. Sci. U.S.A.* **79**, 5862 (1982).
[22] R. Schlegel and M. Wade, *J. Biol. Chem.* **259**, 4691 (1984).

red blood cells in 1 ml of distilled water. The spontaneous hemolysis of the cells in the absence of peptides is also determined and subtracted from the value obtained in the presence of peptides.

Using this method, Schlegel and Wade[22] have found that the 25-amino acid peptide corresponding to the N terminus of the vesicular stomatitis virus envelope glycoprotein (G) causes hemolysis of sheep erythrocytes below pH 6.5. This pH range corresponds to that for induction of the cell–cell fusion[23] and hemolytic[24] activities of the virus, as well as the range where virus–cell fusion is observed.[25]

Conductance Changes in Planar Bilayers

Insertion of peptides into planar bilayers composed of phospholipids leads to conductance fluctuations and an overall increase in conductance.[26] These measurements are recommended only for investigators who have access to instruments already set up in laboratories specialized in planar bilayer studies.

Solvent-free planar membranes are formed across a hole of about 80-μm diameter made in a thin Teflon sheet (25-μm thickness) by electric discharge. Two monolayers are spread on aqueous solutions [100 mM KCl, 10 mM N - 2 - hydroxyethylpiperazine - N' - 2 - ethanesulfonic acid (HEPES), pH 7.0] on both sides of the Teflon septum, and the solvent used for phospholipids (usually hexane) is allowed to evaporate. The levels of the solutions are raised by micrometer-driven syringes, and the two mono-layers are apposed across the hole.[27,28] The membrane conductance is measured by a conventional voltage clamp system. The signal is filtered with a low-pass filter of 2-kHz bandwidth, processed by a Sony (Ridge Park, NJ) PCM-F1, and stored on video recorder tapes. The signal is monitored continuously on an oscilloscope and an oscillographic paper recorder (Hewlett Packard 7402A, Palo Alto, CA).

A synthetic peptide (HA2.7) corresponding to the seven amino acids of the N terminus of HA$_2$ (X-31 strain) was found to induce conductance fluctuations in planar bilayers at neutral pH.[26] Mutant peptides, in which the glycines at the N terminus or the 4-position were replaced with glutamic acid, were considerably less effective in causing conductance changes

[23] J. White, K. Matlin, and A. Helenius, *J. Cell Biol.* **89,** 674 (1981).
[24] K. Mifune, M. Ohuchi, and K. Mannen, *FEBS Lett.* **137,** 293 (1982).
[25] R. Blumenthal, A. Bali-Puri, A. Walter, D. Covell, and O. Eidelman, *J. Biol. Chem.* **262,** 13614 (1987).
[26] N. Düzgüneş and F. Gambale, *FEBS Lett.* **227,** 110 (1988).
[27] M. Montal, this series, Vol. 32, p. 545.
[28] F. Gambale, A. Menini, and G. Rauch, *Eur. Biophys. J.* **14,** 369 (1987).

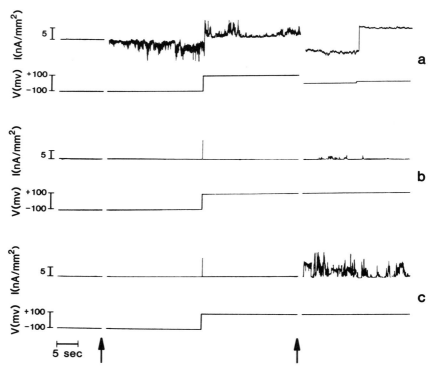

FIG. 1. Typical current measurements of planar asolectin membranes in the presence of influenzavirus hemagglutinin (HA$_2$) N-terminal peptides and its mutants. Peptides are added to the chamber on one side of the membrane (the cis side) at a concentration of 50 μg/ml (arrow on left). (a) Peptide HA2.7 (Gly-Leu-Phe-Gly-Ala-Ile-Cys); (b) peptide HA2.7mu1 (in which the terminal Gly is replaced with Glu); (c) peptide HA2.7mu4 (in which the Gly at position 4 is replaced with Glu). The initial applied voltage, shown as the lower trace in each section, was -100 mV, the sign of the voltage being that of the trans compartment with respect to the cis side. The arrow on the right indicates the trace 8 min (a), 55 min (b), or 11 min (c) after the addition of the peptides. (Reproduced with permission from Düzgüneş and Gambale.[26])

(Fig. 1), analogous to the impaired fusion activity of HA$_2$ molecules with identical mutations.[29] A 22-amino acid synthetic peptide corresponding to the N terminus of the HIV envelope glycoprotein gp41 appeared to form channels in planar bilayers.[30] It is possible that during the fusion of the

[29] M.-J. Gething, R. W. Doms, D. York, and J. M. White, *J. Cell Biol.* **102,** 11 (1986).
[30] V. A. Slepushkin, S. M. Andreev, M. V. Sidorova, G. B. Melikyan, V. B. Grigoriev, V. M. Chumakov, A. E. Grinfeldt, R. A. Manukyan, and E. V. Karamov, *AIDS Res. Hum. Retroviruses* **8,** 9 (1992).

virus with a target membrane, several N termini interact with the membrane to form the beginnings of a fusion pore, as observed during the fusion of influenzavirus hemagglutinin-expressing cells with erythrocytes.[31]

For experiments with planar bilayers, the use of peptides dissolved in certain organic solvents, such as dimethyl formamide or ethanol–guanidine hydrochloride, as stock solutions for subsequent injection into the aqueous compartments is not advisable, because the solvent itself can affect the sensitive conductance measurements. For this reason, the experiments reported above[26] were performed with sonicated dispersions of the peptides (in the same aqueous medium used for the bilayer experiments). Although this method may work for peptides that dissolve in aqueous solutions, it may not be useful for highly hydrophobic peptides that dissolve only in organic solvents. The inability of a 17-amino acid HA_2 peptide (HA2.17) to induce significant conductance changes[26] may have been the result of its poor solubility in the aqueous buffer. In the study of Slepushkin et al.,[30] the use of dimethyl sulfoxide (DMSO) did not appear to affect the conductance of planar bilayers.

Release of Aqueous Contents

The insertion of peptides into phospholipid vesicles (liposomes) can result in the breakdown of the permeability barrier of the membranes. The permeability of low molecular weight solutes across the membrane can be determined conveniently by measuring the release of fluorescence markers encapsulated in the liposomes.[26,32–35] A solution of the fluorophore 1-aminonaphthalene-3,6,8-trisulfonic acid (ANTS; 12.5 mM) and its collisional quencher N,N'-p-xylylenebis(pyridinium bromide) (DPX; 45 mM), buffered to pH 7.4 with 10 mM N-tris(hydroxymethyl)methyl-2-aminoethanesulfonic acid (TES), is coencapsulated in large unilamellar liposomes.[36,37] The osmolality of the solution is adjusted with NaCl to that of the medium buffer by means of a vapor pressure osmometer (Wescor Instruments, Logan, UT). Large unilamellar vesicles (LUVs) are prepared by reversed-phase evaporation, followed by extrusion four times through

[31] A. E. Spruce, A. Iwata, J. M. White, and W. Almers, Nature (London) 342, 555 (1989).

[32] C. Kayalar and N. Düzgüneş, Biochim. Biophys. Acta 860, 51 (1986).

[33] K. Shiffer, S. Hawgood, N. Düzgüneş, and J. Goerke, Biochemistry 27, 2689 (1988).

[34] D. W. Hoyt and L. M. Gierasch, Biochemistry 30, 10155 (1991).

[35] M. Rafalski, A. Ortiz, A. Rockwell, L. C. van Ginkel, J. D. Lear, W. F. DeGrado, and J. Wilschut, Biochemistry 30, 10211 (1991).

[36] H. Ellens, J. Bentz, and F. C. Szoka Biochemistry 23, 1532 (1984).

[37] N. Düzgüneş, R. M. Straubinger, P. A. Baldwin, D. S. Friend, and D. Papahadjopoulos, Biochemistry 24, 3091 (1985).

polycarbonate membranes (Nuclepore, Pleasanton, CA) of 100-nm pore diameter.[38-40] The size distribution of the vesicles is determined by dynamic light scattering in a Coulter (Hialeah, FL) N4 MD submicron particle analyzer. The mean diameter of the vesicles prepared in this way is about 150 nm. Unencapsulated material is separated from the liposomes by gel filtration on a Sephadex G-75 (Pharmacia) column, using medium buffer for elution.

When ANTS and DPX are released into the medium the fluorescence of ANTS increases, because DPX is not effective as a quencher when it is diluted. The excitation monochromator of the fluorometer [e.g., Spex (Edison, NJ) SLM4000 (Urban, IL), or Perkin-Elmer (Norwalk, CT) LS-5B] is set to 360 nm, and the fluorescence above 530 nm is monitored in the emission channel by using a Corning (Corning, NY) 3-68 high-pass filter. The fluorescence can also be followed by setting the monochromator to 530 nm with large slit widths to maximize light intensity. For experiments on peptide-induced leakage or release of contents, the fluorescence scale is set to 100% (F_{max}) by lysing the vesicles with n-dodecyl octaethylene monoether ($C_{12}E_8$) at a final concentration of 0.8 mM. The residual fluorescence of the intact vesicles (which is similar to that of the buffer) is set to 0% fluorescence.

An alternative method for measuring the release of aqueous contents from liposomes entails the use of calcein encapsulated at a self-quenched concentration.[32,35,41] For experiments to be performed in 100 mM NaCl or KCl as the external medium, about 40 mM calcein is used,[35] whereas for experiments at 150 mM salt, either the calcein concentration is raised to 60 mM[41] or the osmolality of the solution is increased by adding NaCl to the solution of calcein. The excitation monochromator is set at 490 nm, and the emission monochromator is set at 520 nm. Alternatively, a high-pass filter, such as the Corning 3-68, can be used in the emission channel.

Phospholipids used for liposome preparations are obtained from Avanti Polar Lipids (Alabaster, AL), and stored in chloroform under argon at $-70°$. The detergent $C_{12}E_8$ is from Calbiochem (San Diego, CA). The fluorophores ANTS and calcein, and DPX are obtained from Molecular Probes (Eugene, OR).

Investigators may have peptides synthesized by any of several companies specializing in peptide synthesis (e.g., Peninsula Laboratories, Belmont, CA; Multiple Peptide Systems, San Diego, CA; Neosystem Labora-

[38] F. Szoka, F. Olson, T. Heath, W. Vail, E. Mayhew, and D. Papahadjopoulos, *Biochim. Biophys. Acta* **601**, 559 (1980).
[39] N. Düzgüneş, J. Wilschut, K. Hong, R. Fraley, C. Perry, D. S. Friend, T. L. James and D. Papahadjopoulos, *Biochim. Biophys. Acta* **732**, 289 (1983).
[40] N. Düzgüneş and J. Wilschut, this series, Vol. 220 [1].
[41] R. M. Straubinger, N. Düzgüneş, and D. Papahadjopoulos, *FEBS Lett.* **179**, 148 (1985).

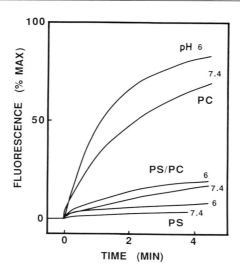

FIG. 2. Influence of lipid composition and pH on the kinetics of HA2.17-induced release of ANTS–DPX from large unilamellar vesicles. The sequence of HA2.17 is Gly-Leu-Phe-Gly-Ala-Ile-Ala-Gly-Phe-Ile-Glu-Asn-Gly-Trp-Glu-Gly-Cys). The peptide concentration is 6.7 μg/ml, and the lipid concentration is 50 μM. PS, Phosphatidylserine; PC, phosphatidylcholine. (Reproduced with permission from Düzgüneş and Shavnin.[19])

toire, Strasbourg, France). The purity of the peptides is assessed by high-performance liquid chromatography (HPLC) analysis. A more thorough analysis of purity involves mass spectrometry. In the case of some hydrophobic peptides, it may be impossible to dissolve the peptide in solvent systems used for HPLC analysis. Amino acid analysis should indicate the expected composition. One of the best solvent systems for hydrophobic peptides is $2\,M$ guanidine hydrochloride/50% (v/v) ethanol.[26] Other solvents that can be used include dimethyl sulfoxide,[42,43] 5 mM formic acid–ammonia buffer,[44] and dimethyl formamide. The stock peptide solution is diluted 100-fold or more when injected into the buffer containing the liposomes in the fluorometer cuvette. Additions of the solvent alone, or solvent plus a hydrophilic control peptide, to the liposomes should not produce any significant leakage from the vesicles.

Figure 2 gives examples of the results obtained with the peptide HA2.17, corresponding to the 17 amino acids of the N terminus of the

[42] J. D. Lear and W. F. De Grado, *J. Biol. Chem.* **262,** 6500 (1987).
[43] K. N. J. Burger, S. A. Wharton, R. A. Demel, and A. Verkleij, *Biochim. Biophys. Acta* **1065,** 121 (1991).
[44] M. Murata, Y. Sugahara, S. Takahashi, and S.-I. Ohnishi, *J. Biochem. (Tokyo)* **102,** 957 (1987).

cleaved hemagglutinin (HA$_2$) of the X-31 strain of influenzavirus.[19] Shorter peptides of 7 or 10 amino acids are not as effective as this peptide. Decreasing the pH to 6 enhances the release of contents slightly. Figure 2 also shows the effect of membrane composition on peptide-induced membrane destabilization. Liposomes composed of phosphatidycholine are most sensitive to the effect of HA2.17, compared to mixed phosphatidylserine–phosphatidylcholine or pure phosphatidylserine liposomes. Leakage induced by N-terminal peptides of the simian immunodeficiency virus gp32 had a different peptide length dependence, decreasing with increasing peptide length.[45] Membrane composition also affected the leakage process in this system, with membranes containing phosphatidylethanolamine displaying the highest extent of leakage. Using a mathematical model for leakage kinetics, Parente *et al.*[46] proposed that a critical number of the synthetic amphipathic peptide GALA assemble into a supramolecular aggregate, forming a transbilayer channel. They suggested that the theoretical analysis may be of general use in defining the state of aggregation of peptides that act in a similar manner.

In many cases, peptide–liposome interactions do not result in the complete release of the internal aqueous contents. One reason for this may be the possibility that several peptides need to form an intermolecular complex before the membrane is destabilized, and that this is a dynamic process.[26,46] Long-term observation is likely to reveal a more extensive release of contents.

Membrane Adhesion and Fusion: Lipid Mixing Assays

Aggregation and fusion of liposomes induced by peptides can be monitored conveniently by an assay measuring the proximity of fluorescent phospholipids.[47,48] This is achieved by measuring resonance energy transfer between N-(7-nitrobenz-2-oxa-1,3-diazol-4-yl)phosphatidylethanolamine (NBD-PE) and N-(lissamine) rhodamine B sulfonyl PE (Rh-PE) incorporated into the liposome membrane, because resonance energy transfer has a sharp dependence ($1/r^6$) on the distance r between the probes.

Aggregation or adhesion of liposomes can be measured sensitively by the "probe mixing" configuration of the assay,[49,50] involving two popula-

[45] I. Martin, F. Defrise-Quertain, V. Mandieau, N. M. Nielsen, T. Saermark, A. Burny, R. Brasseur, J.-M. Ruysschaert, and M. Vandenbranden, *Biochem. Biophys. Res. Commun.* **175,** 872 (1991).

[46] R. A. Parente, S. Nir, and F. C. Szoka, Jr., *Biochemistry* **29,** 8720 (1990).

[47] D. K. Struck, D. Hoekstra, and R. E. Pagano, *Biochemistry* **20,** 4093 (1981).

[48] D. Hoekstra and N. Düzgüneş, this series, Vol. 220 [2].

[49] D. Hoekstra, *Biochim. Biophys. Acta* **692,** 171 (1982).

[50] N. Düzgüneş, T. M. Allen, J. Fedor, and D. Papahadjopoulos, *Biochemistry* **26,** 8435 (1987).

tions of vesicles, one containing 1 mol% of NBD-PE and the other containing 1 mol% of Rh-PE. Interbilayer adhesion results in the close proximity of the probes, and the efficiency of energy transfer increases. If membrane fusion follows the adhesion step, energy transfer will be enhanced, because more of the probes will be intermixed, both in the outer and inner monolayers of the liposome membrane.

To exclusively measure resonance energy transfer resulting from lipid mixing during membrane fusion (and not due to adhesion), the "probe dilution" configuration of the assay is used.[37,47,50,51] Three populations of vesicles are prepared for this assay: (1) labeled liposomes, containing 1 mol% of each of Rh-PE and NBD-PE (Avanti Polar Lipids), (2) unlabeled liposomes, and (3) liposomes to be used for the calibration of fluorescence to 100%, containing 0.2 mol% of each probe. The lipid concentration of the liposome preparation is determined by phosphorus analysis.[52] Labeled vesicles are mixed with unlabeled vesicles at a 1 : 4 ratio, at a total lipid concentration of 50 nmol/ml in 1 ml 100 mM NaCl, 5 mM TES, 5 mM morpholineethanesulfonic acid (MES), 5 mM citrate, pH 7.4 at 25°. If the lipids of the labeled and unlabeled liposomes intermixed completely, a "megavesicle" containing 0.2 mol% of each probe would be produced.[37] Thus, the fluorescence intensity of liposomes (50 nmol lipid/ml) containing 0.2 mol% of each probe is set to 100% (designated F_{max}). The fluorescence intensity of the labeled vesicles is taken as 0%. Dilution of the probe molecules from labeled liposomes into unlabeled liposomes results in a decrease in efficiency of energy transfer from NBD to Rh, resulting in an increase in NBD fluorescence intensity. The probe dilution method is a reliable indicator of lipid mixing during membrane fusion, and is not affected by the aggregation of liposomes.[50] Fluorescence measurements are made in an SLM 4000, Spex, or Perkin Elmer LS-5B fluorometer. The excitation wavelength is set to 450 nm, and the emission wavelength to 520 nm, with the monochromator slits set at 4 nm. The fluorometer cuvette is stirred continuously and thermostatted to the desired temperature.

Using the "probe mixing" assay, the seven amino acid N-terminal peptide of the vesicular stomatitis virus G protein (G.7) was found to decrease the fluorescence of labeled phosphatidylserine liposomes, indicating that the liposomes aggregated and possibly intermixed their lipids.[19] The initial rate and extent of probe mixing increased as the pH was decreased. With the "probe dilution" method, a low level of lipid mixing was observed at pH 7.4 in the presence of G.7, and the initial rate and extent of lipid mixing increased at pH 6.0, with no further increase at

[51] J. Rosenberg, N. Düzgüneş, and C. Kayalar, *Biochim. Biophys. Acta* **735,** 173 (1983).
[52] G. R. Bartlett, *J. Biol. Chem.* **234,** 466 (1959).

pH 5.0. The change in fluorescence intensity at low pH was considerably lower than that observed in the probe mixing assay.

The 20-amino acid peptide (H-20) corresponding to the HA_2 N terminus of influenzavirus (B/Lee/40 strain) induced the fusion of small unilamellar liposomes composed of palmitoyloleoylphosphatidylcholine, as determined by the "probe dilution" assay.[42] However, the fusion activity was independent of pH. A shorter peptide (16 amino acids) did not induce fusion. The 17-amino acid peptide from the HA_2 of the X-31 strain did not induce any fusion of small unilamellar phosphatidylcholine liposomes.[26] Martin et al.[45] have reported that peptides corresponding to the N-terminal sequence of the simian immunodeficiency virus transmembrane protein gp32 induced fusion of small unilamellar liposomes composed of dioleoylphosphatidycholine, or a mixture of phospholipids.

Several studies on the fusion activity of viral peptides have used small unilamellar liposomes, in some cases with different fusion assays than the one described above.[42,44,53] Because of their inherent lipid packing defects, such liposomes are considerably more susceptible to fusion than large unilamellar liposomes.[10,54,55] A 20-amino acid peptide from the X-31 HA_2 N terminus did not induce the fusion of large unilamellar palmitoyloleoylphosphatidylcholine vesicles (unpublished data of Wilschut et al., cited in Rafalski et al.[35]) The HA2.17 peptide described above also did not mediate the fusion of large unilamellar liposomes composed of phosphatidylcholine or phosphatidylserine.[19] On the other hand, large liposomes containing phosphatidylethanolamine are indeed prone to fusion by peptides under conditions that do not induce the fusion of liposomes without the phosphatidylethanolamine[45] (Wilschut et al., quoted in Rafalski et al.[35]). The presence of this lipid in membranes has been shown previously to render the membranes susceptible to fusion by other agents such as calcium, magnesium, and protons.[9,37,56,57]

Membrane Fusion: Intermixing of Aqueous Contents

Intermixing of aqueous contents during membrane fusion is monitored by the ANTS–DPX assay.[37,30,58] One population of liposomes contains

[53] S. A. Wharton, S. R. Martin, R. W. H. Ruigrok, J. J. Skehel, and D. C. Wiley, J. Gen. Virol. 69, 1847 (1988).
[54] J. Wilschut, N. Düzgüneş, and D. Papahadjopoulos, Biochemistry 20, 3126 (1981).
[55] R. A. Parente, S. Nir, and F. C. Szoka, Jr., J. Biol. Chem. 263, 4724 (1988).
[56] N. Düzgüneş, J. Wilschut, R. Fraley, and D. Papahadjopoulos, Biochim. Biophys. Acta 642, 182 (1981).
[57] N. Düzgüneş, K. Hong, P. A. Baldwin, J. Bentz, S. Nir, and D. Papahadjopoulos, in "Cell Fusion" (A. E. Sowers, ed.), p. 241. Plenum, New York, 1987.
[58] H. Ellens, J. Bentz, and F. C. Szoka, Biochemistry 24, 3099 (1985).

25 mM ANTS and the other 90 mM DPX, with 10 mM TES, ph 7.4. The osmolality is adjusted to that of the medium buffer, using NaCl. Unencapsulated material is separated from the liposomes as described above. The two populations are mixed at a 1:1 ratio, to a final lipid concentration of 50 nmol/ml. The ANTS fluorescence is measured using a Corning 3-68 high-pass filter (>530 nm), with the excitation monochromator set at 360 nm. The initial fluorescence of the vesicle mixture is taken as 100% (corresponding to 0% fusion). The fluorescence of vesicles containing coencapsulated ANTS–DPX is set to 0% (corresponding to 100% fusion, a hypothetical situation in which the contents of all the vesicles have intermixed).

This assay has been used to investigate whether the seven-amino acid peptide of vesicular stomatitis virus G protein (G.7), which caused aggregation and lipid mixing between phosphatidylserine liposomes, indeed caused the intermixing of aqueous contents of such liposomes. The results indicate no contents mixing.[19] This observation emphasizes the importance of monitoring contents mixing as well as lipid mixing to establish the occurrence of membrane fusion. Probe dilution may be an indicator of the intermixing of the outer monolayers of labeled and unlabeled liposomes, a process that does not necessarily result in aqueous contents mixing.[51,59] It is also possible, however, that leakage of contents is so rapid that it precludes the observation of intermixing of aqueous contents.[19,60] Nevertheless, some contents mixing has been observed even in systems in which leakage is rapid.[37] In ambiguous cases, it is advisable to perform negative-stain, thin-section, or freeze–fracture electron microscopy.[37,61]

Tryptophan Fluorescence

For peptides containing tryptophan, the interaction of the peptide with membranes can be characterized by measuring the wavelength shift of the tryptophan emission spectrum in the presence of liposomes.[35,42] Liposomes are prepared by the methods described above and elsewhere in this series.[40] Peptides are added from a concentrated stock solution. The peak of tryptophan fluorescence at 346 nm, observed with free peptide in solution,

[59] N. Düzgüneş and J. Bentz, in "Spectroscopic Membrane Probes" (L. M. Loew, ed.), Vol. 1, p. 117. CRC Press, Boca Raton, FL, 1988.

[60] I. Martin, F. Defrise-Quertain, N. M. Nielsen, T. Saermark, A. Burny, R. Brasseur, M. Vandenbranden, and J.-M. Ruysschaert, in "Membrane Interactions of HIV: Implications for Pathogenesis and Therapy in AIDS" (R. C. Aloia and C. C. Curtain, eds.), p. 365. Wiley-Liss, New York, 1992.

[61] C. E. Larsen, S. Nir, D. R. Alford, M. Jennings, K.-D. Lee, and N. Düzgüneş, Biochim. Biophys. Acta, in press.

shifts to lower wavelengths when liposomes are added. By titrating the amount of lipid, it is possible to determine the number of lipids per peptide at saturation, and an apparent dissociation constant.[35,42,62]

Circular Dichroism

Circular dichroism spectra can reveal the conformation of peptides in solution or in association with liposomes. The ellipticity at 222 nm is a measure of the helical content of the peptide.[35,42,63] A flat spectrum between 210 and 250 nm is indicative of the absence of a unique folded conformation.[42] Minima at 222 and 208 nm are observed when the peptide adopts an α-helical conformation.[35,42,63] A 20-amino acid HA$_2$ N-terminal peptide displayed random coil conformation in solution, while adopting a partially α-helical conformation in the presence of small unilamellar liposomes.[35] Light-scattering artifacts are minimized by the use of small liposomes rather than large ones. Solvents used to dissolve peptides may present difficulties, which should be ascertained. For example, in one study, hexafluoroacetone hydrate had to be substituted for DMSO.[42] The use of solvents was eliminated altogether in another investigation, in which small liposomes were prepared from a dried mixture of lipid and peptide (the peptide in hexafluoropropanol was mixed with the lipid in chloroform, the mixture was dried under nitrogen and vacuum, and liposomes were formed by hydrating and sonicating the dry film).[35]

Concluding Remarks

Fusion induced by N-terminal peptides does not necessarily reflect the fusogenic activity of viral fusion proteins. Site-directed mutagenesis studies have shown that the N termini of influenzavirus HA$_2$ and of HIV-1 envelope glycoprotein gp41 are clearly involved in the fusion of these viruses.[29,64] However, the remainder of the protein and the other members of the oligomeric structure formed by the proteins are also likely to be involved in the fusion process. Thus, studies on the membrane-destabilizing effects of the N-terminal peptides, and on the molecular and structural requirements for these effects, appear to be more relevant to understanding how these segments of the viral envelope proteins mediate membrane

[62] J. Dufourcq and J.-F. Faucon, *Biochim. Biophys. Acta* **467**, 1 (1977).

[63] N. K. Subbarao, R. A. Parente, F. C. Szoka, Jr., L. Nadasdi, and K. Pongracz, *Biochemistry* **26**, 2964 (1987).

[64] M. Kowalski, J. Potz, L. Basiripour, T. Dorfman, W. C. Goh, E. Terwilliger, A. Dayton, C. Rosen, W. Haseltine, and J. Sodroski, *Science* **237**, 1351 (1987).

fusion than are investigations of the induction of membrane fusion by these peptides.

Acknowledgment

This work was supported by National Institutes of Health Grant AI-25534.

Section III

Membrane Fusion during Exocytosis

[8] Simultaneous Electrical and Optical Measurements of Individual Membrane Fusion Events during Exocytosis

By JOSHUA ZIMMERBERG

Membrane fusion is characterized by two criteria: (1) the melding of two phospholipid bilayer membranes and (2) the mixing of the two aqueous compartments previously separated by the two membranes. To understand the mechanism of membrane fusion, it is essential to understand that the two phenomena are usually evaluated separately, by distinct techniques. Only by using multiple methods simultaneously can one construct a pathway for intermediates in fusion. We have learned that flaccid vesicles can fuse, that fusion pore widening is not accompanied by morphological changes, and that swelling follows fusion, all by following fusion with more than one technique simultaneously.[1] In this chapter we summarize which steps in fusion are measured by optical, and then by electrical, measurements. We then describe the methods we have used to combine these two types of assays. This chapter concentrates on the techniques required to measure both electrical and optical signals in a coordinated fashion.

Optical Measurements of Membrane Fusion

Optical measurements of membrane fusion have, historically, provided the oldest evidence for fusion. Syncytium formation, fertilization, trichocyst release, and zymogen granule release were appreciated prior to the physiological elucidation of the structure of biological membranes. This most dramatic consequence of fusion consists of changes in the image of the system that reflect the events finalizing fusion: the mixing of enclosed compartments with each other (multinuclear cell formation) or with the extracellular space (secretion). In many cases, swelling of the secretory granule on contact with the external fluid is the first visual indication of exocytosis.[2] It is often followed by movement of granule contents out of a cell. The methods for good visualization of content release are light scattering[3] and differential-interference contrast microscopy (DIC).[4]

---

[1] J. Zimmerberg, *Biosci. Rep.* **7**, 251 (1987).

[2] A. Finkelstein, J. Zimmerberg, and F. S. Cohen, *Annu. Rev. Physiol.* **48**, 163 (1986).

[3] J. Zimmerberg, C. Sardet, and D. Epel, *J. Cell Biol.* **101**, 2398 (1985).

[4] S. Inoué, "Video Microscopy." Plenum, New York, 1986.

METHODS IN ENZYMOLOGY, VOL. 221

Fluorescence microscopy can detect either membrane mixing or content mixing, as we have detailed in another chapter of this volume.[5] In measurements of dequenching, video recordings are often inferior to photomultiplier measurements, which can have better time resolution and lower noise. This is especially true when electrical measurements are compared to indicators of either pH or calcium.[6] Although electron microscopy is valuable for detection of the omega figure in small granules such as synaptic vesicles, and for detection of small pores that can widen to finalize fusion, these still images cannot be correlated to dynamic electrical changes in single cells.

Electrical Measurements of Membrane Fusion

Electrical measurements are of voltage and current. We must interpret these measurements to reflect fusion. Because new membrane is incorporated during fusion, and membranes are semipermeable, changes in ion permeability (conductance) can result from fusion. For example, ionic channels can be incorporated into vesicles, and these vesicles can be fused to planar bilayers. Fusion is then measured as an increase of membrane current as channels are incorporated and membrane conductance increases.[7] Another electrical sequela of new membrane incorporation is an increase in total membrane surface area. Because bilayer membranes have nearly the same thickness, this increase in area results in a proportional increase in membrane capacitance. Fusion is then measured as an increase in current due to an increase in membrane capacitance.[8] New membrane must be charged to the same potential as the original membrane. Fusion is then measured as a current transient: the movement of ionic charge as the new membrane equilibrates.[9,10] Finally, the contents of a vesicle can be oxidized or reduced on contact with an external electrode.[11] Fusion is then measured as the current that results from release of vesicular contents.

Capacitance Measurements

Membrane capacitance can be measured in a number of ways, using the formula $I = C \, dV/dt$, where I is current, C is capacitance, V is voltage, and t is time. By using a voltage clamp, the transmembrane potential can

[5] S. J. Morris, J. Zimmerberg, D. P. Sarkar, and R. Blumenthal, this volume, [4].
[6] E. Neher, *J. Physiol. (London)* **395**, 193 (1988).
[7] F. S. Cohen, J. Zimmerberg, and A. Finkelstein, *J. Gen. Physiol.* **75**, 251 (1980).
[8] L. A. Jaffe, S. Hagiwara, and R. T. Kado, *Dev. Biol.* **67**, 243 (1978).
[9] L. J. Breckenridge and W. Almers, *Nature (London)* **328**, 814 (1987).
[10] A. E. Spruce, L. J. Breckenridge, A. K. Lee, and W. Almers, *Neuron* **4**, 643 (1990).
[11] J. Millar and G. V. Williams, *J. Electroanal. Chem.* **282**, 33 (1990).

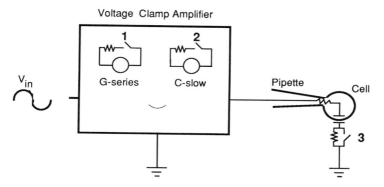

FIG. 1. Modifications to an EPC-7 patch-clamp amplifier (Adams and List, Westbury, NY) that simplify capacitance calibration and phase estimation. The potentiometers for the capacitance and series resistance compensation circuits are shown as dials. A series circuit consisting of a reed relay and a resistor are added in parallel to the potentiometers that control the series resistance (G-series) (1) and capacitance (C-slow) (2) compensations. A 1-MΩ resistor (3) is added in series to the ground electrode for phase tracking. (Modified from Fidler and Fernandez.[15])

be controlled. The capacitive current that results from a step in voltage can be fit to yield capacitance.[12] If a linear ramp of voltage (dV/dt is a constant) is applied across the membrane, a constant current is produced that is proportional to capacitance. However, the most elegant approach to measuring capacitance has been in conjunction with the whole-cell recording technique.[13] Here, a glass pipette is placed on the cell surface, a seal between the glass and the membrane is attained, and then the small patch of membrane in the lumen of the pipette is disrupted to allow fluid continuity between the pipette interior and the cell interior (Fig. 1). If the voltage applied to the pipette is a sine wave, the resulting capacitive current is a cosine wave. In general, membranes have both capacitance and conductance, therefore the response of a cell to a sine wave (of frequency f) also includes a component that follows Ohm's law, $I = G_{ac}V$, where G_{ac} is conductance. Thus $I = G_{ac}\sin(\omega t) + C\cos(\omega t)$ where ω is $2\pi f$. To distinguish between changes in capacitance and conductance, a lock-in amplifier can be used to generate signals that are proportional to *small* changes in G_{ac} and C, if the appropriate phase is determined.[13] By using the whole-cell recording mode, phase is determined by the cell capacitance and the pipette series resistance. In practice, one first achieves the whole-cell configuration, applies a small pulse of voltage, and uses the compensation

[12] M. Landau and E. Neher, *Eur. J. Physiol.* **411**, 137 (1988).
[13] E. Neher and A. Marty, *Proc. Natl. Acad. Sci. U.S.A.* **79**, 6712 (1982).

circuits on the amplifier to measure and null the capacitance and series resistance. A sine wave of voltage is then applied to the pipette. The phase is determined by increasing and decreasing cell capacitance slightly with the capacitance compensation circuit, while simultaneously changing the phase of the lock-in amplifier until there is no change in the sine output with a change in capacitance. Then the cosine output is proportional to capacitance *for small changes in capacitance.* An alternative method for setting the phase can also be used, in which the response of the cosine signal is nulled in response to a change in series resistance compensation.[14] However, there are no significant differences in the methods for changes in capacitance less than 7 pF, which is often the case. Because the series resistance of a pipette can change with time after attaining the whole-cell mode (usually increasing as pipettes clog), phase and calibration must be rechecked after each series of exocytotic events from each cell, and data must be discarded if a significant change is detected.

Two modifications to the method of Neher and Marty[13] can reduce the time for calibration and determination of phase. First, using relays, Fidler and Fernandez[15] modified the compensation circuits to add small, fixed, computer-controlled amounts of capacitance and conductance to the whole-cell capacitance and series conductance compensation circuits (Fig. 1). Second, a relay can be used to switch in and out a 1-Ω resistor in series with the bath reference electrode. From this known change in series resistance, the factor by which calibration is altered during the experiment can be calculated.[15] These modifications are both of great practical value. Zieler has recently published a method for rapid setting of the phase angle.[15a]

Fusion Pore Determination

When patch clamped in the whole-cell recording mode, the equivalent electrical circuit representing the cell during fusion of a single granule with the plasma membrane is shown in Fig. 2A. If the cell is clamped to a sine wave voltage of 300–3200 Hz, current flows predominantly through the capacitive components of the cell and granule membranes.[16] Thus the equivalent circuit reduces to the simplified circuit shown in Fig. 2B, which effectively describes the time course for the conductance of an exocytotic pore. The use of this equivalent circuit allows an analytical solution for the

[14] C. Joshi and J. M. Fernandez, *Biophys. J.* **53**, 885 (1988).
[15] N. H. Fidler and J. M. Fernandez, *Biophys. J.* **56**, 1153 (1989).
[15a] K. Zieler, *Biophys. J.* **63**, 854 (1992).
[16] J. Zimmerberg, M. Curran, F. S. Cohen, and M. Brodwick, *Proc. Natl. Acad. Sci. U.S.A.* **84**, 1585 (1987).

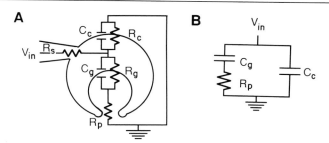

FIG. 2. (A) Diagram of an exocytotic pore that has resulted from fusion, and the appropriate equivalent circuit. R_c and C_c, resistance and capacitance of the cell membrane; R_g and C_g, resistance and capacitance of the granule membrane; R_p, resistance of the exocytotic pore; and R_s, series resistance of the patch electrode. (B) The reduced equivalent circuit. R_c and R_g are not significant for $f > 200$ Hz because $R_{c,g} \ll \frac{1}{2}\pi f C_{c,g}$. The change in admittance as a result of fusion is given in complex notation in text.

As a result of fusion, R_p varies from practically infinite, before fusion, to a small value after fusion. Fusion does not cause any net change in the real part of the admittance, G_{ac}. However, while the fusion pore is small enough to have a significant voltage drop across it, G_{ac} changes — it increases and then decreases. The maximum possible change (due to varying R_p) in the real component of the admittance equals $\omega C_g/2$, which is one-half the final change of the imaginary component, ωC_g. (Reproduced from Zimmerberg et al.[16])

pore conductance as a function of the measured cell current. For circuit analysis, it is convenient to use the terms of impedance and admittance (1/impedance), which are AC analogs of resistance and conductance.

The experimentally obtained electrical current gives the time course for the conductance of the exocytotic pore, as shown below.

The only component of Fig. 2B that changes during exocytosis is the resistance of the pore, R_p, which can be expressed as the conductance of the pore, $g_p = 1/R_p$. C_g, the capacitance of the fused granule, and g_p are in series and constitute a voltage divider. At the beginning of pore formation, g_p is small and the applied clamp potential, V_c, predominantly drops across it. As g_p increases (i.e., the lumen of the pore dilates) the voltage drop across C_g increases and the measured capacitive component of the admittance increases. When g_p is sufficiently large, the entire voltage drop occurs across the capacitor. Quantitatively the change in admittance during granule fusion, ΔY, is given in complex notation by

$$\Delta Y(\omega) = \frac{(\omega C_g/g_p)^2}{1 + (\omega C_g/g_p)^2} + j\frac{\omega C_g}{1 + (\omega C_g/g_p)^2}$$

where $j = -1^{1/2}$. The first term of the above equation, denoted by R, represents G_{ac}, whereas the second term, denoted as I, is the capacitance and is associated with the voltage drop across C_g. Each of these components is measured vs time as the output of a two-channel lock-in amplifier. Note that at the final stages of pore opening g_p is large, R equals 0, and I equals C_g. The conductance of the pore can be calculated from either the measured in-phase R component, denoted g_p^R, or the out-of-phase I component, written as g_p^I, by the equations

$$g_p^R = \frac{2R}{1 + n\,[1 - (2R/\omega C_g)^2]^{1/2}}$$

where when $n = 1$, $I < \omega C_g/2$, and when $n = -1$, $I > \omega C_g/2$, and

$$g_p^I = \frac{\omega C_g}{[(\omega C_g/I) - 1]^{1/2}}$$

Because $g_p = g_p^R = g_p^I$, the pore conductance can be calculated from either the capacitance or G_{ac}. Taking series resistance (R_s) into account, we obtain the equations

$$g_p^R = \frac{2R}{1 + n\,[1 - (2R/\omega C_g)^2]^{1/2} - 2RR_s} \tag{1}$$

where when $n = 1$, $I < \omega C_g/2$, and when $n = -1$, $I > \omega C_g/2$, and

$$g_p^I = \frac{\omega C_g}{[(\omega C_g/I) - 1]^{1/2} - \omega C_g R_s} \tag{2}$$

We now describe the computer programs that derive pore conductance from capacitance and G_{ac}. Capacitance and G_{ac} are digitized and subscripted to correspond to time, and are read into separate arrays, R and I. Each digitized point can be treated as a distinct measurement of admittance and used to calculate a distinct pore conductance by using simple Fortran programs (obtainable on request). The mean and standard deviation of the baseline data are calculated before the onset of fusion as well as after the end of fusion (Fig. 3). Next, using the experimental parameters, C_g is calculated as the difference between these two capacitance baselines.

Six parameters of noise are determined for each fusion event: the two pore conductances, from R and I, that correspond to a 2.34-standard deviation increase above baseline before fusion; the two pore conductances, from R and I, that correspond to a 2.34-standard deviation decrease in the respective baselines after the end of fusion; and two parameters from R equal to 2.34 standard deviations of the baseline noise (1) subtracted from and (2) added to the peak of R ($\frac{1}{2}\,\omega C_g$) (Fig. 3). The last two parameters are calculated because the transfer function for R to g_p^R is steep at the peak of R, so that g_p^R varies greatly with insignificant changes in R. We then reject points that fall within any of these experimental noise

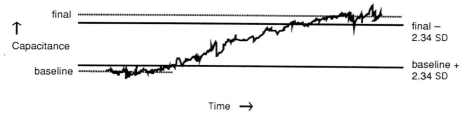

FIG. 3. Diagram of a noisy capacitance change. The baseline and final values are calculated from the mean of the values prior to and after fusion. Only points greater than 2.34 standard deviations (SD) above the baseline, and 2.34 SD below the final value, are used for fusion pore determination.

ranges, corresponding to significance at the 1% probability level, using a single-tailed Gaussian distribution.

Multiple Histograms Needed to Show Probabilities Properly

To study a large variety of pore conductances, one can stimulate cells with command voltages (V_{in}) of differing frequencies, because granule admittance is a function not only of size but of stimulating frequency as well. The experimentally controlled frequency of V_{in}, combined with the natural size distribution of granules, leads to a large range of granule impedances. Only pores comparable in conductance to the granule admittance can be measured. The admittance of each granule is given by $2\pi f C_g$, where f is the frequency of the stimulating voltage and C_g is the granule capacitance. Granules from beige mouse mast cells range in capacitance from 80 to 700 fF. To illustrate the consequences, a 100-fF granule stimulated at 800 Hz has an admittance of 0.5 nS, and may have a detectability range of 0.1 to 2.5 nS, whereas a 500-fF granule stimulated at 3200 Hz has an admittance of 10 nS, and a detectability range of 2 to 50 nS. These two granules would overlap only at 2 to 2.5 nS. The conductance corresponding to the baseline noise sets the limit of detecting small pores. Similarly, noise of the final level sets the limit of detecting large pores. However, because the pore conductance measurement is bounded by baseline noise as well as noise at the end of fusion, for each fusion event pore conductances can be detected only within a unique range, smaller than the range of all possible pore conductances. This heterogeneity could also bias histograms by creating artificial peaks. Data histograms based on model experiments in which pore conductance was linear with random noise indicate some of these biases. The individual histograms are flat, consistent with a flat distribution (Fig. 4a–e); however, each histogram is within different limits (x axis, Fig. 4a–e). When all of the data are combined, peaks emerge

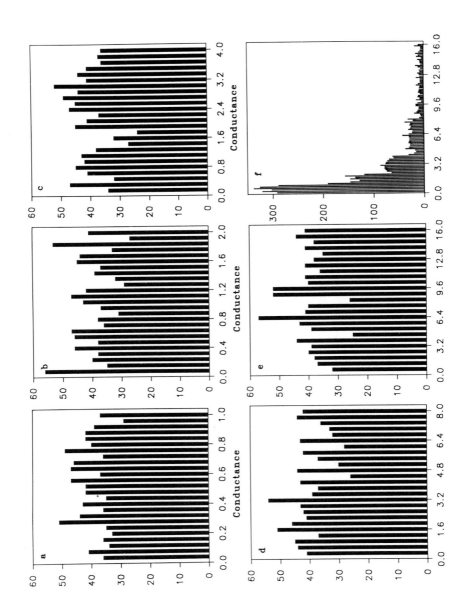

in the resulting histogram, although clearly they are artifacts (Fig. 4f). To avoid such spurious peaks, we prepare multiple histograms, each within a narrow detectability range, to avoid bias due to detection abilities[16a] (Fig. 5). Each histogram is generated over a 10-fold range (e.g., 0.4–4 nS and 0.8–8 nS), using only data from granules whose range of detectability encompasses the range of the histogram. In this way, each histogram reflects the probability of pore conductance, rather than, for example, the distribution of factors determining the noise. Significant peaks are seen in a broad distribution. Further filtering does not remove these peaks (Fig. 6). A new method for preparing composite histograms has recently been described.[16a]

Simultaneous Measurements

Equipment

The microscope must be fitted with suitable equipment for recording images. Usually research microscopes have options available for a C-mount, the usual video camera mounting adapter. For differential-interference contrast (Nomarski), a high-resolution, low light level video or "scientific" CCD camera is employed. For fluorescent work, we use a microchannel plate intensifier followed by a nuvicon–tube camera. It is important to choose a suitable optical device for the experiment. For example, if high spatial resolution is needed, then time resolution is sacrificed. If high time resolution is needed, then time resolution is sacrificed and a photomultiplier used instead of a camera. A new use for the CCD camera has been described that allows software control of the balance between time and spatial resolution.[17]

The microscope should also be equipped with a suitable manipulator for positioning of pipettes for the electrical recording. A variety of manipulators and mounting devices exist; because the whole-cell configuration

[16a] M. Curran, F. S. Cohen, D. E. Chandler, P. J. Munson, and J. Zimmerberg, *J. Membr. Biol.*, **133**(1), 61 (1993).
[17] N. Lasser-Ross, H. Miyakawa, V. Lev-Ram, S. R. Young, and W. N. Ross, *J. Neurosci. Methods* **36**, 253 (1991).

FIG. 4. A random number generator (MATLAB; The MathWorks, Inc., South Natick, MA) was used to create 5 arrays of 1000 numbers each, between 0 and 1.0 (a), 2.0 (b), 4.0 (c), 8.0 (d), and 16.0 (e). From these arrays, histograms were generated with 25 bins (a–e), to simulate the histogram expected from a flat distribution of fusion pore conductances, as one may expect from a fusion pore that grew at a rate linear in conductance. These arrays were combined to form one large array, the histogram shown in (f). As can be seen, this resulting histogram contains spurious peaks and is weighted to smaller values. To avoid this bias, one may combine measurements only within the same range of conductance values.

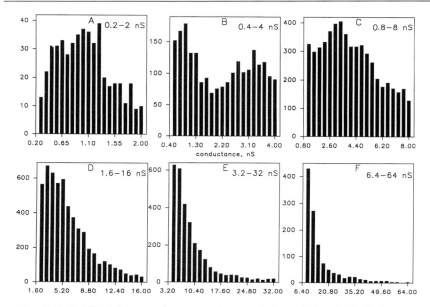

FIG. 5. Probability histogram for exocytotic pores between 0.2 and 64 nS. Pores were calculated from capacitance, as described in the text, from 224 fusion events in 65 beige mouse mast cells.[15a] Multiple histograms are prepared in a series of detectability ranges (windows) to avoid bias, as described in the text. The windows are given by the abscissa labels. Within a broad distribution between 1 and 5 nS there are broad peaks at 1 and 3–5 nS.

must be maintained for the duration of the experiment, one should select manipulators with minimal drift. It is useful to mount the manipulator and cell chamber together and separate them from the microscope. In this way, focusing of the microscope does not disturb the relative positions of the cell and pipette, which vibrate together. The entire rig should rest on a vibration isolation table. Care is taken to avoid oscillations induced by high center of gravity. This can be compensated for by either a greater distance between legs or by adding lead bricks to the table surface.

Recording Data

Two cables now lead from the rig: one containing electrical information and one containing images. How does one record and synchronize data? Although good computer software is available for on-line recording and analysis of either membrane currents or images, there are few hardware/ software combinations that integrate both kinds of signals. Accordingly, we have devised one effective solution (described below). There are undoubtedly other solutions, and the problems with the present design will be discussed. Most current trends in computing are toward virtual memory

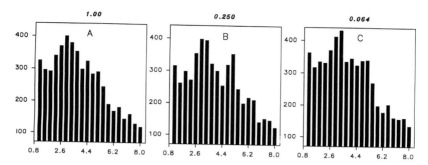

FIG. 6. Effect of filtering on pore conductance histograms. To determine if noise in these components was obscuring peaks in the pore conductance histograms, we digitally filtered both capacitance and G_{ac}, using an eight-pole butterworth filter (The Math Works, South Natick, MA). We digitally filtered at various fractions (A) 1.0, (B) 0.25, and (C) 0.064 of the original analog filtering frequency. Since many different frequencies were employed and granule size varied, histograms were constructed within limits determined by the baseline and final level noise, as described in the text. Only capacitance in the detectability range from 0.8 to 8 nS is shown here, but all ranges between 0.2 and 128 nS were calculated. In all cases, varying the digital filter frequency did not affect the shape of the histogram, did not make peaks in the histogram significantly more apparent, and did not alter the number of peaks. Thus the wide peaks and broad distributions that we found are not due to insufficient filtering.

operating systems, such as Unix; therefore real-time computing is poorly supported by commercial vendors.

We use tape recorders to record signals, because there is no way of knowing *a priori* when a single fusion event will occur. Although one may synchronize a population of cells to secrete via the addition of a secretagogue, the aim of single-event studies is to isolate the fusion of one granule from the others. Because the electrical measurements may last up to 15 sec, and the morphological events follow, we adjust the concentration of secretagogue to separate each event by minutes. This results in large amounts of data that can be accommodated only with tape media.

Synchronization of Electrical and Optical Data

We use two methods to align the two sets of data. First, an oscilloscope screen displaying the electrical data is recorded optically with a video camera. This image is then inserted into the optical video signal with a Burle "splitter/inserter" (Burle Industries, Lancaster, PA). It is important that the two video cameras be "genlocked" together, that is, begin each frame at the same instant. This can be accomplished by driving the "synch" pulses of the second video with those of the first, through an external connector on the splitter/inserter.

A video field has a duration of 17 msec, and consists of 256 lines, each lasting 63.5 μsec. If one uses an oscilloscope on which the trace persists for at least 17 msec, one can see the electrical data up to a time resolution limited by the electrical apparatus and the horizontal resolution of the video recorder. In practice, the electrical recordings of capacitance are filtered at 200 Hz. The morphological changes are usually limited by the video field, so that images 17 msec apart can be compared. In principle, one could use faster video cameras or scan one line over and over, to achieve better visual time resolution.

A second method of synchronizing data is to display a digitized voltage on the video screen. This is accomplished by constraining the date digits of a time/date generator to read the output of an analog/digital converter. The voltage acquisition is triggered at the beginning of each video field.

Results of Studies

We began these studies to learn if membrane stretching was required for biological membrane fusion. By choosing appropriate biological material, we could improve both the relative spatial resolution of an image and the electrical resolution of the capacitance measurement. A large secretory granule was needed for a good signal-to-noise ratio. The beige mouse mast cell is an animal model for the Chédiak-Higashi syndrome. It is a cell with granules whose diameters, 2–6 μm, are half that of the cell, 10 μm. In a series of experiments, we measured the DIC image of a beige mouse mast cell in the whole-cell recording configuration, and recorded both the image and the capacitance simultaneously.[16] The capacitance changes always preceded the secretory granule swelling; therefore stretching of the secretory granule membrane is not causal to fusion. In a control experiment, vesicles were made flaccid with hypertonic solution, and then the dual measurements were made again. Flaccidity was assured with measured reswelling of granules after hyperosmotic shrinkage. Once again, the capacitance changes preceded the swelling.

As an example, a simultaneous measurement of capacitance and DIC image of a beige mouse mast cell is seen in Fig. 7. The image of the patch pipette is not seen, because DIC images only a thin optical section. As can be seen, there are no morphological sequelae of a transient opening of the fusion pore, seen as an increase followed by a decrease in capacitance in the oscilloscope recording.

Conclusions

Simultaneous measurements can reveal the relative timing of different events in fusion, so that we can test sequential hypotheses of cause and effect. Electrical measurements are the first to change, followed by lipid

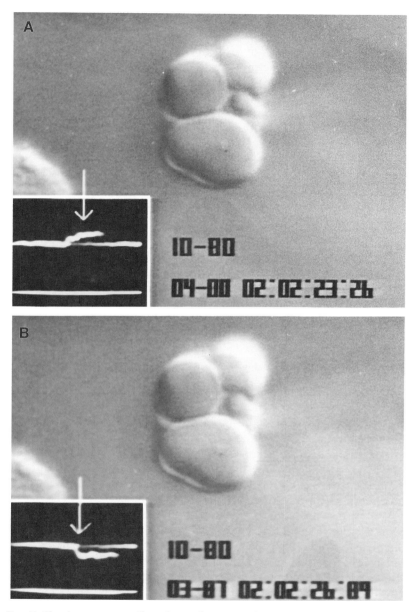

FIG. 7. Simultaneous recording of capacitance, conductance, and DIC image. A beige mouse mast cell is internally dialyzed with 150 mM potassium glutamate, 6 mM MgCl$_2$, 0.1 mM EGTA, 10 mM HEPES, 0.5 mM inosine triphosphate, and 2 μM GTPγS, pH 7.2. From left to right, one sees the oscilloscope tracing of capacitance (top), the digitized values of G_{ac} and capacitance, and the time. At 0.2:02:23:26, an increase in capacitance is seen (arrow). Capacitance returns to baseline 3.63 sec later (arrow). No morphological changes are seen, despite this transitory opening of the fusion pore. (Reproduced from J. Zimmerberg, S. S. Vogel, and L. V. Chernomordik, *Annu. Rev. Biophys. Biomol. Struct.* **22,** 433 (1993).)

diffusion, and then swelling of secretory granule contents. It is a powerful approach, but difficult. The more techniques one is using at one time, the greater the opportunity for an aborted experiment due to equipment failure or other technical difficulties. However, the information may not yet be available in any other way. Perhaps automation of some of the experimental manipulations or reduction of instruments to a single computer–user interface may enhance success.

Acknowledgments

I would like to thank Nancy Fidler Lim and Ongun Onaran for stimulating conversations about capacitance measurements, and Michael J. Curran for help with the flicker experiments.

[9] Visualization of Exocytosis by Quick Freezing and Freeze–Fracture

By Carrie J. Merkle and Douglas E. Chandler

Quick-Freezing Techniques

Exocytosis occurs extremely rapidly. Quick freezing in combination with freeze–fracture offers advantages for visualizing exocytosis in that it captures events occurring within milliseconds after stimulation and reveals panoramic views of membranes. Cells frozen ultrarapidly, that is, by removing heat so fast that ice crystals are undetectable by electron microscopy (EM),[1] can yield superbly preserved ultrastructure without membrane fusion-mimicking artifacts resulting from slow, selective chemical fixatives and membrane-disrupting cryoprotectants.

Cold metal block freezing or "slam" freezing is one method with proven success in capturing membrane fusion. Introduced in the 1940s,[2] its first use in preserving cells for electron microscopy came in 1964 when Van Harreveld froze liver on a silver block cooled to $-207°$ with N_2 slush.[3] However, the first truly successful cold metal block machine was developed in the late 1970s by Heuser and Reese.[4] This machine (Fig. 1) relies on

---

[1] J. C. Gilkey and L. A. Staehelin, *J. Electron Microsc. Tech.* **3**, 177 (1986).

[2] W. L. Simpson, *Anat. Rec.* **80**, 173 (1941).

[3] A. Van Harreveld and J. Crowell, *Anat. Rec.* **149**, 381 (1964).

[4] J. E. Heuser, T. S. Reese, M. J. Dennis, Y. Jan, L. Jan, and L. Evans, *J. Cell Biol.* **81**, 275 (1979).

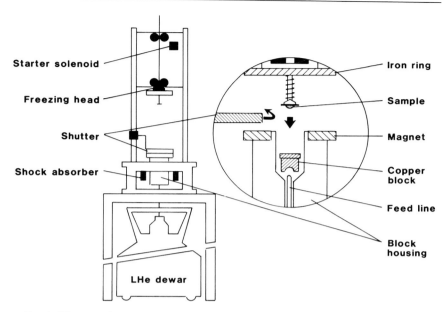

FIG. 1. Diagram of the cold metal block freezing machine designed by Heuser and Reese.[4] Exploded view shows sample on freezing head falling toward the cold copper block.

gravity to bring the sample held by a telescopic freezing head into firm contact with a helium-cooled copper block. As the block is cooled by the flow of boiling cryogen from a pressurized Dewar flask, the sample is loaded onto the pedestal in the center of the freezing head. On release of the freezing head, a shutter covering the block opens, allowing the sample to contact the cold block. High-quality freezing (to depths of 10 to 20 μm from the surface contacting the block) has been achieved in cells and tissues and in studies requiring precise timing with electrical[4,5] and chemical stimulation.[6-8]

Propane jet freezing (Fig. 2) is a second method that freezes tissues well. Quick freezing occurs when liquid N_2-cooled propane sprays both sides of a sample "sandwich" positioned between two jets emerging from a pressurized propane cylinder. Commercial models of the device allow samples to be frozen between two gold alloy, hat-type specimen holders separated

[5] J. E. Heuser and T. S. Reese, *J. Cell Biol.* **88,** 564 (1981).

[6] C. J. Kazilek, C. J. Merkle, and D. E. Chandler, *Am. J. Physiol.* **254,** C709 (1988).

[7] M. Curran, F. Cohen, D. Chandler, P. Munson, and J. Zimmerberg, *J. Membr. Biol.* (in press).

[8] C. J. Merkle and D. E. Chandler, *J. Membr. Biol.* **112,** 223 (1989).

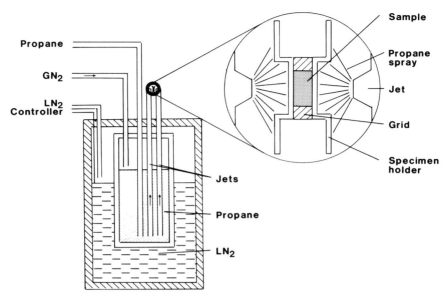

FIG. 2. Diagram of the propane jet freezing machine adapted from Gilkey and Staehelin[1] and Sitte *et al.*[14] The enlarged portion shows the sample "sandwich" positioned between the propane jets.

by electron microscope grids with punched-out centers to eliminate sample loss and damage. Although impractical for thicker tissues, this method produces excellent results in isolated cells and thin tissues.[9,10] The sandwich configuration offers the advantage that the sample is cooled on two surfaces simultaneously, allowing specimens as thick as 30 μm to be well frozen. The sample is then fractured in a double-replica device that eliminates knife marks and increases fractured membrane surface area.

Larger samples can be ultrarapidly frozen by high-pressure freezing, a method introduced by Moor 1968.[11] Here the sample is subjected to a pressure of 2100 atm immediately before being frozen by a jet of liquid N_2. Under high pressure, the freezing point of water is decreased, thereby reducing ice crystal formation. The main advantage to high-pressure freezing is that cells or tissue blocks as large as 0.6 mm in diameter[12] can be ultrarapidly frozen with excellent preservation throughout. Membrane fu-

[9] G. H. Haggis, *J. Microsc.* **143**, 275 (1986).
[10] J. C. Gilkey and L. A. Staehelin, *Planta* **178**, 425 (1989).
[11] H. Moor, *in* "Cryotechniques in Biological Electron Microscopy" (R. A. Steinbrecht and K. Zierold, eds.), p. 175. Springer-Verlag, Berlin, 1987.
[12] S. Craig and L. A. Staehelin, *Eur. J. Cell Biol.* **46**, 80 (1988).

sion events can be captured using this technique.[12,13] Potential drawbacks are sample damage due to longer preparation time and exposure to high pressure.[1,11,12]

Further information on rapid freezing techniques and their application to specific tissues can be found in several reviews.[1,11,14-18]

Technique of Cold Metal Block Freezing

Preparation

Preparation of specimen stages for cold metal block freezing can be done in advance. We generally use a basic stage (Fig. 3A–D), consisting of an aluminum planchet to which a disk of Whatman (Clifton, NJ) filter paper (made with a standard paper punch) is glued to the center using "5-min" epoxy. Next, a plastic ring is glued to the planchet with two drops of cyanoacrylate glue. After lightly imprinting notations to identify the sample on the bottom of the planchet, a piece of double-stick tape is placed on the bottom of the stage.

If samples are to be frozen on a cushion of liver or lung (see below), this tissue must be fixed, sliced, and washed prior to quick freezing. Rabbit lung, for example, can be fixed with 2% glutaraldehyde/10% ethanol, encased in 4% (w/v) agar, sliced thinly (0.4 to 0.6 mm) with a homemade razor blade microtome, and washed thoroughly in the buffer to be used during quick freezing.

Shortly before a quick-freezing session, the surface of the copper block is thoroughly cleaned with metal polish and the block is sonicated in acetone, then methanol. To set up the quick-freezing machine, one first passes a straight, double-walled feed line into the helium Dewar flask, then centers and levels the machine over the feed line. A constant positive pressure of 5 cm of water is kept on the Dewar flask at all times to avoid ice buildup in the feed line.

To prepare isolated cells, it is important to establish suitable conditions for maintaining normal cell physiology, while manipulating the cells for

[13] B. Draznin, R. Dahl, N. Sherman, K. E. Sussman, and L. A. Staehelin, *J. Clin. Invest.* **81**, 1042 (1988).

[14] H. Sitte, L. Edelmann, and K. Neumann, *in* "Cryotechniques in Biological Electron Microscopy" (R. A. Steinbrecht and K. Zierold, eds.), p. 87. Springer-Verlag, Berlin, 1987.

[15] B. P. M. Menco, *J. Electron Microsc. Tech.* **4**, 177 (1986).

[16] D. E. Chandler, *Curr. Top. Membr. Transp.* **32**, 169 (1988).

[17] R. Dahl and L. A. Staehelin, *J. Electron Microsc. Tech.* **13**, 165 (1989).

[18] J. Heuser, *J. Electron Microsc. Tech.* **13**, 244 (1989).

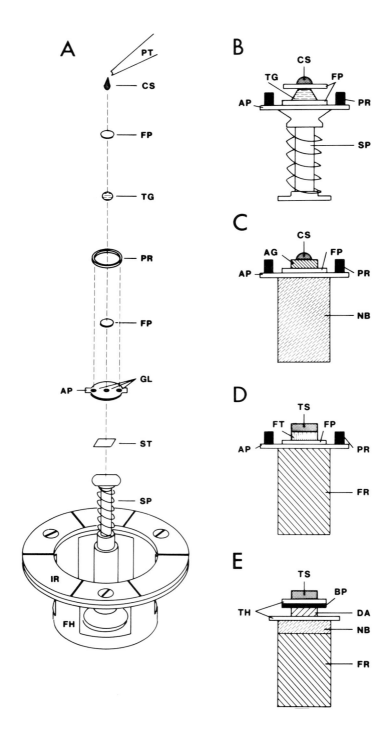

freezing. We suggest incubating cells in 1.5-ml microfuge tubes in medium that is pH regulated with a nongaseous buffer, for example, N-2-hydrox-yethylpiperazine-N'-2-ethane sulfonic acid (HEPES). To keep cells at the proper temperature, cells should be incubated in a water bath until just before freezing.

To prepare tissues for cold metal block freezing, one must use careful dissection to expose the cell layers of interest so that these contact the block first and are well frozen. The tissue block must be trimmed small enough to fit on the specimen stage and thick enough to extend above the plastic ring.

Loading

As illustrated in Fig. 3, there are several methods for loading cells onto the freezing head. One possibility (Fig. 3A and B) is to place an aliquot of concentrated cells on a filter paper disk, which can be dipped into stimulant. The sample is then placed on the specimen stage, which adheres by double-stick tape to the central pedestal of the freezing head, which can be a spring-type plunger (Fig. 3A and B), a Neoprene block (Fig. 3C), a block of foam rubber (Fig. 3D), or a combination of Neoprene and foam rubber (Fig. 3E). Between the sample and stage is a small cushion of ultrasound transmission gel (Fig. 3A and B), fixed liver or lung tissue (Fig. 3D), or agar (Fig. 3C and E), which serves to raise the level of the sample above the plastic ring and to soften its impact as the sample hits the block. In some cases, tissue can be placed directly on the specimen disk with no intervening cushion; uncushioned specimens have been used with success in studies on amphibian eggs.[19]

[19] C. A. Larabell and D. E. Chandler, *J. Electron Microsc. Tech.* **13**, 228 (1989).

FIG. 3. Specimen stage configurations for cold metal block freezing. (A) Exploded view of freezing head and specimen stage, consisting of a plastic ring and filter paper disk glued to an aluminum planchet. A drop of ultrasound transmission gel is placed between the specimen stage and filter paper disk supporting the sample. A piece of double-stick tape secures the stage to a spring-type plunger in the center of the freezing head. (B) Cutaway view of (A). (C) An alternative mount using agar instead of transmission gel to support cells. Cells may be placed directly or indirectly, using filter paper, onto the agar. Neoprene serves as an alternative to the spring-type plunger in the center of the freezing head. (D) An alternative mount using a slice of fixed liver or lung to support tissues. This mount may also be used to freeze cells. Foam rubber serves as an alternative to the spring-type plunger in the center of the freezing head. (E) An alternative mount in which tissue is placed on Thermanox covering a brass plate, which is supported by 2.5% agar on Thermanox. The plunger is made of double-density foam. AG, Agar; AP, aluminum planchet; BP, brass plate; CS, cell suspension; DA, 2.5% agar; FH, freezing head; FP, filter paper; FR, foam rubber; FT, fixed tissue; GL, glue; IR, iron ring; NB, Neoprene block; PR, plastic ring; PT, pipette tip; SP, spring-type plunger; ST, double-stick tape; TG, ultrasonic transmission gel; TH, Thermanox; TS, tissue.

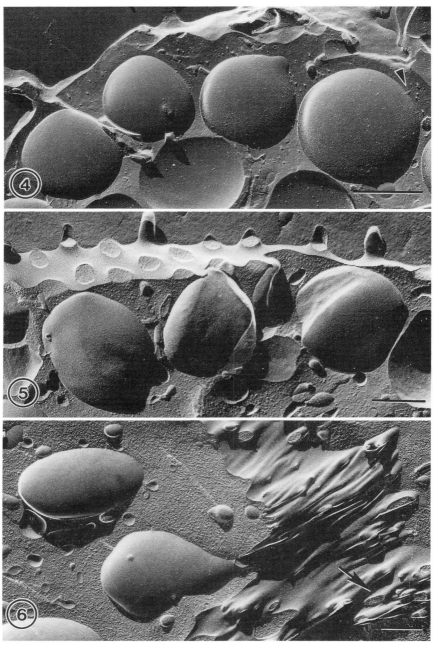

Fig. 4. Histamine-containing granules in a rat peritoneal mast cell that was quick frozen. Note filaments (arrows). This and subsequent figures have been photographically reversed; platinum deposits appear white. Bar: 0.5 μm.

Freezing and Storage

At the start of each run the liquid helium Dewar flask is pressurized to 45 to 50 g/cm² to initiate cryogen flow. After loading the sample, the freezing head is snapped into place on the raised plunger. Once the copper block is cold (detected by a thermocouple) Dewar flask pressure is dropped to background level, the plunger drop mechanism is activated, and the shutter opens to expose the cold copper block to the falling specimen. It is possible to stimulate tissue electrically as a specimen drops by electrodes built into the freezing head.[4] One then releases the electromagnetic catch, transfers the freezing head to liquid N_2, and removes the specimen stage for long-term storage in liquid N_2. Specimens are stored in plastic scintillation vials with perforated walls and caps.

Freeze–fracture

The aluminum planchet to which specimens are frozen is designed to fit directly onto the Balzers (Nashua, NH) 301 specimen table and is secured with a screw cap. Alternatively, tables fitted with a pair of flange clamps can be custom designed for many freeze–fracture units, including the Balzers 400 model. It is crucial that the sample be fractured with one or two passes of a microtome knife so as to just graze the surface of the flattened specimen. The knife edge must be lined up parallel to the plane of the table before the run. After normal platinum–carbon replication and cleaning procedures, replica pieces are picked up on grids such that the edges of the grazed area (containing the best frozen cells) are positioned in the center of the grid.

Assessment of Freeze–Fracture Replicas of Quick-Frozen Cells

Quick freezing on a cold metal block is characterized by a gradient of ice crystal formation in the sample. This ranges from near-vitreous ice at the surface of the sample that contacts the block to extremely large and disruptive crystals deeper in the sample. This gradient is easily seen in thin sections of specimens that have been freeze-substituted, embedded, and cut perpendicularly to the plane of block contact.[1,20] Crystal size increases with

---

[20] R. L. Ornberg and T. S. Reese, *in* "Freeze Fracture: Methods, Artifacts, and Interpretations" (J. E. Rash and C. S. Hudson, eds.), p. 89. Raven Press, New York, 1979.

FIG. 5. Cortex of a quick-frozen sea urchin egg, showing a single row of cortical granules just beneath the plasma membrane. Bar: 0.5 μm.

FIG. 6. Sea urchin egg cortex quick frozen just after exocytosis. Note the plasma membrane and stretched microvilli (arrow) that have undergone shearing forces. Bar: 0.5 μm.

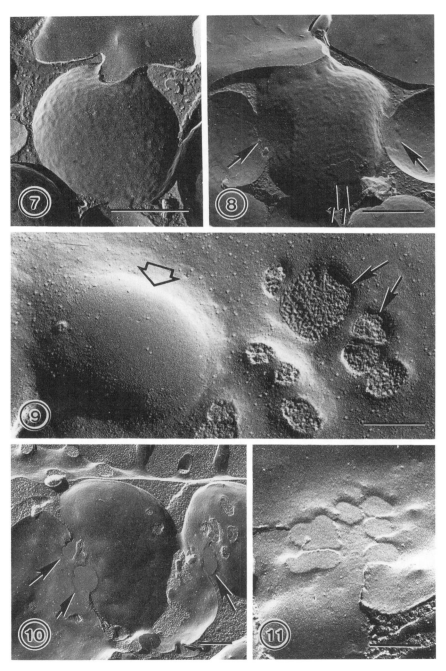

FIG. 7. Poor freezing in this quick-frozen mast cell is indicated by mottled granule membranes. Bar: 0.5 μm.

depth, usually becoming unacceptable at 20 to 25 μm from the sample surface.

Fracturing a quick-frozen sample in a freeze–fracture unit is an imprecise process. One attempts to graze the surface of the sample with a shallow cut (see above), but on most occasions the razor blade cuts more deeply into the sample at the center than at the periphery. Thus, when viewing replicas of quick-frozen cells, it is commonplace to find large areas of poorly frozen tissue (anywhere from 50 to 90% of total replica area) combined with smaller areas of well-preserved material. Clearly, one accepts data from the highest quality freezing, which represents preservation far superior to that found in chemically fixed tissues.

Well-frozen cells contain organelles having extremely smooth membranes and an appearance of turgidity (Figs. 4 and 5). All organelles are distinctly separated (Figs. 5 and 6). Of particular significance is the fact that the extracellular space shows little or no sign of ice crystal formation, as indicated by a honeycomb or irregular latticed appearance (Fig. 5).

The most sensitive indicator of a poorly frozen cell is a bumpy appearance to organellar membranes — a distortion caused by formation of small ice crystals. As shown in Figs. 7 and 8, secretory granule membranes are particularly susceptible. In mast cells this is often accompanied by characteristic appositions between membranes of granules that appear to have been forced together (arrows, Fig. 8).

A second artifact, specific to cold metal block freezing, is that created by mechanical compression and shearing forces as the sample is literally squashed against the block. These forces seldom cause distortions in small cells, but commonly do in large cells. In the sea urchin egg, distortion at the EM level is seen as parallel wrinkles in the plasma membrane and microvilli that appear to have been stretched along the same axis (arrow, Fig. 6). Despite this potential problem, the large majority of cells in any given sample do not show evidence of distortion; in fact, *Xenopus laevis* eggs,

---

FIG. 8. Contacts between neighboring mast cell granules are artifacts of poor freezing (arrows). Bar: 0.5 μm.

FIG. 9. Plasma membrane of a fertilized sea urchin egg subsequently fixed with glutaraldehyde and glycerinated. An IMP-free area is seen where the underlying granule bulges against the plasma membrane (large arrow); multiple etched pores join the plasma membrane with an underlying cortical granule (small arrows). Bar: 0.25 μm.

FIG. 10. Single bilayer diaphragms (arrows) between cortical granule membranes in a glutaraldehyde-fixed and glycerinated sea urchin egg. Bar, 0.5 μm.

FIG. 11. Single bilayer diaphragms are IMP free and are continuous with both neighboring granule membranes. Bar: 0.25 μm.

which are macroscopically flattened where they have contacted the block, show few if any signs of shearing when viewed at the EM level.

Given the problems of poorly frozen tissue and mechanical stress that must be avoided, one may wonder why quick freezing is the method of choice for studying membrane fusion events. The fact is that the alternative, chemical fixation, produces artifacts that are more serious in the sense that they can easily be mistaken for structures considered to represent intermediates in the membrane fusion process. First, in many cells chemically fixed during exocytosis, granules bulge against the plasma membrane, frequently producing intramembrane particle (IMP)-cleared areas such as that seen in Fig. 9 (large arrow). Intramembrane-free areas are thought to be equivalent to pentalaminar structures in thin sections and have been seen in a number of secretory cells.[16] Second, the granule membranes of some cells become joined by single bilayer diaphragms (arrows, Fig. 10). At higher magnification (Fig. 11) diaphragms are seen to be the equivalent of trilaminar structures as visualized in thin sections of many secretory cells.[16] These structures, in freeze–fracture at least, are usually present only in chemically fixed and glycerinated cells, suggesting that they may be either fixation or glycerination artifacts. We have previously shown that plasma and granule membranes in sea urchin eggs can artifactually fuse during glycerination despite the fact that they are already aldehyde fixed.[21] Such fusions lead to multiple etchable pores joining the plasma membrane with the granule below (arrows, Fig. 9). Finally, aldehyde fixation can lead to formation of plasma membrane blebs, as documented previously in tissue culture cells.[22]

Thus it is clear that routine preparation of secretory cells for freeze–fracture (aldehyde fixation and glycerination) can lead to artifacts that are remarkably similar to proposed intermediates in the membrane fusion process. For this reason, quick freezing has proved remarkably useful in differentiating between artifactual and physiologically important features. Our present hypothesis is that these artifacts arise from actual membrane fusion intermediates but become enlarged or altered during the relatively slow, nonphysiological process of chemical fixation and dehydration. Quick-frozen specimens reveal that the initial fusion event takes place over a small domain, and electrophysiological techniques suggest that the initial pore within this domain may be as small as 1 or 2 nm.[7] Future studies using cryofixation will be required to determine whether pore formation is preceded by such traditional structures as the pentalaminar and trilaminar structures, albeit over a domain that is restricted in size.

[21] D. E. Chandler, *J. Cell Biol.* **83,** 91 (1979).
[22] D. L. Hasty and E. D. Hay, *J. Cell Biol.* **78,** 756 (1978).

Acknowledgments

We thank Steven Pfeiffer of LifeCell Corporation (Woodlands, TX) for helpful discussions concerning specimen stage configurations for cold metal block freezing machines, Lisa Moore for bibliographic assistance, and Charles Kazilek for photographic assistance. Research from our laboratory described herein was supported by National Science Foundation Grant IBN-9117509.

[10] Electropermeabilized Platelets: A Preparation to Study Exocytosis

By Derek E. Knight and Michael C. Scrutton

Exocytosis is a process whereby an intracellular vesicle fuses with the inner surface of the plasma membrane. Exocytosis not only alters the composition of the cell surface membrane by the addition of vesicle membrane but also exports molecules previously trapped within the vesicles into the extracellular medium. Enzymes, hormones, and neurotransmitters are secreted in this manner. The mechanism of exocytosis has been examined by gaining access to the interior of the cell in order to define the chemical environment at the site of the exocytotic event. Definition of this chemical environment can be achieved by use of the patch pipette[1,2] and by permeabilizing the plasma membrane so that extracellular solutes can diffuse into the cell and equilibrate with the cytosol. Permeabilization of the plasma membrane can be achieved by treatment with detergents or toxins and by application of high-voltage electric fields (electropermeabilization or electroporation). These methods have been applied to analyze a number of secretory systems.[3,4] Comparisons of techniques and preparations have been well documented elsewhere in this series,[5,6] as well as in other sources.[3,7,8] One preparation that has generated considerable interest because of the central role of the cell in many pathological processes involves the permeabilized blood platelet.[9]

[1] E. Neher and A. Marty, *Proc. Natl. Acad. Sci. U. S. A.* **79,** 6712 (1982).

[2] J. M. Fernandez, E. Neher, and B. D. Gomperts, *Nature (London)* **312,** 453 (1984).

[3] D. E. Knight and M. C. Scrutton, *Biochem. J.* **234,** 497 (1986).

[4] D. E. Knight, H. Grafenstein, and C. M. Athayde, *Trends Neurosci.* **12,** 451 (1989).

[5] P. F. Baker and D. E. Knight, this series, Vol. 98, p. 28.

[6] P. F. Baker and D. E. Knight, this series, Vol. 171, p. 817.

[7] B. D. Gomperts and J. M. Fernandez, *Trends Biochem. Sci.* **10,** 414 (1985).

[8] P. F. Baker and D. E. Knight, *in* "In Vitro Methods for Studying Secretion" (Poisner and Trifaro, eds.), p. 223. Elsevier, New York, 1987.

[9] D. E. Knight and M. C. Scrutton, *in* "Platelets in Biology and Pathology III" (J. Gordon Macintyre, ed.). 1987.

Platelet activation, caused by many agonists, includes the secretion, by exocytosis, of various factors from this cell. Because platelet secretion is triggered by a range of secretagogues it has been suggested that multiple, and possibly independent, pathways may be involved in the exocytotic mechanism. Electroporation of the platelet plasma membrane has provided new insight into this area. This chapter describes the procedures necessary for successful preparation of platelets rendered permeable by a high-voltage electric field, the potential and limitations of this experimental system, and some of the results that have been obtained from experiments in which it has been used.

Electroporation Preferred to Other Methods

Several methods have been used to permeabilize platelets, including exposure to detergents such as digitonin or saponin,[10,11] to organic solvents such as dimethyl sulfoxide, or to divalent ion chelation.[12] Detergent-induced permeabilization has been widely employed and depends on a more or less selective removal of cholesterol from the membrane, hence causing loss of structural integrity. Preferential disruption of the plasma membrane is achieved because this membrane has a three- to fourfold higher content of cholesterol as compared to the boundary membranes of intracellular organelles.[13]

Detergent-induced permeabilization appears simple, but requires careful control of experimental variables such as incubation time and temperature in order to achieve a selective effect on the plasma membrane.[14] The required conditions must be established empirically and it appears difficult to obtain a homogeneous preparation in which lesions are restricted to the plasma membrane.[10] Furthermore the effect of the detergent is neither reversible nor localized. Divalent ion chelation has been used less frequently to permeabilize platelets, and there is little understanding of the mechanisms by which this treatment destabilizes membrane structure although the effect does appear to be reversible.[12] In contrast electroporation of the plasma membrane is reasonably well understood, being primarily a function of the relative radii of the various membrane-bound compartments that are exposed to the field. The technique therefore takes advan-

[10] A. D. Purdon, J. L. Daniel, G. J. Stewart, and H. Holmsen, *Biochim. Biophys. Acta* **800,** 178 (1984).

[11] K. Authi, B. J. Evenden, and N. Crawford, *Biochem. J.* **233,** 709 (1986).

[12] P. C. Johnson, J. A. Ware, P. B. Clivenden, M. Smith, A. M. Dvorak, and E. W. Salzman, *J. Biol. Chem.* **260,** 2069 (1984).

[13] S. Menashi, H. Weintroub, and N. Crawford, *J. Biol. Chem.* **256,** 4095 (1981).

[14] P. F. Zuurendonk and J. M. Tager, *Biochim. Biophys. Acta* **333,** 393 (1974).

tage of a clear-cut difference in cellular anatomy. The studies performed thus far with this method suggest that selective permeabilization of the plasma membrane can be achieved with minimal damage to other parts of the platelet.[15,16]

Theory of Electroporation

When a cell or organelle is placed in an electric field a potential (V) develops across its membrane. If the cell is approximately spherical, then the maximum potential develops across the membrane at the two points in line with the applied field. Furthermore, if the conductivity of the fluid inside the cell is approximately the same as that of the extracellular fluid and is much greater than the conductivity of the membrane itself, then the magnitude of this potential is given by the equation $V = 1.5Er$, in volts, where E is the magnitude (in volts per centimeter) of the applied field, and r is the radius (in centimeters) of the cell or organelle.[5,6,17,18] Hence, the smaller the radius *(r)* of the cell or organelle, the smaller will be the potential difference *(V)* imposed across its limiting membrane when exposed to an electric field. The membrane breaks down and becomes leaky to extracellular solutes when the potential developed across it is approximately 1.1 V. However, in many cells when this voltage is imposed briefly across the membrane, only a transient leakiness develops as the membrane reseals within a short time. If a potential difference in excess of 3 V is imposed briefly across the membrane, it appears that the membrane is rendered fully permeable and cannot effectively reseal. Platelets are on average about 2 μm in diameter and so, from the above equation, an applied field strength of 20,000 V/cm would be expected to induce a potential difference of 3 V across the membrane. The diameter of organelles within the platelet, such as lysosomes and the protein and amine storage granules, is about 0.2 μm; so, even if they were exposed to this same brief field strength, the maximal potential difference generated across their limiting membranes would be approximately 0.3 V. As this is insufficient to render them leaky, an applied field strength of 20 kV/cm would therefore allow the selective permeabilization of the plasma membrane. More comprehensive discussion of the theoretical basis of this method can be found elsewhere.[19]

[15] D. E. Knight and M. C. Scrutton, *Thromb. Res.* **20,** 437 (1981).
[16] R. J. Haslam and M. M. L. Davidson, *Biochem. J.* **222,** 351 (1984).
[17] U. Zimmerman, G. Pilwat, and F. Riemann, *Biophys. J.* **14,** 881 (1974).
[18] D. E. Knight, *in* "Techniques in Cellular Physiology" (P. F. Baker, ed.), p. 113. Elsevier/North-Holland, Amsterdam, 1981.
[19] E. Neumann, A. E. Sowers, and C. A. Jordan, eds., "Electroporation and Electrofusion in Cell Biology." Plenum, New York, 1989.

The principle of the technique is to place the platelet suspension between two electrodes and to discharge a capacitor through it, the time course of the electric field decaying exponentially, with a time constant (seconds) given by the produce of the capacitance (farads) and electrical resistance of the suspension between the electrodes (ohms).

Apparatus Required for Electroporation

Figure 1 shows the electroporation equipment presently in use in our laboratories. It consists of a 0- to 6-kV high-voltage power supply (A), one 2.5-μF capacitor (B) plus two 1-μF capacitors (C) connected in parallel, a switch (D) to permit discharge of the capacitors through the platelet suspension, and a chamber (E) into which the platelet suspension is placed. This chamber is shown diagrammatically in Fig. 2A. The electrodes of the chamber are two 5.2 × 3.6 cm stainless steel plates, each inlaid into the upright section of an L-shaped piece of Perspex, and separated by a U-shaped Perspex spacer 1 mm in thickness.

The cell is made watertight by *lightly* greasing the apposed faces of the L pieces and the spacer, and the component parts are held together using two large bulldog clips with the jaws placed approximately 0.5 to 1 cm from the outside edge of the cell (Fig. 1). Care should be taken not to allow grease on

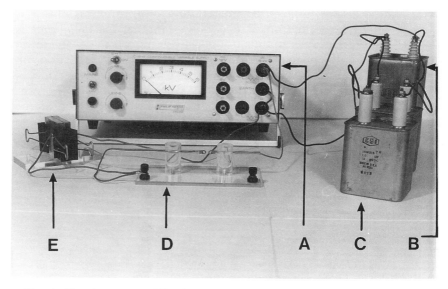

FIG. 1. The electropermeabilization apparatus as used in studies on platelets; details described in text.

the metal electrodes. Although its design is simple, this chamber offers two valuable features. First, the required field strength of 20 kV/cm necessary to permeabilize the platelets can be achieved by discharging 2 kV across the stainless steel electrodes (1 mm apart). An increased spacing between the electrodes would require a larger voltage difference between the electrodes to achieve the same applied field. This increase would give rise to significant heating of the platelet suspension and also run the risk of the platelet suspension being blown out of the chamber during the discharge. Second, the chamber design allows easy dismantling to clean the electrodes. The apparatus as described here makes no provision for temperature control or for variation of capacitance. Although both of these features may be desirable and could be incorporated, we have not found them to be necessary for successful permeabilization of the platelet when using this chamber design. Despite the high voltages employed only a minimal 2 to 3° rise in temperature occurs during permeabilization. Alteration of the capacitance would permit variation of the duration of the exposure of the platelets to the electric field. The use of three smaller capacitors rather than a single one of 4.5 μF, as used elsewhere,[16] allows some variation in capacitance. Control of field duration can, however, also be achieved by variation of the ionic strength of the suspending medium.[3,18,20]

Experimental

The chamber is assembled and completely filled with the platelet suspending medium in order to check for leaks. After removal of this medium, the platelet suspension (approximately 1.2 ml) is loaded into the chamber, taking care not to introduce air bubbles and ensuring that the level of fluid in the chamber is adequate to cover the electrodes completely. The total absence of trapped air bubbles from the chamber and the complete covering of the electrodes are important, as they reduce the risk that the current might pass nonuniformly between the electrodes, causing a localized heating effect, rapid boiling, and hence explosive ejection of the platelet suspension from the chamber.[3] Air bubbles can be removed most simply by passing a fine piece of flexible tubing through the cell suspension once it is loaded into the chamber.

The capacitors are charged to 2 kV by using the high-voltage supply and then discharged through the platelet suspension by closing the switch. This process is repeated as desired to achieve the required permeabilization. Because it is unlikely that a given platelet will maintain the same orientation with respect to the applied field over the time period required

[20] D. E. Knight, V. Niggli, and M. C. Scrutton, *Eur. J. Biochem.* **143**, 437 (1984).

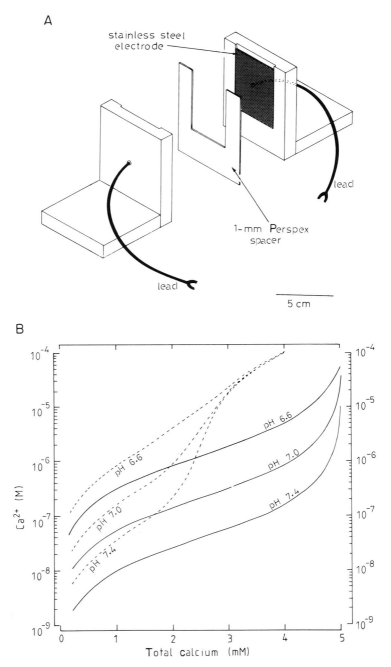

FIG. 2. (A) View of the disassembled cell suspension chamber. (B) Calcium ion titration curves for two calcium buffers at different pH values. Calcium ion concentrations were

to repeat the charge/discharge cycle (approximately 4 sec), unique areas of localized membrane breakdown ("holes") are created as a consequence of each discharge. To achieve a permanently permeabilized preparation we routinely apply 10 discharges, each of 2 kV and corresponding to a field of 20 kV/cm. It is advisable occasionally to reverse the polarity of the electrodes during a longer succession of discharges (approximately 20) as this reduces the possibility of gas bubbles forming in the chamber and hence creating an explosive condition.

After permeabilization, the platelet suspension is carefully removed from the chamber by using a 2-ml plastic syringe attached to a fine (0.5 mm) plastic tube. This preparation is either used directly for subsequent studies,[3] or can be subjected to gel filtration on Sepharose CL-4B in order to remove the low molecular weight components released as a consequence of permeabilization.[16]

Preparation of Platelets for Electropermeabilization

The medium in which the platelets are suspended for exposure to the capacitor discharge needs to be (1) electrically conducting and (2) to be of such a composition as not to alter the response of the platelet when diffused into the cytosol. For permeabilization the platelets are usually suspended at a density of approximately 10^9 cells/ml. Three different suspending media that fulfill these criteria have been used successfully in electropermeabilization studies.

Medium I: 150 mM Potassium glutamate, 20 mM K$^+$–piperazine-N,N'-bis(2-ethanesulfonic acid) (PIPES), 5 mM Na$_2$ATP, 7 mM magnesium diacetate, 5 mM glucose, 1 mM ethylene glycol-bis (β-aminoethyl ether)-N,N,N',N'-tetraacetic acid (EGTA) {or BAPTA [bis(o-aminophenoxyethane-$N,N,N'N'$-tetraacetic acid)]}. The pH is adjusted to 6.6 or 7.4 by addition to KOH[15,21,22]

Medium II: As for medium I, but containing 280 mM glycine and

[21] K. S. Authi, K. Hughes, and N. Crawford, *FEBS Lett.* **254**, 52 (1989).
[22] K. Hughes and N. Crawford, *Biochim. Biophys. Acta* **981**, 277 (1989).

calculated at the various pH values indicated, using the dissociation constants given in Table I, and for a total ligand concentration of 5 mM. The ordinate indicates the concentration of Ca^{2+} obtained when the total calcium concentration indicated on the abscissa is added to buffers containing 5 mM EGTA, 7 mM Mg, and 5 mM ATP (solid lines) or 2.5 mM EDTA, 2.5 mM EGTA, 12.5 mM Mg, and 5 mM ATP (dashed lines). The Ca^{2+} titration curve for a buffer containing 5 mM BAPTA, 7 mM Mg, and 5 mM ATP is not shown, but follows closely that for EGTA at pH 7.0 and is relatively insensitive to variation of pH over the range shown.

10 mM potassium glutamate in place of 150 mM potassium glutamate[20]

Medium III (a modified Ca^{2+}-free Tyrode's solution): 154 mM NaCl, 2.7 mM KCl, 1 mM $MgCl_2$, 5.6 mM glucose, 7 mM $NaHCO_3$, 0.6 mM NaH_2PO_4, 5 mM Na–PIPES (pH 6.5), 0.35% (w/v) bovine serum albumin (fraction V), 5 mM K^+–EGTA (pH 6.5)[23]

Media I and II contain low concentrations of Na^+ and no added Cl^- and hence, in this respect, at least resemble the natural chemical environment. In some permeabilized cells, addition of a high concentration of Cl^- causes irreversible inhibition of secretion.[24] In medium I, the ionic strength is high and hence the electrical resistance is low, limiting the duration of the intense electric field to which the platelet is exposed on discharge of the capacitors (see below). Replacement of most of the potassium glutamate by glycine (as in medium II) increases the electrical resistance and hence ensures that the duration of exposure to the electric field is sufficient to cause lasting localized plasma membrane breakdown. With the setup as described, the time constant of the electric field when using medium II is close to 30 μsec.

If medium III is used, the permeabilized platelets are immediately transferred by gel filtration on Sepharose CL-4B at 4° into medium IV, containing a high concentration of K^+. The composition of medium IV is as follows: 160 mM potassium glutamate, 20 mM K^+–HEPES (pH 7.4), 2.5 mM ethylenediaminetetraacetic acid (EDTA), 2.5 mM EGTA, 12.5 mM $MgCl_2$. The pH is adjusted to 7.4 at 4° with KOH.[16,23]

The Sepharose CL-4B column is washed before use with 0.15 vol of medium IV containing 10% (w/v) bovine serum albumin, followed by 4 vol of medium IV lacking albumin. After elution in medium IV the platelets are diluted to the required platelet count with medium IV containing ATP to give a final ATP concentration of 4–5 mM. The platelet suspension is then stored at 2° until used.

Criteria for Permeabilization

The criteria used to establish successful permeabilization must be related to the size of the molecule(s) that are to be introduced into the platelet. For studies involving control of Ca^{2+} concentration, using EGTA as the ligand, or of nucleotide concentration, it is necessary to achieve permeabilization to solutes having a molecular weight of approximately

[23] R. J. Haslam and M. M. L. Davidson, *J. Recept. Res.* **4**, 605 (1984).
[24] D. E. Knight and P. F. Baker, *J. Membr. Biol.* **68**, 107 (1982).

500, while preferably allowing retention within the cytosol of proteins and other macromolecules. The criteria cited below are those we have used to define such a platelet preparation.

1. For platelets incubated in a medium containing ^3H_2O and $^{45}Ca-$EGTA permeabilization followed by isolation of the platelets by centrifugation through silicone oil should lead to a rapid increase in the space accessible to $^{45}Ca-EGTA$. In our hands the ^{45}Ca-EGTA space is approximately 20% of the ^3H_2O space in control cells and increases to approximately 70% within 5 min after permeabilization in medium I or II.[3,15] The relatively high ^{45}Ca-EGTA space observed for control cells presumably reflects the presence of extracellular fluid, which is carried through the silicone oil in association with the platelets due to trapping in the extensive invaginations of the plasma membrane characteristic of this cell.

Platelets permeabilized in medium III and then isolated by gel filtration before storage at $0°$ show an extent of equilibration with ^{45}Ca-EGTA that depends both on the number of exposures to a 15-kV/cm field and on the time of incubation at $0°$. The time course of equilibration with ^{45}Ca-EGTA is biphasic, with the slower component being attributed to uptake of $^{45}Ca^{2+}$ into an organelle compartment.[23]

It is particularly important to note that release of $^{86}Rb^+$ from platelets that had previously been loaded with this marker is not a reliable index of permeabilization when access of the preparation to larger molecules is required. In platelets release of Rb^+ can be observed on exposure to a field that fails to permit access of the intracellular environment to ^{45}Ca-EGTA.[3,15]

2. Immediate release of approximately 70% of total cellular ATP should be observed on permeabilization in a medium containing less than $0.01 \mu M$ Ca^{2+}, without release of a significant amount of lactate dehydrogenase.[25] The release of ATP under these conditions should be incomplete because a significant proportion of total cellular ATP is sequestered in the amine storage granules. This fraction varies in platelets obtained from different mammalian species, and hence the quantitative extent of release of ATP resulting from permeabilization will also vary.[26]

3. Platelets, when electroporated, increase in volume. If they reseal, they recover their original volume. Sizing of platelets using a Coulter (Hialeah, FL) counter may therefore give a measure of the extent of resealing after electroporation.[21,22]

[25] D. E. Knight, T. J. Hallam, and M. C. Scrutton, *Nature (London)* **296**, 256 (1982).
[26] K. Ugurbil and H. Holmsen, *in* "Platelets in Biology and Pathology II" (J. L. Gordon, ed.), p. 147. Elsevier/North-Holland, Amsterdam, 1981.

Size of Areas of Localized Damage in Platelet Caused by
  Electroporation

Platelets permeabilized by 10 exposures to a 20-kV/cm field show
near-instantaneous release of small cytosolic markers, for example, ATP,
but do not release significant amounts of several cytosolic proteins ranging
in molecular weight from 17,500 (calmodulin) to 125,000 (lactate dehydrogenase).[25] Furthermore, the areas of damage cannot be identified in
transmission electron micrographs of platelets permeabilized in this way.[27]
Such observations suggest that the areas of damage are relatively small, but
their dimensions cannot be defined at present. For bovine adrenal medullary cells, also permeabilized by imposing 3 V across the membrane, the
effective diameter of the "holes" has been estimated as 2 to 4 nm.[28]

Calcium Buffers Used

Exocytosis in electropermeabilized platelets is triggered by exposure to
Ca^{2+} in the micromolar range. It is necessary to define such Ca^{2+} concentrations by using a Ca^{2+} buffer system. We have used either EGTA or
EGTA/EDTA as the buffer. Figure 2B summarizes the properties of these
two buffer systems. The Ca–EGTA system gives the most effective buffering over the physiological range of cytosolic Ca^{2+} concentration (approximately 0.1 to 10 μM)[29,30] but only if the studies are performed at pH 6.6.
The pH must be strictly controlled, because the K_D for the Ca–EGTA
varies sharply with pH in the physiological range (Fig. 2B, Table I).[30a]
Decreased sensitivity to changes in pH can be obtained by using a Ca–
BAPTA buffer, but the K_D for the Ca–BAPTA complex (approximately
0.1 μM) makes it impossible to achieve effective control of Ca^{2+} concentration at concentrations above 2 μM because the ligand is then almost
fully saturated with this cation. Buffering of Ca^{2+} concentrations over the
physiological range and at pH 7.4 can be obtained by using a Ca–EGTA/
EDTA system in the presence of excess Mg^{2+}. However, this system is not
an effective buffer over the physiological range of Ca^{2+} concentration (Fig.
2B), and must therefore be used with caution. For example, when, as
shown in Fig. 2B, 2.5 mM EDTA and 2.5 mM EGTA are added together,
an increase in total calcium from 2 to 3 mM will cause the Ca^{2+} to change
from 0.2 to 17 μM. Furthermore, because the K_D for Ca–EDTA approxi-

[27] D. E. Knight, V. Niggli, and M. C. Scrutton, *Adv. Biol. Med.* **192**, 171 (1985).
[28] D. E. Knight and P. F. Baker, *J. Membr. Biol.* **68**, 107 (1982).
[29] T. J. Rink, S. W. Smith, and R. Y. Tsien, *FEBS Lett.* **148**, 21 (1982).
[30] T. J. Rink and T. J. Hallam, *Trends Biochem. Sci.* **12**, 215 (1984).
[30a] A. E. Martell and L. G. Sillén, *Spec. Publ.—Chem. Soc.* **17**, 634, 651, 697 (1964).

TABLE I
DISSOCIATION CONSTANTS FOR VARIOUS LIGANDS AND METAL IONS[a]

Metal ligand	pK_a at pH				
	6.6	6.8	7.0	7.2	7.4
Ca–EGTA	5.9277	6.326	6.723	7.118	7.512
Mg–EGTA	1.25	1.47	1.70	1.94	2.19
CaATP	3.437	3.555	3.650	3.727	3.783
MgATP	3.574	3.693	3.790	3.866	3.922
Ca–EDTA	6.796	7.040	7.271	7.492	7.705
Mg–EDTA	4.90	5.14	5.37	5.59	5.81

[a] Expressed as pK_a at different pH values, the dissociation constants were calculated using equilibrium constants as given by Martell and Sillén.[30a]

mates that for Mg–EDTA (Table I) addition of calcium will displace Mg from EDTA. When this system is titrated with calcium, therefore, the level of Mg^{2+} increases from 4 mM to in excess of 7 mM. In contrast, the calcium titration curves of the EGTA or BAPTA buffers are rather insensitive to the variation in magnesium concentration.

Effect of Conditions of Electroporation on Secretory Properties of Platelets

After 10 exposures to a 20-kV/cm field in medium II the platelets are stable in the permeabilized state for at least 60 min at temperatures between 2 and 25°, as indicated by the ability of 10 μM Ca^{2+} to release 90 to 100% of an amine storage granule marker ([14C]serotonin). A decrease in the number of exposures at this field strength or the use of a lower field strength (10 kV/cm) even with repeated exposures yields a preparation in which responsiveness to added Ca^{2+} declines slowly (Fig. 3a). The progressive decline in responsiveness after exposure to the field is probably due to resealing of the membrane. The conditions necessary for complete resealing of electropermeabilized platelets are described elsewhere.[21,22]

Exposure to more intense fields than 20 kV/cm, or an increased number of exposures at 20 kV/cm, also leads to a decrease in the responsiveness of the preparation (Fig. 3b). This effect is not due, however, to resealing of the membrane, but rather is caused either by the destruction of some element of the intracellular machinery or by the loss of an essential factor. Loss of responsiveness under these conditions is not accompanied by an increased extent of lactate dehydrogenase release.

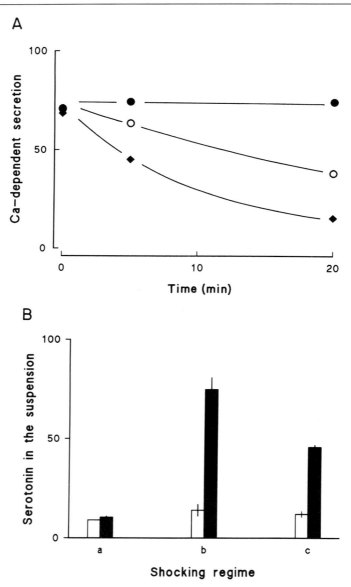

FIG. 3. The extent of the Ca²⁺-dependent secretory response observed at different times after exposing platelets to various electropermeabilization regimes. (A) Platelets loaded with [¹⁴C]serotonin and suspended in the glycine-based medium (medium II) containing 0.5 mM EGTA were permeabilized by 10 exposures to 20 kV/cm (●), 1 exposure to 20 kV/cm (○), and 10 exposures to 10 kV/cm (◆). At various times after permeabilization, as indicated,

Studies on Exocytosis Using Electropermeabilized Platelet

Intracellular Factors Controlling Secretion

Serotonin is secreted from electropermeabilized platelets on exposure to micromolar levels of buffered Ca^{2+}, with half-maximal secretion occurring at about 2 μM. Optimal secretion requires addition of approximately 10 μM Ca^{2+}. The exocytotic machinery operates equally well at pH 6.6 and 7.4 using the two buffer systems, and the dose–response curve for Ca^{2+} is independent of the calcium buffer capacity. This finding indicates that the secretory system is indeed triggered by micromolar levels of Ca^{2+} and not by transient exposure to higher concentrations of this cation that might result from Ca^{2+} release from intracellular stores. In the presence of suboptimal Ca^{2+} concentration the extent rather than the rate of secretion is reduced.[31] Secretion of serotonin can also be induced by addition of micromolar levels of strontium, but other divalent cations are ineffective. Secretion of serotonin also requires the presence of millimolar concentrations of a Mg–nucleoside triphosphate complex. MgATP is the most effective nucleotide although other nucleotides, for example MgCTP, are also active. Nonhydrolyzable analogs of ATP do not support secretion, suggesting a role for Ca^{2+}-driven phosphorylation or for a Ca^{2+}-driven ATPase. A model that may explain these data implicates a cycle of phosphorylation and dephosphorylation as an essential part of the exocytotic mechanism.[28,32]

At Ca^{2+} concentrations in excess of 10 μM, exocytosis in the electropermeabilized platelet is inhibited due at least in part to activation of a

[31] D. E. Knight and M. C. Scrutton, *FEBS Lett.* **223**, 47 (1987).
[32] D. E. Knight, *Biosci. Rep.* **7**, 355 (1987).

aliquots of the permeabilized suspension were challenged with 10 mM Ca–EGTA buffers corresponding to 0.01 and 10 μM Ca^{2+}. The cell suspension was incubated for 4 min at 20° and the cells then removed by centrifugation (8000 g for 2 min). The [^{14}C]serotonin secreted in response to the 10 μM Ca^{2+} challenge relative to the level secreted by 0.01 μM Ca^{2+} is shown and is expressed as a percentage of the total cellular content. The figure shows that responsiveness to addition of Ca^{2+} decays with time after permeabilization if either the field strength or the number of discharges is markedly decreased from the standard regime (10 exposures to 20 kV/cm). (B) Platelets loaded with [^{14}C]serotonin and suspended in medium II were either not exposed to any electric field (a), or were exposed 10 times to 20 kV/cm (b), or 20 times to 20 kV/cm (c). After 10 min at 20° aliquots were added to 10 mM Ca–EGTA buffers corresponding to 0.01 μM Ca^{2+} (open bars) and 10 μM Ca^{2+} (closed bars). After incubation for a further 5 min at 20° the cells were removed by centrifugation and the [^{14}C]serotonin content of the supernatant determined. This amount is expressed as a percentage of the total in the suspension.

Ca^{2+}-dependent protease.[33,34] It is, however, unclear whether this effect occurs in the intact cell. Although an increase in $[Ca^{2+}]$ to the micromolar range can clearly drive secretion in the electropermeabilized platelet other factors are also involved. For example, measurement of cytosolic $[Ca^{2+}]$ in intact platelets has shown that exocytosis of serotonin occurs at $[Ca^{2+}]$ levels much lower than those needed to drive this process in the electropermeabilized platelet. This finding, together with other data, has suggested the possibility of another exocytotic pathway.[30] Electropermeabilized platelets do, however, respond to thrombin and in the presence of this agonist the Ca^{2+} dose–response curve for serotonin secretion shifts markedly to the left such that the exocytotic machinery is triggered by levels of Ca^{2+} present in the resting platelet (0.05 to 0.1 μM).[30,35,36] This enhancement of the Ca^{2+} sensitivity of the exocytotic mechanism is probably mediated by the thrombin-induced formation of 1,2-diacylglycerol and consequential activation of protein kinase C. Thus, thrombin causes formation of 1,2-diacylglycerol from electropermeabilized cells,[23] and addition of a synthetic 1,2-diacylglycerol to the electropermeabilized platelet mimicks the effect of thrombin on Ca^{2+}-dependent secretion.[35,36] Furthermore, other activators of protein kinase C, for example, tetradecanoylphorbol acetate (TPA), mimic the effects of 1,2-diacylglycerol and thrombin on Ca^{2+}-dependent serotonin release.[35,36]

In the case of thrombin, the enhanced Ca^{2+} sensitivity appears entirely dependent on the presence of GTP, suggesting the involvement of a GTP-binding protein in the stimulus–secretion pathway. This postulate is supported by the finding that the nonhydrolyzable analogs of GTP, for example, GTPγS, both enhance the effect of thrombin on the release of endogenous 1,2-diacylglycerol and on the Ca^{2+} dependence of secretion[16,23,31,35,36] and can also produce these effects in the absence of the agonist.

Although these data seem to implicate protein kinase C in the exocytotic mechanism,[38,39] it is uncertain whether activation of this kinase is an integral and essential part of the mechanism, or whether Ca^{2+} and the other secretagogues can also act via different pathways. For example, Ca^{2+} alone might act via activation of a calmodulin-dependent kinase, or GTPγS might act directly on a GTP-binding protein at or near the site of exocyto-

---

[33] C. M. Athayde and M. C. Scrutton, *Eur. J. Biochem.* **189,** 647 (1990).

[34] T. Morimoto, C. Oho, M. Ueda, S. Ogihara, and H. Takisawa, *J. Biochem. (Tokyo)* **108,** 311 (1990).

[35] R. J. Haslam and M. M. L. Davidson, *FEBS Lett.* **174,** 90 (1984).

[36] D. E. Knight and M. C. Scrutton, *Nature (London)* **309,** 66 (1984).

[37] D. E. Knight and M. C. Scrutton, *Eur. J. Biochem.* **106,** 183 (1986).

[38] Y. Nishizuka, *Nature (London)* **308,** 693 (1984).

[39] Y. Nishizuka, *Nature (London)* **334,** 661 (1988).

sis. Because the effect of thrombin is GTP dependent this latter postulate might also explain the action of this agonist. Tetradecanoylphorbate acetate might enhance the Ca^{2+} sensitivity of the secretory system by recruiting a protein kinase C component that enhances, but is not essential for, secretion.

The possible contribution of such differing pathways has been examined pharmacologically in the electropermeabilized platelet by addition of protein kinase C inhibitors such as staurosporine[40] and the protein kinase C pseudosubstrate.[41]

Staurosporine inhibits secretion evoked by Ca^{2+} alone, and secretion evoked by GTPγS, TPA, or thrombin together with GTP. Half-maximal inhibition (IC_{50}) for these responses induced by all these additions occurs at a similar concentration of staurosporine (30 nM). The protein kinase C pseudosubstrate also inhibits secretion induced by all these secretagogues with an IC_{50}, in each case, of close to 5 μM.[42]

Such data strongly support the postulate that activation of protein kinase C is necessary for secretion, with the different secretagogues all acting via a common final pathway in which activation of protein kinase C is an integral part.

Differential Secretion

Many cells contain several types of secretory vesicles. Release of the contents of one population of secretory vesicles can often be triggered without the release of the contents of the other population. Such differential secretion is exhibited by the intact platelet. For example, serotonin release from the amine storage granules is triggered by low concentrations of thrombin, whereas secretion of lysosomal acid hydrolases occurs only when higher concentrations of this agonist are used.[25] Other agonists, for example, thromboxane A_2, are only capable of causing amine storage granule secretion. The electropermeabilized platelet preparation allows examination of the intracellular factors controlling this phenomenon. The secretory systems for acid hydrolases and for serotonin have approximately the same requirement for MgATP and are sensitive to the same range of $[Ca^{2+}]$ when secretion is triggered by Ca^{2+} alone. In the presence of thrombin, however, the affinity for Ca^{2+} increases dramatically in the case of serotonin secretion, whereas this secretagogue enhances acid hydrolase secretion by increasing the extent of the response without appreciably altering the affinity for Ca^{2+} (Fig. 4). Qualitatively similar effects on acid hydrolase secretion are seen on addition either of GTPγS or of activators of

[40] S. P. Watson, J. McNally, L. J. Shipman, and P. P. Godfrey, *Biochem. J.* **249**, 345 (1988).
[41] C. House and B. E. Kemp, *Science* **238**, 1726 (1987).
[42] C. M. Athayde and D. E. Knight, *J. Physiol. (London)* **426**, 78P (1990).

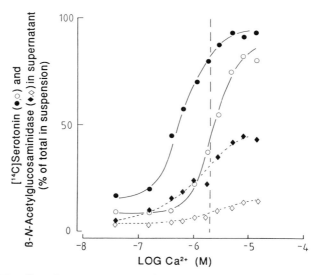

FIG. 4. The effect of thrombin on the Ca^{2+} dependence of amine storage granules (O, ●) and lysosomal (◇, ◆) secretion from electropermeabilized human platelets induced by Ca^{2+} in the absence (O, ◇) and the presence (●, ◆) of thrombin. The Ca^{2+} concentration, when added alone, needed to activate half-maximal secretion of both lysosomal and amine secretion is close to 2 μM and is shown by the dashed line. In the presence of thrombin the Ca^{2+} sensitivity for lysosomal secretion does not alter appreciably, unlike that for amine release. (From Knight et al.[27])

protein kinase C (TPA and 1,2-diacylglycerol).[27,33] These results can explain the differential response seen in the intact cell because at low levels of thrombin concentration the small increase in cytosolic $[Ca^{2+}]$ will permit secretion only of serotonin. Addition of higher thrombin concentrations cause a greater increase in cytosolic $[Ca^{2+}]$ and thus trigger acid hydrolase release. The exocytotic machinery of yet another secretory vesicle in the platelet (the protein storage granule) seems to respond differently to agents such as GTPγS.[43] If, therefore, activation of protein kinase C is required for all these secretory responses, we must consider whether isoenzymes of this kinase present in the same cell can exert different effects on the exocytotic machinery and thereby confer differential properties on the cellular secretory system.[39,44-46]

[43] K. Peltola and M. C. Scrutton, Biochem. Soc. Trans. 18, 466 (1990).
[44] P. F. Baker and D. E. Knight, Br. Med. Bull. 42, 399 (1986).
[45] J. L. Knopf, M. H. Lee, L. A. Sultzman, R. W. Kriz, C. R. Loomis, R. M. Hewick, and R. M. Bell, Cell (Cambridge, Mass.) 46, 491 (1986).
[46] D. E. Knight, in "Electroporation and Electrofusion in Cell Biology" (E. Neumann, A. E. Sowers, and C. A. Jordan, eds.), Chapter 18, p. 283. Plenum, New York, 1989.

[11] Exocytotic Membrane Fusion as Studied in Toxin-Permeabilized Cells

By GUDRUN AHNERT-HILGER, BRIGITTE STECHER, CORDIAN BEYER, and MANFRED GRATZL

Introduction

Permeabilized cells have been widely used in the analysis of exocytotic membrane fusion or intracellular Ca^{2+} regulation. They allow the study of the function of intracellular organelles *in situ* under conditions that are close to the physiological situation in intact cells.

High-voltage discharges[1,2,2a] or detergents such as digitonin or saponin[3-5] have been applied to permeabilize secretory cells. The pitfalls of these techniques, such as the resealing of pores or disintegration of intracellular membranes, have been discussed.[2,6]

To overcome some of the problems inherent in the techniques mentioned above, we developed an approach that makes use of the well-defined pores generated by pore-forming toxins: Alpha-toxin from *Staphylococcus aureus* yields only small pores. Streptolysin O (SLO) from *β*-hemolytic streptococci yields large pores and therefore allows the diffusion of large molecules into and out of secretory cells.[6,7]

Permeabilized rat pheochromocytoma cells (PC-12)[8-11] and bovine adrenal chromaffin cells kept in culture for a short time[12-15] have been

[1] P. Baker and D. Knight, *Nature (London)* **276**, 620 (1978).

[2] D. Knight and M. Scrutton, *Biochem. J.* **234**, 497 (1986).

[2a] D. E. Knight and M. Scrutton, this volume [10].

[3] S. Wilson and N. Kirshner, *J. Biol. Chem.* **258**, 4989 (1983).

[4] L. Dunn and R. Holz, *J. Biol. Chem.* **258**, 4989 (1983).

[5] J. Brooks and S. Treml, *J. Neurochem.* **40**, 468 (1983).

[6] G. Ahnert-Hilger, W. Mach, K. J. Föhr, and M. Gratzl, *Methods Cell Biol.* **31**, 63 (1989).

[7] S. Bhakdi and J. Tranum-Jensen, *Rev. Physiol. Biochem. Pharmacol.* **107**, 147 (1987).

[8] G. Ahnert-Hilger, S. Bhakdi, and M. Gratzl, *J. Biol. Chem.* **260**, 12730 (1985).

[9] G. Ahnert-Hilger, M.-F. Bader, S. Bhakdi, and M. Gratzl, *J. Neurochem.* **52**, 1751 (1989).

[10] G. Ahnert-Hilger, M. Bräutigam, and M. Gratzl, *Biochemistry* **26**, 7842 (1987).

[11] G. Ahnert-Hilger and M. Gratzl, *J. Neurochem.* **49**, 764 (1987).

[12] M.-F. Bader, D. Thierse, D. Aunis, G. Ahnert-Hilger, and M. Gratzl, *J. Biol. Chem.* **261**, 5777 (1986).

[13] G. Ahnert-Hilger, U. Weller, M. E. Dauzenroth, E. Habermann, and M. Gratzl, *FEBS Lett.* **242**, 245 (1989).

[14] B. Stecher, M. Gratzl, and G. Ahnert-Hilger, *FEBS Lett.* **248**, 23 (1989).

[15] B. Stecher, U. Weller, E. Habermann, M. Gratzl, and G. Ahnert-Hilger, *FEBS Lett.* **255**, 391 (1989).

preferentially used to analyze exocytotic membrane fusion. In addition, alpha-toxin as well as SLO have been successfully applied to other secretory systems such as cytotoxic T lymphocytes,[16,17] mast cells,[18,19] or cortical synaptosomes.[20] Toxin-permeabilized preparations have also been instrumental in analyzing the intracellular glucose metabolism in hepatocytes,[21] the chain of events leading to smooth muscle contraction,[22] and the regulation of intracellular Ca^{2+} sequestration[23] (see also [12] in this volume).

Materials and Methods

Alpha-toxin is prepared as described[24] from the culture supernatant of *S. aureus* strain wood 46 (kindly provided by S. Bhakdi, Mainz, Germany). The purified toxin is dialyzed against KG buffer (see Table I),[24a] lyophilized, and stored at $-20°$ for several months. Streptolysin O is purified as in Bhakdi *et al.*[25] and is kindly provided by S. Bhakdi; it can also be dialyzed against KG buffer and stored at $-20°$ without loss of activity for several months. The activity of both pore-forming toxins is determined using 2.5% (v/v) rabbit erythrocytes and is given in hemolytic units (HU)/ml.[6,24] Rat pheochromocytoma cells (PC-12) (kindly provided by H. Thoenen, Max Planck Institut für Psychiatrie, Martinsried, Germany) are cultivated as described earlier.[8] Bovine adrenal chromaffin cells are prepared and kept in short-term cultures.[14,26]

Properties of Alpha-Toxin- and Streptolysin O-Permeabilized Cells

Alpha-toxin permeabilizes cells only for small molecules (up to 1000 Da), as tested by measuring the escape of Rb^+ or ATP, whereas cytoplasmic lactate dehydrogenase remains within the cells.[8,12,24] The free

[16] H. Schrezenmeier, G. Ahnert-Hilger, and B. Fleischer, *J. Exp. Med.* **168,** 817 (1988).
[17] H. Schrezenmeier, G. Ahnert-Hilger, and B. Fleischer, *J. Immunol.* **141,** 3785 (1988).
[18] T. Howell and B. Gomperts, *Biochim. Biophys. Acta* **927,** 177 (1987).
[19] T. Howell, S. Cockcroft, and B. Gomperts *J. Cell Biol.* **105,** 191 (1987).
[20] L. Decker, P. DeGraan, B. Oestreicher, D. Versteeg, and W. Gispen, *Nature (London)* **342,** 74 (1989).
[21] B. F. McEwen and W. J. Arion, *J. Cell Biol.* **100,** 1922 (1985).
[22] T. Kitazawa, S. Kobayashi, K. Horiuti, A. V. Somlyo, and A. P. Somlyo, *J. Biol. Chem.* **264,** 5339 (1989).
[23] K. J. Föhr, J. Scott, G. Ahnert-Hilger, and M. Gratzl, *Biochem. J.* **262,** 83 (1989).
[24] I. Lind, G. Ahnert-Hilger, G. Fuchs, and M. Gratzl, *Anal. Biochem.* **164,** 84 (1987).
[24a] M. Bräutigam, R. Dreesen, and A. Herken, *Naunyn-Schmiedeberg's Arch. Pharmacol.* **320,** 85 (1982).
[25] S. Bhakdi, M. Roth, A. Sziegoleit, and J. Tranum-Jensen, *Infect. Immun.* **46,** 394 (1984).
[26] B. Livett, *Physiol. Rev.* **64,** 1103 (1984).

TABLE I
ASSAY FOR EXOCYTOSIS FROM PC-12 CELLS[a]

1. Load cells with labeled dopamine (noradrenaline) for 1–2 hr in serum-free culture medium supplemented with 1 mM ascorbic acid
2. Wash the cells with Ca^{2+}-free balanced salt solutions several times
3. Suspend the cells in KG buffer (add KG buffer to cells on plates) containing 150 mM potassium glutamate, 10 mM PIPES, 5 mM NTA, 0.5 mM EGTA, pH 7.2 (plus 2 mM Mg^{2+}–ATP and 1 mM free Mg^{2+}). In some experiments 1 mM free Mg^{2+} was added. For permeabilization, 120 HU/ml (60 Hu/ml) alpha-toxin or 120 HU/ml (60 HU/ml) SLO corresponding to 300–500 HU/10^7 cells were used
4. Incubate the cells with pore-forming toxins diluted in KG buffer containing 0.1 BSA: alpha-toxin (20–30 min at 25, 30, or 37°) (same conditions) or SLO (5 min at 0°; 1 or 2 min at 25, 30, or 37°). For permeabilization with SLO the addition of DTT (1 mM) is necessary.
5. Centrifuge (3000 g, 30 s) and remove supernatant (removal of supernatant)
6. If desired, perform a further incubation with substances to be tested dissolved in KG buffer (plus Mg^{2+}–ATP and 1 mM free Mg^{2+}) for 20 to 40 min at 25, 30, or 37°
7. Repeat step 5
8. Stimulate with micromolar amounts of free Ca^{2+} in KG buffer (plus Mg^{2+}–ATP and 1 mM) free Mg^{2+}) for 10 min at 25, 30, or 37°
9. Centrifuge and count released catecholamines in the supernatant or perform HPLC analysis of catecholamines and their metabolites in the supernatant, extracted with 100 mM $HClO_4$
10. Solubilize the cells with 0.2% (w/v) sodium dodecyl sulfate (SDS) and count the catecholamines remaining in the cells. For HPLC analysis, extract cells with 100 mM $HClO_4$
11. Prior to HPLC analysis, dilute the supernatant or the cell extract with 50 mM $HClO_4$. HPLC separation and electrochemical detection (octadecylsilane, 5 μm, oxidation potential +700 mV) is performed as described,[24a] with some modifications: for separation of the catecholamines and the dopamine metabolites a mobile phase is used containing 50 mM sodium acetate, 20 mM citric acid, 2.8 mM octanesulfonic acid, 0.001 mM EDTA, 1 mM di-n-butylamine, pH 4.5, supplemented with 5% (v/v) methanol. The flow rate is usually 0.8 ml and 20-μl samples are injected

[a] Procedures for chromaffin cells in primary culture are given in parentheses.

passage of Ca^{2+}, which allowed a careful analysis of intracellular Ca^{2+} sequestration (see Ref. 23, and [12] in this volume), is also an excellent indicator of sufficient permeability. However, the pores formed in the plasma membrane by hexamerization of alpha-toxin monomers are too small to allow the free passage of toxin monomers.[7,27] Thus the attack of the toxin is restricted to the plasma membrane.

Streptolysin O enables not only small molecules but also proteins to escape from or enter cells. This has been demonstrated by measuring the release of lactate dehydrogenase[8] or the access of antibodies to intracellular

[27] R. Füssle, S. Bhakdi, A. Sziegoleit, J. Tranum-Jensen, T. Kranz, H.-J., and Wellensiek, *J. Cell Biol.* **91,** 83 (1981).

proteins.[9] The large SLO pores also allow the study of the intracellular action of clostridial neurotoxins and their active fragments on exocytosis.[9,11,13–15,28,29]

Damage of intracellular membranes by SLO can be avoided in two ways: by a short incubation of the cells with SLO (1–2 min) at 25, 30, or 37°, or by an incubation at 0°, a condition under which all the SLO monomers present bind to the plasma membrane, followed by warming to trigger pore formation.[7,30] In contrast to the action of SLO,[30] membrane permeabilization by digitonin is insensitive to temperature and therefore is more difficult to control.[6]

Assay for Exocytosis in Permeabilized PC-12 or Bovine Adrenal Chromaffin Cells

In most of the studies dealing with exocytosis from permeabilized cells, an "intracellular medium" containing potassium as a main cation and glutamate as an anion[1] (see also Table I) was used. Because the free Ca^{2+} concentration within the cells under resting conditions, as well as during stimulation, is in the micromolar range, this ion must be carefully controlled in the buffers used. A combination of chelators for divalent cations is suitable to buffer the free Ca^{2+} concentration from 0.1 to 100 μM under experimental conditions. Thus a typical buffer for permeabilization contains 150 mM potassium glutamate, 0.5 mM ethylene glycol-bis(β-aminoethyl ether)-N,N,N',N'-tetraacetic acid (EGTA), 5 mM ethylene diamine tetraacetic acid (EDTA), 10 mM piperazine-N,N'-bis(2-ethanesulfonic acid) (PIPES), pH 7.2 (no differences were found using pH between 6.6 and 7.2). Added Mg^{2+} and ATP, as well as the pH of the medium, must be carefully considered because they alter the equilibrium between Ca^{2+} and the chelators present. The free Ca^{2+} and Mg^{2+} concentrations are calculated by a computer program and controlled by Ca^{2+}- and Mg^{2+}-specific electrodes (see [12] in this volume). Each Ca^{2+} buffer is prepared separately from stock solutions with a final check of pH, pCa, or pMg.[31] Buffers can be stored at −20° but should be thawed only once because decomposition of ATP may occur.

[28] B. Stecher, G. Ahnert-Hilger, U. Weller, T. P. Kemmer, and M. Gratzl, *Biochem. J.* **283**, 899 (1992).

[29] B. Stecher, J. Hens, U. Weller, M. Gratzl, W. H. Gispen, and P. De Graan, *FEBS Lett.* **312**, 192 (1992).

[30] F. Hugo, J. Reichweiss, M. Arvand, S. Krämer, and S. Bhakdi, *Infect. Immun.* **54**, 641 (1986).

[31] U. Wegenhorst, M. Gratzl, K. J. Föhr, and G. Ahnert-Hilger, *Neurosci. Lett.* **106**, 300 (1989).

Adrenal chromaffin cells in culture[26] or rat pheochromocytoma cells[32] take up labeled catecholamines and store them within secretory vesicles from which they can be released on stimulation. The released catecholamines can be detected either directly by high-performance liquid chromatography (HPLC) or the intracellular stores can be labeled with tritiated dopamine or noradrenaline. Although the different labeled catecholamines can be taken up by both types of cells, we generally used [^{3}H]dopamine to preload PC-12 cells and [^{3}H]noradrenaline to label bovine adrenal chromaffin cells. After permeabilization of the plasma membrane, release of stored labeled or endogenous catecholamines can be triggered by micromolar concentrations of Ca^{2+}. Table I summarizes the assay of exocytosis for both PC-12 and adrenal chromaffin cells, using alpha-toxin or SLO to permeabilize the plasma membrane.

Proof of Exocytotic Release of Catecholamines from Toxin-Permeabilized PC-12 or Adrenal Chromaffin Cells

The observed release of secretory product from permeabilized cells may occur by exocytosis or may be due to an unspecific leakiness of secretory vesicles. Even the loss of intact vesicles from the cells was observed in digitonin-permeabilized adrenal chromaffin cells.[12] The parallel release of low and high molecular weight secretory products can be taken as a proof for an exocytotic event, if under the same conditions large cytoplasmic constituents remain within the cells. Parallel release of catecholamines and dopamine β-hydroxylase (dopamine β-monooxygenase) from electrically permeabilized chromaffin cells,[33] or of catecholamines and chromogranin A from alpha-toxin-permeabilized chromaffin cells,[12] has been reported. Similarly, the release of vesicular serine esterase from alpha-toxin-permeabilized cytotoxic T lymphocytes was not accompanied by leakage of lactate dehydrogenase.[16,17]

Another approach to distinguish between unspecific release and exocytosis requires a precise analysis of the endogenous catecholamines and their metabolites in permeabilized PC-12 cells. The discharge of vesicular dopamine into the cytoplasm (e.g., by nigericin) results in its enzymatic oxidation, mainly to 3,4-dihydroxyphenylacetic acid (DOPAC). By contrast the direct exocytotic release of catecholamines avoids the cytoplasm and, therefore, metabolic oxidation. The pattern of catecholamines and their metabolites released by alpha-toxin- or SLO-permeabilized PC-12 cells is in accordance with these predictions[10] (Figs. 1 and 2). Calcium ions

[32] L. Greene and A. Tischler, *Adv. Cell. Neurobiol.* **3**, 373 (1982).
[33] D. Knight and P. Baker, *J. Membr. Biol.* **68**, 107 (1982).

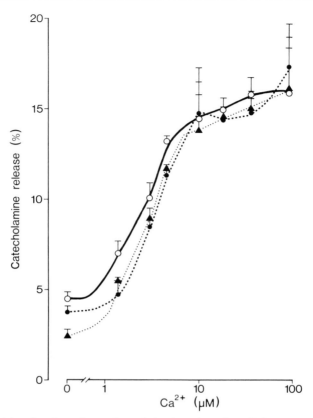

Fig. 1. Calcium ion dependence of catecholamine release from SLO-permeabilized PC-12 cells. Prelabeled ([³H]dopamine) (O) PC-12 cells were washed three times with a Ca^{2+}-free balanced salt solution. The washed cells were suspended in ice-cold ATP-free KG buffer as described in Table I, but containing 1 mM EGTA, 1 mM HEDTA, and 1 mM NTA (instead of 0.5 mM EGTA and 5 mM NTA) adjusted to pH 7.0, for optimal buffering of the free Ca^{2+} concentration between 1 and 10 μM (see also K. J. Föhr, W. Warchol, and M. Gratzl, this volume [12]), and SLO (120 HU/ml), as outlined in Table I. The cells were incubated for a further 25 min at 25° before they were stimulated for 10 min with the various free Ca^{2+} concentrations given on the abscissa. The supernatant was collected. The released radioactive and endogenous catecholamines were detected by β counting and HPLC, respectively. The total (100%) endogenous content of catecholamines was 0.57 ± 0.1 μg dopamine and 0.43 ± 0.1 μg noradrenaline per 10^6 cells ($n = 24$, SD). Each sample contained about 2.5 × 10^5 cells.

result in the release of dopamine and noradrenaline that parallels that of [³H]dopamine (Fig. 1). In contrast, the release of DOPAC is not stimulated. However, DOPAC is released in considerable amounts from permeabilized PC-12 cells on treatment with nigericin, which discharges the vesicular content directly into the cytoplasm (Fig. 2).[10] Besides Ca^{2+},

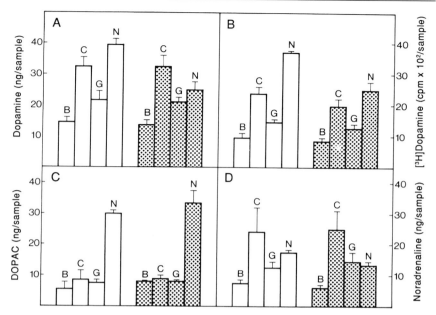

FIG. 2. Basal and stimulated release of catecholamines and DOPAC from SLO-permeabilized PC-12 cells. Prelabeled PC-12 cells were treated with SLO as described in Table I. After removal of the supernatant, the cells were suspended in fresh KG buffer containing either no additive (B), 14 μM free Ca^{2+} (C), 200 μM GMPPNHP (G), or 2 μM nigericin (N). After 20 min at 25°, the released catecholamines [(A) dopamine, (B) [^{3}H]dopamine, (C) DOPAC, and (D) noradrenaline] were determined in the supernatant either by their radioactivity or by HPLC. Cells were incubated in KG buffer without ATP (open bars) or in KG buffer containing 2 mM ATP (dotted bars). The free Mg^{2+} concentration in all buffers was 1 mM. Note that ATP does not substantially alter exocytosis. An increased release of DOPAC was detected only after nigericin treatment, as previously found for alpha-toxin-permeabilized PC-12 cells.[10] The endogenous content of catecholamines was as follows: 0.44 ± 0.09 μg dopamine and 0.35 ± 0.06 μg noradrenaline per 10^6 cells in the absence of ATP, and 0.42 ± 0.04 μg dopamine and 0.26 ± 0.05 μg noradrenaline per 10^6 cells in the presence of ATP ($n = 12$, SD). Each sample contained 2 × 10^5 cells.

GMPPNHP (an activator of G proteins)[34] also triggers release of catecholamines by these cells. The unchanged values of DOPAC indicate an exocytotic event (Fig. 2). Thus PC-12 cells permeabilized either with alphatoxin or with SLO release their catecholamines by exocytosis when stimulated with Ca^{2+} or GMPPNHP. These data also demonstrate that toxin-permeabilized PC-12 cells are able to metabolize catecholamines and thus are suitable to study the metabolism of catecholamines under well-defined conditions. Also, in adrenal chromaffin cells Ca^{2+} and GMPPNHP

[34] A. Gilman, *Annu. Rev. Biochem.* **56,** 615 (1987).

cause a parallel release of noradrenaline and adrenaline, which parallels the release observed with [³H]noradrenaline (Fig. 3). Metabolites of catecholamines after nigericin treatment cannot, however, be distinguished clearly from the large amounts of noradrenaline and adrenaline present in these cells (see captions to Figs. 1–3).

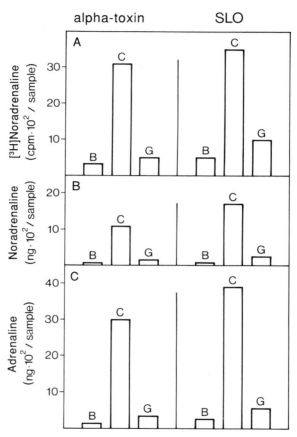

FIG. 3. Basal and stimulated release of catecholamines from alpha-toxin- and SLO-permeabilized adrenal chromaffin cells. Preloaded bovine adrenal chromaffin cells were treated either with alpha-toxin or with SLO as described in Table I. After removal of the supernatant, the cells were suspended in fresh KG buffer supplemented with 1 mM free Mg²⁺ and 2 mM Mg²⁺–ATP containing no additive (B), 14 μM Ca²⁺ (C), or 100 μM GMPPNHP (G). After 20 min at 25° the released catecholamines [(A) [³H]noradrenaline, (B) noradrenaline, and (C) adrenaline] were detected in the supernatant by determining the radioactivity or by HPLC. The endogenous content of catecholamines was 3.6 ± 0.5 μg noradrenaline and 11.5 ± 1.5 μg adrenaline per 10⁶ cells ($n = 24$, SD). Each sample contained 8 × 10⁵ cells.

TABLE II
EFFECT OF ATP ON EXOCYTOSIS FROM
STREPTOLYSIN O-PERMEABILIZED ADRENAL
CHROMAFFIN CELLS[a]

Incubation time (min)	Ca^{2+}-stimulated [^{3}H]noradrenaline release (%)	
	−ATP	+ATP
2	3.4	13.1
25	0.8	4.5

[a] Bovine adrenal chromaffin cells were permeabilized with SLO for 2 min at 37° in KG buffer containing 1 mM free Mg^{2+}, and either no ATP or 2 mM ATP (see Table I). After removal of the supernatant the cells were incubated for the times indicated in Ca^{2+}-free KG buffer with or without ATP. Stimulation was performed for 10 min in KG buffer with or without ATP and supplemented with 20 μM free Ca^{2+}. Basal release under both conditions was subtracted [minus ATP (3.3 ± 0.3) and plus ATP (2.3 ± 0.2), $n = 6$, SD]. Values represent the mean of two samples.

Exocytotic Membrane Fusion: An ATP-Dependent Process

The two types of chromaffin cells used in the authors' laboratory differ in their molecular requirements for exocytosis.

In PC-12 cells permeabilized with alpha-toxin or SLO, Ca^{2+} alone is sufficient to release the stored catecholamines (see also Figs. 1 and 2).[9–11] In contrast, bovine adrenal chromaffin cells require additional Mg^{2+}–ATP[1,12,31] (see also Fig. 3 and Tables II and III). Provided that the molecular mechanism of exocytotic membrane fusion as analyzed in various secretory cells is a common process, it should occur under the same conditions. Thus in permeabilized adrenal chromaffin cells an ATP-dependent step, which is probably responsible for the vesicle transport to the plasma membrane, operates in addition to the exocytotic membrane fusion between the vesicular and the plasma membrane. The former is presumably not necessary within PC-12 cells because most of the vesicles are already located near the plasma membrane.[35] In permeabilized adrenal chromaffin cells, a small fraction of chromaffin vesicles is also near the

[35] O. Watanabe, M. Torda, and J. Meldolesi, *Neuroscience* 10, 1011 (1983).

TABLE III

MAGNESIUM ION REQUIREMENT OF ATP-DRIVEN
STEP DURING EXOCYTOSIS FROM
STREPTOLYSIN O-PERMEABILIZED ADRENAL
CHROMAFFIN CELLS[a]

Condition	Ca^{2+}-stimulated [^{3}H]noradrenaline release (%)	
	$-$ATP	$+$ATP
No Mg^{2+}	3.4	3
Free Mg^{2+} (1 mM)	1.8	13.6

[a] The experimental procedure followed the protocol given in Table II. The Mg^{2+}-free KG buffer contained either no ATP or 2 mM ATP. The basal release was subtracted: no addition (4.1%), ATP alone (4.9%), Mg^{2+} alone (3.9%), Mg^{2+} plus ATP (3.2%). Values are the mean of two determinations.

plasma membrane and can be released without additional ATP. Indeed, after permeabilization with SLO, Ca^{2+} alone causes some catecholamine release, which amounts to roughly one fourth of that observed in the presence of ATP (Table II). Table III demonstrates that not ATP alone, but ATP in combination with Mg^{2+}, is required for the energy-consuming step during exocytosis. Similar results have been obtained for digitonin-permeabilized adrenal chromaffin cells.[36,37] Exocytosis as an ATP-independent step, which is inhibited by the light chains of tetanus toxin and botulinum A toxin, was also observed in permeabilized neurosecretosomes[38-41] and is thus not unique in PC-12 cells.

[36] T. Schäfer, U. Karli, E. Gratwohle, F. Schweizer, and M. Burger, *J. Neurochem.* **49,** 1697 (1987).
[37] R. Holz, M. Bittner, S. Peppers, R. Senter, and D. Eberhard, *J. Biol. Chem.* **264,** 5412 (1989).
[38] M. Cazalis, G. Dayanithi, and J. Nordmann, *J. Physiol. (London)* **390,** 71 (1987).
[39] G. Dayanithi and J. Nordmann, *Neurosci. Lett.* **106,** 305 (1989).
[40] G. Daganithi, G. Ahnert-Hilger, U. Weller, J. J. Nordmann, and M. Gratzl, *Neuroscience* **39,** 711 (1990).
[41] G. Daganithi, U. Weller, G. Ahnert-Hilger, H. Link, J. J. Nordmann, and M. Gratzl, *Neuroscience* **46,** 489 (1992).

Acknowledgment

This work and the publications cited from the authors' laboratory were supported by the Deutsche Forschungsgemeinschaft (Gr 681), the State of Baden Württemberg, and the University of Ulm.

[12] Calculation and Control of Free Divalent Cations in Solutions Used for Membrane Fusion Studies

By KARL J. FÖHR, WOJCIECH WARCHOL, and MANFRED GRATZL

Introduction

The investigation of intracellular processes requires aqueous media that mimic the intracellular fluid. The adjustment of the free Ca^{2+} concentration in these media is of critical importance because of the essential role of Ca^{2+} in the control of exocytotic membrane fusion (see, e.g., [11] in this volume). It is evident that precisely defined free Ca^{2+} concentrations in the submicromolar range cannot be easily obtained by adding the salts to a solution, because laboratory equipment, distilled water, and chemicals are contaminated with Ca^{2+}. In addition, Ca^{2+} binding to cellular constituents and membranes as well as active sequestration by cellular organelles must be taken into account. Therefore divalent cations must be buffered, as is routinely done for protons. One common problem is the choice of appropriate ligands to buffer free Ca^{2+} at a given value, and the calculation of complex media that contain more than one ligand and more than one metal ion. In this chapter a computer program is described that allows the calculation of multiple equilibria between different ligands and metal ions. Although the media prepared according to the calculation generally give concentrations in good agreement with measured values, the calculation should always be controlled. For this purpose, an easy and inexpensive procedure for the preparation of ion-selective electrodes is then described.

Calculation of Ligand–Metal Equilibria

Special ligands have been developed for buffering metal ions.[1] The cation-binding sites of these ligands also bind protons. Thus the addition of acid to aqueous solutions of metal–ligand complexes leads to an increase of free metal ions. Conversely, alkalinization results in stronger binding

---

[1] G. Schwarzenbach, H. Senn, and G. Anderegg, *Helv. Chim. Acta* **40,** 1886 (1957).

between ligands and metal ions and, consequently, in a decrease in the free metal ion concentration. This competition between protons and metal ions is not a serious problem because the pH is generally buffered to a fixed value that can be used for the purposes of calculation as a constant. To simplify the mathematical calculation, so-called "apparent association constants" with different definitions have been introduced.[2-5] The term *apparent association constants* is used in this chapter for recalculation of absolute metal–ligand association constants for a fixed pH value.[2]

The absolute association constants required for the calculation of metal buffers were originally determined at an ionic strength I of 0.1 and a temperature T of 20°.[1,3,6] Biological experiments are often carried out at different temperatures and ionic strengths. Mathematical procedures have been proposed to adjust the absolute association constants for the desired conditions.[7,8] It should be remembered that absolute association constants are listed in terms of concentrations, whereas pH measured with a glass electrode is determined in terms of activity.[6,9] To obtain the same units, either the proton activity can be converted to concentration[10] or metal association constants can be expressed in terms of activities.[11] Alternatively, mixed binding constants may be used.[4,6,9]

It is often necessary to buffer Ca^{2+} in the submicromolar range and Mg^{2+} in the millimolar range, that is, at concentrations occurring in the cytosol of living cells. In Fig. 1 a computer program is described to calculate such a metal buffer with ethylene glycol-bis(β-amino ethyl ether)-N,N,N',N'-tetraacetic acid (EGTA) as a ligand. Briefly, in Part I of Fig. 1 the absolute association constants (for $I = 0.1$ and $T = 20°$)[6] and the considered equilibria are listed. Part II (Fig. 1) contains the input of the final parameters (free divalent metal ions and total amount of ligands), including conversion of proton activity to proton concentration.[10,11] Part III (Fig. 1) calculates the apparent association constants (according to Ref. 2) followed by that of the free ligand concentration. From the free ligand

[2] H. Portzehl, P. C. Caldwell, and J. C. Rüegg, *Biochim. Biophys. Acta* **79**, 581 (1964).

[3] A. Fabiato and F. Fabiato, *J. Physiol. (London)* **75**, 463 (1979).

[4] J. R. Blinks, W. G. Wier, P. Hess, and F. G. Prendergast, *Prog. Biophys. Mol. Biol.* **40**, 1 (1982).

[5] N. Stockbridge, *Comput. Biol. Med.* **17**, 299 (1987).

[6] A. E. Martell and R. M. Smith, "Critical Stability Constants," Vol. 1, Plenum, New York, 1974.

[7] O. Scharf, *Anal. Chim. Acta* **109**, 291 (1979).

[8] S. M. Harrison and D. M. Bers, *J. Am. Physiol.* **256**, C1250 (1989).

[9] R. Y. Tsien and T. J. Rink, *Biochim. Biophys. Acta* **599**, 623 (1980).

[10] D. Ammann, T. Bührer, U. Schefer, M. Müller, and W. Simon, *Pflügers Arch.* **409**, 223 (1987).

[11] A. C. H. Durham, *Cell Calcium* **4**, 33 (1983).

concentration and the apparent association constants the concentration of each metal–ligand complex is calculated. Finally, the sum of each metal species (free and complexed forms) is calculated to give the total amount of required metals to prepare the medium of interest. This procedure forms the basis for calculating complex media and has the advantage over other published programs[3,5,13] that no iterative calculation is required.

The example in Fig. 1 shows the computation to buffer free Ca^{2+} in the submicromolar range. To investigate the Ca^{2+} dependency of biological processes appropriate buffers in the range between 0.1 and 100 μM are required. Calcium ion buffering in the higher micromolar range can be achieved by lowering the pH (Fig. 2a). Alternatively, if the pH must be kept constant, other ligands must be selected. For this purpose, ligands like HEDTA or nitrilotriacetic acid (NTA) are applicable[9] although these ligands, like ethylenediaminetetraacetic acid (EDTA), do not discriminate as well as EGTA between Ca^{2+} and Mg^{2+} (Fig. 2b). Furthermore, in some experiments, naturally occurring ligands such as ATP or GTP must also be taken into account. To follow these requirements the program must be enlarged. This can be done by a simple routine that calculates the additional apparent association constants of the new ligand metal complexes. Thereafter, the concentration of the new ligand–metal complexes can be calculated and summarized as described above.

The computer program developed by the authors considers nine different ligands (EDTA, EGTA, HEDTA, NTA, ATP, ADP, GTP, phosphate, and creatine phosphate), and corrections for temperature and ionic strength. The program calculates either the total amount of metals to give the desired free metal concentrations (Ca^{2+}, Mg^{2+}) or, in the reversed mode, it calculates the free metal concentration for a given total amount of metals and the selected mixture of ligands. Furthermore, an option exists for calculating the apparent association constants under different conditions (pH, T, I), in order to choose the appropriate ligands for the experimental purposes. A further option illustrates the complex situation by drawing buffer curves (see Fig. 2b). In addition, absolute association constants, enthalpy values for temperature correction, and Debye–Hückel parameters for correction of ionic strength can be changed and saved as a separate file. (The program may be obtained from the authors on request.)

Despite the sophisticated calculation of metal buffers as described above, the media prepared do not necessarily have the desired free divalent metal concentrations. Apart from uncertainties in the absolute association

---

[12] P. C. Meier, D. Ammann, W. E. Morf, and W. Simon, *in* "Medical and Biological Applications of Electrochemical Devices" (J. Koryta, ed.), p. 13. 1980.
[13] A. Fabiato, this series, Vol. 157, p. 378.

```
rem   Part I
rem   Main forms present in a solution containing:
rem   the Ligand EGTA (L), Ca and Mg:  L_free + Ligand-Metal-Complexes + Metal_free

rem   Ligand_free              : L, HL, H2L, H3L, H4L        (L = EGTA^4-)
rem   Ligand-Metal-Complexes   : CaL, CaHL, MgL, MgHL
rem   Metal_free               : Ca_free, Mg_free

rem   absolute association constants (log K values for
rem   T = 20°C, I = 0.1); for other conditions the
rem   absolute association constants must be recalculated
```

equilibria

```
rem:  H  +  L    --  HL
rem:  H  +  HL   --  H2L
rem:  H  +  H2L  --  H3L
rem:  H  +  H3L  --  H4L
rem:  Ca +  L    --  CaL
rem:  H  +  CaL  --  CaHL
rem:  Mg +  L    --  MgL
rem:  H  +  MgL  --  MgHL
```

$$K_{H1} = 10^{9.47}$$
$$K_{H2} = 10^{8.85}$$
$$K_{H3} = 10^{2.66}$$
$$K_{H4} = 10^{2.0}$$
$$K_{Ca1} = 10^{10.97}$$
$$K_{Ca2} = 10^{3.79}$$
$$K_{Mg1} = 10^{5.21}$$
$$K_{Mg2} = 10^{7.62}$$

```
rem   Part II
rem   Input: desired conditions: Metal_free, Ligand_total, pH
```

$$Ca_{free} = 10^{-7}$$
$$Mg_{free} = 10^{-3}$$
$$L_{total} = 10^{-2}$$

$$pH = 7.2$$
$$H = 10^{-7.089}$$

```
rem:  Ca = Ca^2+      rem:  (activity)
rem:  Mg = Mg^2+      rem:  (concentration)
```

```
rem  Part III
rem  Calculation of apparent association constants (depending on H)          rem: Calculated values

Sum = 1 + KH1*H + KH1*KH2*H^2 + KH1*KH2*KH3*H^3 + KH1*KH2*KH3*KH4*H^4         rem: 14108
                                                                                              log Kapp
KappCa1 = KCa1 / Sum                                                         rem: 6614605.5    6.82
KappCa2 = KCa1 * KCa2 * H / Sum                                              rem: 3322.8       3.52

KappMg1 = KMg1 / Sum                                                         rem: 11.5         1.06
KappMg2 = KMg1 * KMg2 * H / Sum                                             rem: 39           1.59

rem  Calculation of free Ligand concentration (Lfree)

Lfree = Ltotal / ( 1 + KappCa1 * Cafree + KappCa2 * Cafree                   rem: 5.84^-3   (Mol/l)
                       KappMg1 * Mgfree + KappMg2 * Mgfree )

rem  Calculation of Ligand-Metal-Complexes (CaL, CaHL, MgL, MgHL)

CaL  = Lfree * KappCa1 * Cafree                                             rem: 3.863^-3  (Mol/l)
CaHL = Lfree * KappCa2 * Cafree                                             rem: 1.941^-6  (Mol/l)

MgL  = Lfree * KappMg1 * Mgfree                                            rem: 6.713^-5  (Mol/l)
MgHL = Lfree * KappMg2 * Mgfree                                            rem: 2.280^-4  (Mol/l)

rem  Output: required total metal concentrations to achieve the desired free metal concentrations

Catotal = Cafree + CaL + CaHL                                               rem: 3.864^-3  (Mol/l)
Mgtotal = Mgfree + MgL + MgHL                                               rem: 1.295^-3  (Mol/l)
```

FIG. 1. Program to calculate metal buffers.

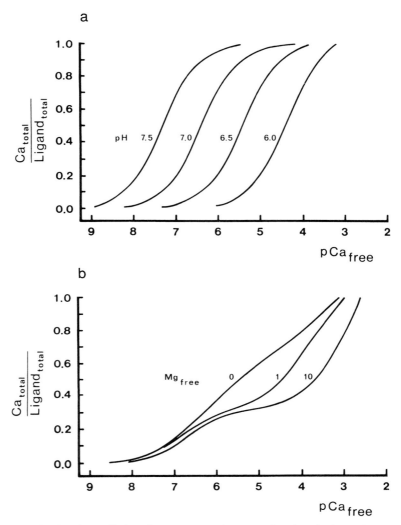

FIG. 2. Calcium ion buffering with selected media. (a) Calcium ion binding to ligands such as EGTA strongly depends on pH. The optimal buffer capacity for a given pH value occurs at a ratio (Ca_{total} to $Ligand_{total}$) of 0.5, which corresponds to the apparent association constant. From these curves it can be deduced that Ca^{2+} buffering is assured at ratios between 0.2 and 0.8. (b) Calcium ion buffer curve for more complex media (5 mM EGTA, 5 mM HEDTA, 5 mM NTA; pH 7.2) to cover a wide range of free Ca^{2+} concentrations at different Mg^{2+} concentrations (0, 1, or 10 mM). In the absence of Mg^{2+} almost constant Ca^{2+} buffering capacity can be obtained between pCa 6.8 and pCa 4. In the presence of increasing amounts of free Mg^{2+} progressively reduced Ca^{2+} buffering exists, as indicated by the shoulder around 10 μM free Ca^{2+}.

constants (as documented by differences in the published constants), the purity of the ligands[14] and errors in pH measurements[15] may contribute to a difference between calculated and actual free metal ion concentrations. Therefore the prepared media should be checked by ion-specific electrodes.

Estimation of Free Metals with Ion-Selective Electrodes

Ion-selective electrodes are valuable tools to measure the activity of ions. To become familiar with the chemical, physical, and mathematical background of ion-selective electrodes the reader should consult previous publications.[16-18]

Ion-selective electrodes may be obtained from different suppliers or can be made in a laboratory workshop. The equipment for ion-selective electrodes is analogous to that of a pH electrode: it consists of ion-selective and reference half-cells connected to a recording system (conventional pH meter).

The most important part of the ion-selective half-cell is the ion-selective membrane. The Ca^{2+}-selective polyvinyl chloride (PVC) membranes are made according to Schefer et al.[19] and Mg^{2+}-selective membranes are made according to Hu et al.[20] For the preparation of Ca^{2+}-selective membranes the neutral ligand ETH129 was chosen because of its low detection limit and high selectivity over other ions.[10,19] All chemicals necessary for the preparation of ion-selective membranes are commercially available from Fluka (Buchs, Switzerland). The Ca^{2+}-selective (Mg^{2+}-selective: values in parentheses) membranes are made by dissolving 102.1 mg (120.1 mg) polyvinyl chloride, 204.1 mg (238 mg) o-nitrophenyloctyl ether, 1.75 mg (2.25 mg) potassium tetrakis(4-chlorophenyl)borate, and 3.1 mg ETH129 (3.64 mg ETH5124) in 5 ml tetrahydrofuran. When fully dissolved the fluid is poured into an appropriate glass petri dish, 3 cm in diameter, which should be partly covered to assure slow evaporation of the solvent overnight. The remaining PVC membrane can be stored in the

[14] D. J. Miller and G. L. Smith, J. Am. Physiol. **246**, C160 (1984).
[15] J. A. Illingworth, Biochem. J. **195**, 259 (1981).
[16] W. E. Morf and W. Simon, in "Ion-selective Electrodes in Analytical Chemistry" (H. Freiser, ed.), Vol. 1, p. 211. Plenum, New York, 1978.
[17] A. K. Covington, "Ion-Selective Electrode Methodology," Vols. 1,2. CRC Press, Boca Raton, FL, 1979.
[18] K. Cammann, "Working with Ion-Selective Electrodes." Springer-Verlag, Berlin and New York, 1979.
[19] U. Schefer, D. Ammann, E. Pretsch, U. Oesch, and W. Simon, Anal. Chem. **58**, 2282 (1986).
[20] Z. Hu, T. Bührer, M. Müller, B. Rusterholz, M. Rouilly, and W. Simon, Anal. Chem. **61**, 574 (1989).

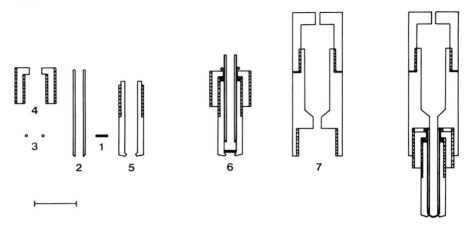

FIG. 3. Design of the ion-selective half-cell. The ion-selective membrane (1) is inserted into the electrode body as described in text. (3) indicates sections through O rings used for sealing. Hatched areas are threads. Bar: 1 cm.

refrigerator at 4° for about 1 year. The major problem in the construction of the electrode body is to separate the sample solution completely from the internal filling solution by the ion-selective membrane. This can be done by using an O ring,[17] by using adhesives,[21] or by mechanical clamping.[16] The electrode body developed and used in the authors' laboratory is shown in Fig. 3. The ion-selective membrane [(1), Fig. 3] is inserted in the tip of the electrode body [(5), Fig. 3] and gently squeezed with piece (2) by screwing piece (6) into piece (7). Afterward the ion-selective half-cell is filled with a syringe with either 10 mM CaCl$_2$ or 10 mM MgCl$_2$.

Tubing or a glass capillary with a salt bridge [1% (w/v) agar in 3 M KCl] at the tip may be used as a reference electrode. The reference electrode is filled with 3 M KCl (saturated with AgCl). Commercially available silver wires chlorinated electrically (1.5 V for 30 min as an anode in a solution containing about 100 mM Cl$^-$) are placed in the electrode filling solutions and connected with the pH (voltage) meter.

Half-cells prepared in this way are stored in the filling solution overnight. Prior to use, the electrodes should be equilibrated for about 2 hr in the experimental medium. The electrode may be checked rapidly by several changes of the experimental solutions containing 0 Ca^{2+} (medium containing 1 mM EGTA) and 1 mM Ca^{2+}. Calibration of the electrodes between pCa 2 and pCa 5 dilutions of neutral CaCl$_2$ (e.g., Orion, Lorch,

[21] H. Affolter and E. Sigel, *Anal. Biochem.* **97**, 315 (1979).

Germany) or $MgCl_2$ in experimental solution without any ligands is suitable. For lower metal concentrations the calibration curve (plotted in a semilogarithmic way) may be extended by extrapolation. Alternatively, Ca^{2+} buffers in experimental media with EGTA as the only ligand are suitable.[9,10,22] Then the electrodes can be used to control complex media as calculated at the beginning of this chapter. The electrodes can also be applied for the analysis of intracellular Ca^{2+} uptake and Ca^{2+} release from permeabilized cells.[23-26]

Acknowledgment

The authors thank Drs. S. Galler (Konstanz), O. Scharf, T. Saermark, M. Treiman (Copenhagen), and D. Sherman (Paris) for their help and suggestions prior to and during preparation of this manuscript. The authors are also indebted to W. Simon (Zurich) for generous gifts of ligands as well as for valuable advice. This work was supported by the Deutsche Forschungsgemeinschaft (Gr 681).

[22] D. M. Bers, *J. Am. Physiol.* **242**, C404 (1982).
[23] K. J. Föhr, J. Scott, G. Ahnert-Hilger, and M. Gratzl, *Biochem. J.* **262**, 83 (1989).
[24] K. J. Föhr, G. Ahnert-Hilger, B. Stecher, J. Scott, and M. Gratzl, *J. Neurochem.* **56**, 665 (1991).
[25] R. Engling, K. J. Föhr, T. P. Kemmer, and M. Gratzl, *Cell Calcium* **12**, 1 (1991).
[26] K. J. Föhr, Y. Wahl, R. Engling, T. P. Kemmer, and M. Gratzl, *Cell Calcium* **12**, 735 (1991).

[13] Manipulation of Cytosolic Free Calcium Transients during Exocytosis in Intact Human Neutrophils

By DANIEL P. LEW, MARISA JACONI, and TULLIO POZZAN

Introduction

The term *exocytosis* is commonly used to indicate the process by which hydrophilic cellular products (e.g., proteins, peptides, and neurotransmitters) segregated within intracellular vesicles are discharged into the extracellular fluid. This process is triggered by stimulation of the cell and consists of several discrete events, including the movement of the storage organelles to the plasma membrane, the fusion between the two membranes, and the liberation of the stored contents.

More than 20 years ago, it was discovered that Ca^{2+} plays a pivotal role in regulated exocytosis.[1] In particular it has been demonstrated that exocy-

[1] A. K. Campbell, "Intracellular Calcium: Its Universal Role as Regulator." Wiley, New York, 1983.

tosis often depends on the presence of Ca^{2+} in the extracellular medium and that it can be induced by ionophores that selectively transport Ca^{2+} across natural and artificial membranes.[2,3] Additional support for the involvement of Ca^{2+} has been provided by the discovery of the inhibitory action of Ca^{2+} antagonists and the role of Ca^{2+}-regulated proteins.[3-5] Two technical developments have led to important progress in this field, namely, the introduction of permeabilized cells that retain their capacity to secrete[6-8] and the use of fluorescent Ca^{2+} chelators for manipulating and measuring cytosolic free calcium concentration $[Ca^{2+}]_i$.[9,10] The quantitative relationship between $[Ca^{2+}]_i$ and secretion has been investigated in permeabilized cells and, more recently, validation of such data has been performed in our laboratory in intact cells.[10-12]

In this chapter we present a summary of the methodology we employed for the investigation of the $[Ca^{2+}]_i$ dependence of exocytosis in intact human neutrophils. Basically, two different methodologies were used to manipulate $[Ca^{2+}]_i$ while assessing exocytosis in parallel:

1. Introduction of high-affinity Ca^{2+} chelators into intact cells (such as quin2 or MAPT) to buffer and control Ca^{2+} transients; in addition, this technique may be used under appropriate experimental conditions to deplete intracellular Ca^{2+} stores and to decrease basal $[Ca^{2+}]_i$ to very low levels.
2. Establishment of a $[Ca^{2+}]_i$ steady state with the Ca^{2+} ionophore ionomycin; for this purpose a strict experimental protocol using intracellular Ca^{2+} chelators and varying the extracellular concentration of Ca^{2+} is necessary.

All the experiments described below have been performed in populations of human neutrophils in suspension. Some caution must be used in extrapolating the methodology described in this paper to other experimental conditions (e.g., neutrophils adhering to a substratum) or to other cell types. For example, adhering neutrophils exhibit different Ca^{2+} homeo-

[2] V. L. Lew and J. Garcia Sancho, *Cell Calcium,* **6,** 15 (1985).
[3] D. Romeo, G. Zabucchi, N. Miani, and F. Rossi, *Nature (London)* **253,** 542 (1975).
[4] W. Y. Cheung, *Science* **207,** 19 (1980).
[5] F. S. Southwick and T. P. Stossel, *Semin. Hematol.* **20,** 305 (1983).
[6] P. F. Baker, D. E. Knight, and J. A. Umbach, *Cell Calcium* **6,** 5 (1985).
[7] B. D. Gomperts, *Nature (London)* **306,** 64 (1983).
[8] S. P. Nilson and N. Kishner, *J. Biol. Chem.* **258,** 4994 (1983).
[9] T. Pozzan, P. D. Lew, C. B. Wollheim, and R. Y. Tsien, *Science* **221,** 1413 (1983).
[10] T. J. Rink, S. W. Smith, and R. Y. Tsien, *FEBS Lett.* **148,** 21 (1982).
[11] P. D. Lew, C. B. Wollheim, F. A. Waldvogel, and T. Pozzan, *J. Cell Biol.* **99,** 1212 (1984).
[12] P. D. Lew, A. Monod, F. A. Waldvogel, B. Dewald, M. Baggiolini, and T. Pozzan, *J. Cell Biol.* **102,** 2197 (1986).

static properties when compared to the same cells in suspension; alternatively, whereas neutrophils maintain a similar $[Ca^{2+}]_i$ in the presence or absence of extracellular Ca^{2+}, in bovine glomerulosa cells there is a marked drop in $[Ca^{2+}]_i$ upon chelation of extracellular Ca^{2+} or after addition of the Ca^{2+} channel blocker nifedipine.[13]

Receptors and $[Ca^{2+}]_i$ in Human Neutrophils

Before dealing with the specific methodology for $[Ca^{2+}]_i$ measurement, it seems appropriate to summarize briefly some of the basic properties of the neutrophil receptors involved in modulation of $[Ca^{2+}]_i$ homeostasis.

The main function of human neutrophils is to sense, approach, and destroy invading microorganisms, in particular pyogenic bacteria. It is possible to divide the various neutrophil receptors into five groups according to their main function: (1) adherence receptors, (2) chemotactic receptors, (3) phagocytic receptors, (4) cytokine receptors, and (5) receptors of unknown function.

The coupling to $[Ca^{2+}]_i$ increases is firmly established only for chemotactic and phagocytic receptors. Recently, it has been demonstrated that adherence receptors can also modulate $[Ca^{2+}]_i$, although the mechanism of this effect has not yet been clarified completely. On the other hand, in neutrophils cytokine receptors have not been studied in detailed, as far as their coupling to $[Ca^{2+}]_i$ homeostasis is concerned. However, extrapolating from other systems, it can be predicted that cytokine receptors will not directly affect $[Ca^{2+}]_i$, although an indirect modulation of other receptors cannot be excluded. This chapter discusses primarily the effects on secretion of chemotactic receptors, with special reference to those for the chemotactic peptide formylmethionylleucylphenylalanine (fMLP). This peptide has been used during the past few years mainly as a model agonist, mimicking chemotactic factors produced by bacteria. Upon stimulation with fMLP neutrophils are known to secrete into the medium the content of three types of granules: primary or azurophil, secondary or specific, and secretory vesicles. Convenient markers for the contents of these three types of granules are glucuronidase (azurophil), vitamin B12-binding protein (secondary), and gelatinase (secretory vesicles).

Introduction of High-Affinity Ca^{2+} Chelators into Intact Cells to Modulate Cytosolic Ca^{2+}-Buffering Capacity

The introduction of a new generation of Ca^{2+} indicators of high affinity and selectivity, the tetracarboxylates, and of their intracellularly hydrolyzable alkyl esters, has made possible the measurement of $[Ca^{2+}]_i$ in small intact mammalian cells. In most cases it has been used either nonquantita-

[13] A. M. Capponi, P. D. Lew, and M. B. Vallotton, *Biochem. J.* **247,** 335 (1987).

tively or simply to demonstrate that the $[Ca^{2+}]_i$ rises as a consequence of certain stimuli.[10,14-16] Then quin2 was used as a high-affinity Ca^{2+} chelator to buffer and control intracellular Ca^{2+} transients while simultaneously monitoring the actual values of $[Ca^{2+}]_i$.[17-19] Some of the more representative experiments performed in human neutrophils are described below.

Effect of Intracellular Quin2, [Quin2]_i, on [Ca²⁺]_i Transients Induced by Chemotactic Peptide fMLP

In Fig. 1A–D,[11] the $[Ca^{2+}]_i$ changes induced by fMLP in the same batch of neutrophils containing low or high intracellular concentrations of quin2 ([quin2]_i) are shown. In Ca^{2+} medium, at low [quin2]_i, there is a rapid increase in $[Ca^{2+}]_i$ that reaches micromolar levels, followed by a slow decrease to basal levels. At high [quin2]_i the amplitude of the $[Ca^{2+}]_i$ increase is markedly reduced, reaching ~300 nM at its maximum. The kinetics of the $[Ca^{2+}]_i$ transients are different at high or low quin2 loadings. At low loading, on addition of fMLP, there is a rapid increase that is completed in < 10 sec, followed by a slow return to the basal level. At high loading, the rapid phase is drastically reduced in amplitude and a slow phase of $[Ca^{2+}]_i$ increase is now observed that lasts for > 4 min. The fast component is attributed to the release of Ca^{2+} from intracellular stores, because it is also observed in the presence of ethylene glycol-bis(β-aminoethyl ether)-N,N,N',N'-tetraacetic acid (EGTA) in the medium, whereas the slow component is probably due to an increased influx from the extracellular medium because it is abolished when external Ca^{2+} is removed.

At low [quin2]_i loadings in Ca^{2+}-free medium, an important $[Ca^{2+}]_i$ increase can still be measured (up to 850 nM); at high loadings the $[Ca^{2+}]_i$ increase is barely significant, from 90 to 115 nM. The differences in $[Ca^{2+}]_i$ increase at high and low loading in Ca^{2+}-free medium should be attributed to the Ca^{2+}-buffering capacity provided by quin2, because the amount of Ca^{2+} released in the two cases is rather similar (see below).

[14] T. R. Hesketh, G. A. Smith, J. P. Moore, M. V. Taylor, and J. C. Metcalfe, *J. Biol. Chem.* **258**, 4876 (1983).
[15] J. R. White, P. H. Naccache, T. F. P. Molski, P. Borgeat, and R. I. Sha'afi, *Biochem. Biophys. Res. Commun.* **113**, 44 (1983).
[16] C. B. Wollheim and T. Pozzan, *J. Biol. Chem.* **259**, 2262 (1984).
[17] V. L. Lew, R. Y. Tsien, C. Minex, and R. M. Bookchin, *Nature (London)* **298**, 478 (1982).
[18] T. Pozzan, P. Arslan, R. Y. Tsien, and T. J. Rink, *J. Cell Biol.* **94**, 335 (1982).
[19] R. Y. Tsien and T. J. Rink, *Curr. Methods Cell. Neurobiol.* **3**, (1982).
[20] R. B. Zurier, B. S. Hoffstein, and G. Weissmann, *Proc. Natl. Acad. Sci. U.S.A.* **70**, 844 (1973).
[20a] R. Gennaro, T. Pozzan, and D. Romeo, *Proc. Natl. Acad. Sci. U.S.A.* **81**, 1416 (1984).

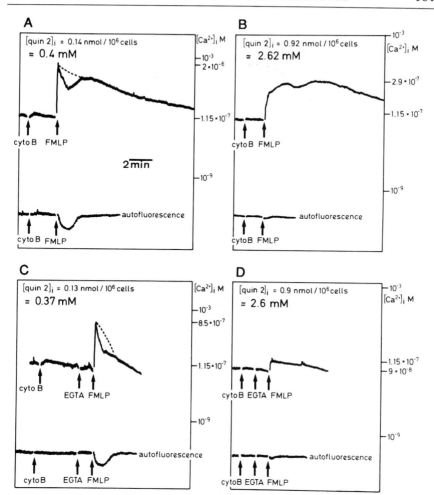

FIG. 1. Effect of intracellular quin2 on $[Ca^{2+}]_i$ transients. $[Ca^{2+}]_i$ changes in neutrophils containing low (A and C) or high (B and D) quin2 concentrations induced by fMLP in Ca^{2+} medium (A and B) or Ca^{2+}-free medium (C and D). In each panel the changes in autofluorescence in control cells are indicated. The dashed lines represent the graphic correction of the $[Ca^{2+}]_i$ transient, taking into account the changes in autofluorescence. This correction factor is significant at low quin2 loading but not at high loading because of the difference in signal intensity. The intracellular quin2 concentration was calculated taking into account a volume of 0.35 μl free water per 10^6 neutrophils.[20a] (Reproduced from the *Journal of Cell Biology*, 1984, **99**, 1212–1220 by copyright permission of the Rockefeller University Press.)

The amount of Ca^{2+} released from intracellular pools can be calculated[18]: quin2 binds Ca^{2+} with a stoichiometry of 1:1; thus the initial amount of Ca^{2+} liberated from the stores is the product of the percentage change in quin2–Ca^{2+} complex times the total quin2 intracellular content. For example, in Fig. 1, the amount of Ca^{2+} released at low loading in Ca^{2+}-free medium was 53 pmol $Ca^{2+}/10^6$ cells (41% × 130 pmol quin2/10^6 cells) whereas at high loading it was 49 pmol $Ca^{2+}/10^6$ cells (5.5% × 900 pmol quin2/10^6 cells). In this calculation the assumption is made[18] that the endogenous Ca^{2+} buffering is negligible compared to that provided by quin2. Thus only the lower limit of the amount of Ca^{2+} released from stores can be inferred from this approach, the error being larger at lower $[quin2]_i$. As discussed in detail by Tsien and Rink[19] and shown by von Tscharner et al.[21] from a series of experiments of this kind, the endogenous Ca^{2+}-buffering capacity can also be calculated.

Effect of [Quin2]$_i$ on Granule Content Release Induced by fMLP

Figure 2A and B shows the time course of release of β-glucuronidase (a marker of primary granules) and vitamin B_{12}-binding protein (a marker of secondary granules) in the same batch of neutrophils loaded with various concentrations of quin2 as shown in Fig. 1, and stimulated by 10^{-7} M fMLP. Increasing quin2 concentration decreases the amount of primary and secondary granule exocytosis in the Ca^{2+} medium. There is a less drastic decrease in the rate of granule release; however, measurements < 15 sec were not performed. The inhibitory effect on granule content release is much more pronounced in Ca^{2+}-free medium even in cells loaded with the lowest quin2 concentrations. Figure 2C indicates the maximal $[Ca^{2+}]_i$ reached in the same batch of neutrophils in Ca^{2+} or Ca^{2+}-free medium at various quin2 loadings. The maximal $[Ca^{2+}]_i$ increase reached correlated well with the maximal extent of granule content released under these various conditions (Fig. 2D and E).

Effect of Intracellular Quin2 on [Ca^{2+}]$_i$ Transients and Granule Content Release Induced by the Ca^{2+} Ionophore, Ionomycin

Figure 3 shows that quin2 affects the $[Ca^{2+}]_i$ increase induced by the Ca^{2+} ionophore, ionomycin, in a predictable way: in Ca^{2+} medium, increasing quin2 loadings decreases the rate of $[Ca^{2+}]_i$ elevation without affecting its final extent (Fig. 3A and B). At a loading of 0.22 nmol quin2/10^6 cells the $[Ca^{2+}]_i$ reaches maximal detectable levels at ~ 10 sec, whereas at 0.9 nmol/10^6 cells the $[Ca^{2+}]_i$ reaches maximal detectable levels at 2 min. In contrast, the rate of Ca^{2+} increase in Ca^{2+}-free medium is

[21] V. von Tscharner, D. A. Deranleau, and M. Baggiolini, *J. Biol. Chem.* **261**, 10163 (1986).

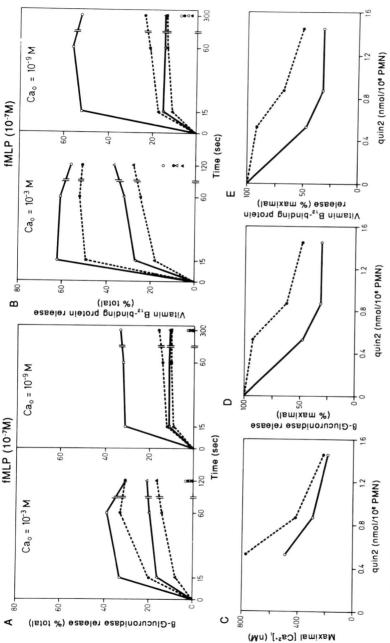

FIG. 2. Effect of intracellular quin2 on granule exocytosis and [Ca²⁺]ᵢ changes induced by fMLP. Neutrophils were incubated as described in Materials and Methods in the presence of 0, 30, 60, and 100 μM quin2/AM and then tested in parallel. (A and B) Time course of β-glucuronidase (A) and vitamin B₁₂-binding protein release (B) in Ca²⁺ or Ca²⁺-free medium. ○, No quin2; ●, 0.4 nmol quin2/10⁶ cells; Δ, 0.8 nmol quin2/10⁶ cells; ▲, 1.2 nmol quin2/10⁶ cells. (C–E) Maximal [Ca²⁺]ᵢ levels, maximal β-glucuronidase release, and maximal vitamin B₁₂-binding protein release as a function of intracellular quin2 in Ca²⁺ (●) or Ca²⁺-free medium (○). The values in (D) and (E) represent the maximal release values from (A) and (B), and are shown for comparison with (C). All these experiments were performed on the same batch of neutrophils. (Reproduced from the *Journal of Cell Biology*, 1984, **99**, 1212–1220 by copyright permission of the Rockefeller University Press.)

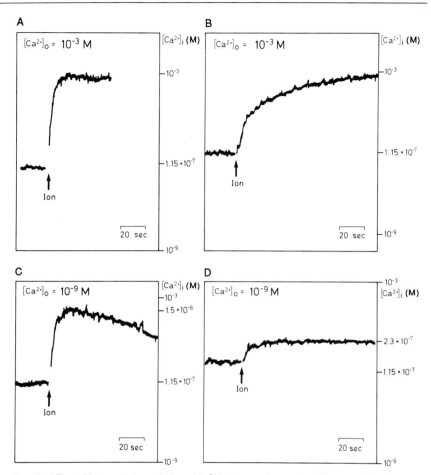

FIG. 3. Effect of intracellular quin2 on $[Ca^{2+}]_i$ changes in neutrophils exposed to ionomy-cin (Ion., 500 nM) in Ca^{2+} or Ca^{2+}-free medium. (A and C) $[quin2]_i = 0.22$ nmol quin2/10^6 cells (0.6 mM); (B and D) $[quin2]_i = 0.9$ nmol quin2/10^6 cells (2.6 mM). (Reproduced from the *Journal of Cell Biology*, 1984, **99**, 1212–1220 by copyright permission of the Rockefeller University Press.)

practically unaffected by the extent of intracellular quin2 whereas the magnitude of the $[Ca^{2+}]_i$ rise is dramatically decreased from 1500 to 230 nM under these two conditions (Fig. 3C and D). Again the difference must be ascribed to the extra cytosolic Ca^{2+}-buffering capacity provided by quin2, because the amount of Ca^{2+} released is similar both at high and low loading, that is, 108 and 95 pmol Ca^{2+}/10^6 cells, respectively. Figure 4 shows the effect of $[quin2]_i$ on the rate and extent of β-glucuronidase, N-acetyl-β-glucosaminidase, and vitamin B_{12}-binding protein release from

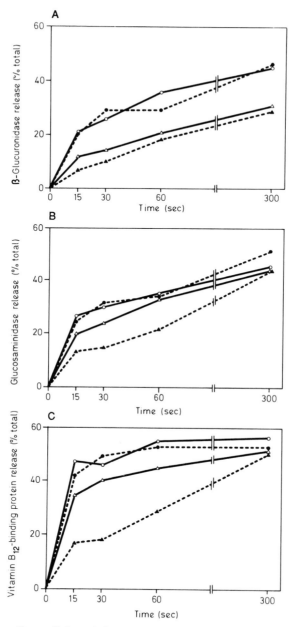

FIG. 4. Effect of intracellular quin2 on granule exocytosis induced by ionomycin (1 μM) as a function of time. Extracellular Ca^{2+} was $10^{-3}\ M$ throughout. (A) β-Glucuronidase release; (B) glucosaminidase release; (C) vitamin B_{12}-binding protein release. ○, 0 nmol quin2/10^6 cells; ●, 0.37 nmol quin2/10^6 cells; △, 0.72 nmol quin2/10^6 cells; ▲, 1.1 nmol quin2/10^6 cells. (Reproduced from the *Journal of Cell Biology*, 1984, **99**, 1212–1220 by copyright permission of the Rockefeller University Press.)

neutrophils treated with ionomycin. Paralleling the changes in $[Ca^{2+}]_i$, increasing $[quin2]_i$ decreases the rate of protein release, while the final extent of secretion eventually approaches the same level, although at different rates for various granule populations.

Establishment of $[Ca^{2+}]_i$ Steady State with the Ca^{2+} Ionophore, Ionomycin, in Ca^{2+}-Buffered Cells

The experimental protocol is as follows: neutrophils are loaded with quin2, washed and resuspended in the fluorimeter cuvette; a fixed concentration of ionomycin (500 nM) is added at different concentration of extracellular Ca^{2+}, $[Ca^{2+}]_o$; secretion of the granule content is measured from aliquots of the cell suspension of the cuvette.

As shown in Fig. 5, ionomycin induces a rapid increase in $[Ca^{2+}]_i$ up to a plateau that depends on the $[Ca^{2+}]_o$. The plateau remains constant for 10 min, indicating that steady state levels ranging from 120 to > 2000 nM can be established by the experimental protocol adopted. Due to the poor indicator sensitivity of quin2 at $[Ca^{2+}]_i$ levels above 2 μM, calibration at higher values has not been attempted.

In Figure 6 the relationship between $[Ca^{2+}]_i$ and the extent of exocytosis is shown. Minimal enzyme release is observed when the $[Ca^{2+}]_i$ is $< 200-250$ nM. Above this level release occurs, albeit to different extents, from all three storage compartments. The $[Ca^{2+}]_i$ thresholds, defined as the intercept of the extrapolated slope of the $[Ca^{2+}]_i$ dependence curve with the abscissa, were determined from three to six experiments of this type. For exocytosis from the specific granules and the secretory vesicles similar threshold $[Ca^{2+}]_i$ values are obtained that range between 190 and 240 nM, independent of cytochalasin B pretreatment of the cells. For primary (azurophil) granule exocytosis, the threshold concentration in the absence of cytochalasin B cannot be determined because the amounts released at $[Ca^{2+}]_i$ levels below 2000 nM are too low for accurate calculation. In the presence of cytochalasin B the threshold is 280 nM, a value slightly higher but not statistically different from that obtained for the specific granules and the secretory vesicles. In view of the well-established fact that cytochalasin B enhances the rate and extent of exocytosis,[22] it is interesting to note that pretreatment with cytochalasin B does not appreciably affect these $[Ca^{2+}]_i$ threshold levels. From the same experiments, the $[Ca^{2+}]_i$ that gives half-maximal release (EC_{50}) of the three markers can be calculated (Fig. 8). In neutrophils pretreated with cytochalasin B, these values are 610 and

[22] J. P. Bennett, S. Cockcroft, and B. D. Gomperts, *Biochim. Biophys. Acta* **601**, 584 (1980).

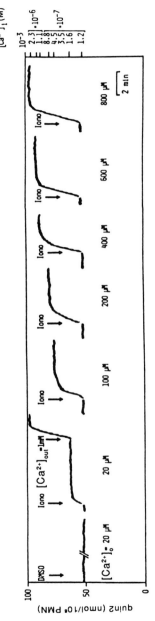

FIG. 5. Establishment of steady state cytosolic free calcium, $[Ca^{2+}]_i$, with ionomycin. Neutrophils (10^7) loaded with quin2 (0.8 nmol quin2/10^6 cells) were suspended in 3.2 ml buffer with increasing concentrations of $CaCl_2$ as indicated. Ionomycin (Iono; 500 nM) was added as indicated, and fluorescence tracings are shown. In neutrophils pretreated with cytochalasin B virtually identical results were obtained. (Reproduced from the *Journal of Cell Biology*, 1986, **102**, 2197–2204 by copyright permission of the Rockefeller University Press.)

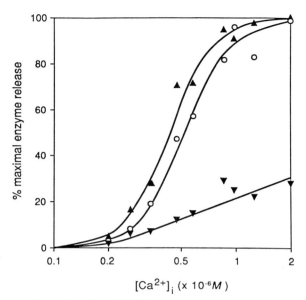

FIG. 6. Exocytosis as a function of $[Ca^{2+}]_i$ steady state. Experimental conditions were as described in Fig. 1. The cells were pretreated with cytochalasin B (5 μg/ml) for 5 min. Samples for the determination of marker release were withdrawn from the fluorimeter cuvette 5 min after addition of ionomycin (500 nM).[12] ▲, Secondary granules; ○, secretory vesicles; ▼, primary granules.

650 nM for vitamin B_{12}-binding protein and gelatinase, respectively (no statistical difference). The EC_{50} for β-glucuronidase release is 2600 nM, much higher than the EC_{50} for release from specific granules and secretory vesicles ($p < 0.001$). The $[Ca^{2+}]_i$ threshold and EC_{50} for exocytosis from specific granules and secretory vesicles are similar. The rate of release, however, is higher for the gelatinase-containing organelles (not shown). In addition, a nearly complete release of gelatinase is observed, whereas maximum release of vitamin B_{12}-binding protein does not exceed 50% of the total cellular content.

Effect of Receptor Activation on $[Ca^{2+}]_i$ Dependence of Exocytosis

Figure 7[12] shows the protocol used in an attempt to estimate the $[Ca^{2+}]_i$ threshold and EC_{50} of fMLP-induced exocytosis. The protocol described in Fig. 1 for ionomycin was not applicable for the following two reasons: (1) the $[Ca^{2+}]_i$ increase induced by fMLP is transient and therefore no steady state $[Ca^{2+}]_i$ can be obtained. The interpretation of an effect of fMLP at steady state $[Ca^{2+}]_i$ levels established with ionomycin in the presence of extracellular Ca^{2+} appeared impossible in view of the stimulatory effect of

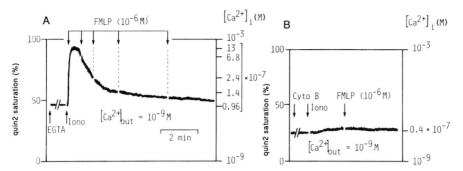

FIG. 7. Experimental protocol for studying the $[Ca^{2+}]_i$ dependence of exocytosis induced by fMLP. (A) Cells (10^7) were loaded with a low concentration of quin2 in Ca^{2+} medium (0.12 nmol quin2/10^6 cells) and then resuspended in Ca^{2+}-free medium. EGTA (1 mM) and ionomycin (Iono; 500 nM) were added where indicated. Cytochalasin B (5 μg/ml) was added 3 min before ionomycin. The arrows on top of the graph designate the time points corresponding to different $[Ca^{2+}]_i$ levels at which 1 μM fMLP was added to parallel samples (i.e., each sample received fMLP at a different time point). In cells not pretreated with ionomycin, fMLP increased the $[Ca^{2+}]_i$ to approximately the same level as the ionophore, whereas fMLP added after ionomycin did not modify the $[Ca^{2+}]_i$ transient.[20,22a] (B) Conditions were as in (A), except that a higher $[quin2]_i$ was used (0.6–1 nmol quin2/10^6 cells) and both loading and the subsequent incubation were performed in Ca^{2+}-free medium containing 1 mM EGTA as described.[19] (Reproduced from the *Journal of Cell Biology,* 1986, **102**, 2197–2204 by copyright permission of the Rockefeller University Press.)

the ionophore alone. We thus applied ionomycin in $[Ca^{2+}]_i$-free medium. Under these conditions the ionophore does not appreciably stimulate exocytosis[9] but releases Ca^{2+} from internal stores, leading to a transient elevation of $[Ca^{2+}]_i$. We then added fMLP at different times after ionomycin and determined the extent of release of granule markers. fMLP added after the ionophore does not modify $[Ca^{2+}]_i$ but is a potent stimulator of exocytosis.[9,11] The protocol of this experiment is illustrated in Fig. 7A, and curves relating fMLP-dependent exocytosis to $[Ca^{2+}]_i$ are shown in Fig. 8. After stimulation with the chemotactic peptide, exocytosis is rapid and reaches completion within 30 sec. It appears therefore justified to calculate $[Ca^{2+}]_i$ thresholds and EC_{50} values on the basis of the $[Ca^{2+}]_i$ levels determined at the time of fMLP addition. Note that high $[Ca^{2+}]_i$ elevations are transient (Fig. 7A), and consequently the exocytosis responses at these levels are less reproducible than at lower $[Ca^{2+}]_i$. Levels of $[Ca^{2+}]_i$ below the resting level are obtained by loading the cells with quin2 in Ca^{2+}-free medium.[23] Under these conditions the Ca^{2+} stores are depleted, and the

[22a] H. Lagast, T. Pozzan, F. A. Waldvogel, and P. D. Lew, *J. Clin. Invest.* **73**, 878 (1984).
[23] F. Di Virgilio, P. D. Lew, and T. Pozzan, *Nature (London)* **310**, 691 (1984).

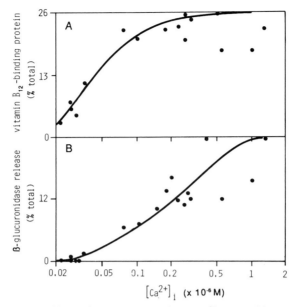

FIG. 8. Dependence of fMLP-induced exocytosis on $[Ca^{2+}]_i$. Conditions were as described in Fig. 5. Cytochalasin B (5 μg/ml) was added 3 min before ionomycin (500 nM). The $[Ca^{2+}]_i$ is the value measured at the time of fMLP addition. Samples for the determination of marker release were collected from the fluorimeter cuvette 5 min after fMLP was added. Release is expressed as the percentage of the initial cellular content minus the value obtained for ionomycin alone. Results are from two experiments. The respective values for ionomycin alone were 14 and 13% for vitamin B_{12}-binding protein, and 5 and 4% for β-glucuronidase. The corresponding values for unstimulated cells were 8 and 7% for vitamin B_{12}-binding protein, and 3% for β-glucuronidase. (Reproduced from the *Journal of Cell Biology,* 1986, **102,** 2197–2204 by copyright permission of the Rockefeller University Press.)

basal $[Ca^{2+}]_i$ is decreased 3- to 10-fold below the normal resting level, depending on the quin2 concentration used (Fig. 7B). Figure 8 shows the results of experiments performed according to the protocols described in Fig. 7A and B. In the presence of cytochalasin B the $[Ca^{2+}]_i$ threshold for fMLP-induced release of vitamin B_{12}-binding protein was found to be <20 nM ($n = 12$) and <50 nM ($n = 5$) for β-glucuronidase. These values are about one order of magnitude lower than those obtained with ionomycin alone. An EC_{50} of 200 ± 60 nM ($n = 3$) was calculated for azurophil and an EC_{50} of 60 ± 20 nM ($n = 3$) ($p < 0.001$) was determined for specific granule exocytosis. Due to the transient nature of the $[Ca^{2+}]_i$ elevation under these conditions reliable values for the threshold and EC_{50} in the absence of cytochalasin B could not be determined.

Usefulness of This Approach for Study of Exocytosis in Neutrophil Suspensions

We have addressed the question of whether Ca^{2+} plays a role in neutrophil activation by using two different methodologies.

The first involves trapping into the cytoplasm of neutrophils various concentrations of the Ca^{2+} indicator quin2, which binds Ca^{2+} with high affinity. This allowed us to increase the cytosolic Ca^{2+}-buffering capacity progressively and subsequently to monitor the kinetics of $[Ca^{2+}]_i$ rises and functional response induced by fMLP or ionomycin.

In the case of exocytosis there was a clear-cut correlation between the final extent of $[Ca^{2+}]_i$ increase and exocytosis of the contents of both primary (β-glucuronidase) and secondary (vitamin B_{12}-binding protein) granules. Increasing $[quin2]_i$ decreased the maximal $[Ca^{2+}]_i$ increases and the extent of exocytosis in response to fMLP and these decreases were even more pronounced in Ca^{2+}-free buffer, in which Ca^{2+} influx from the extracellular medium is prevented and Ca^{2+} originates only from intracellular pools. This approach allowed us to demonstrate clearly that there is a Ca^{2+} dependence for exocytosis in human neutrophils. The second method is to use a Ca^{2+} ionophore to obtain different $[Ca^{2+}]_i$ steady states and thus quantify more accurately the Ca^{2+} dependence of secretion.

The "ionophore clamp" approach overcomes some of the problems found in the studies with permeabilized cells because it manipulates the $[Ca^{2+}]_i$ while leaving all other cellular parameters intact. However, several factors, that is, Ca^{2+} release from intracellular stores, Ca^{2+} buffering, and the effect of pH gradients across the membrane, must be taken into consideration and reliable information can be obtained only if secretion is measured in the very same cells used for $[Ca^{2+}]_i$ monitoring.

The ionophore clamp of $[Ca^{2+}]_i$ has been used previously by other groups, with the assumption that the ionophore can completely overcome the Ca^{2+} homeostatic mechanisms of the cells, so that $pCa^{2+} = 2$ pH. We observed that with $0.5-1$ μM ionomycint the $[Ca^{2+}]_o$ was still 1000-fold higher than $[Ca^{2+}]_i$, and that the equilibrium condition was not reached even at ionomycin concentrations as high as 10 μM. A similar discrepancy between the assumed and measured value for $[Ca^{2+}]_i$ has been shown in adrenal glomerulosa cells.[13,24]

The approach used in this study is related to the "ionophore clamp" method. The high-affinity Ca^{2+} chelator quin2 served two purposes, that is, it allows manipulation of $[Ca^{2+}]_i$ and monitoring of these levels. Human

[24] I. Kojima, I. Lippes, K. Kojima, and H. Rasmussen, *Biochem. Biophys. Res. Commun.* **116**, 555 (1983).

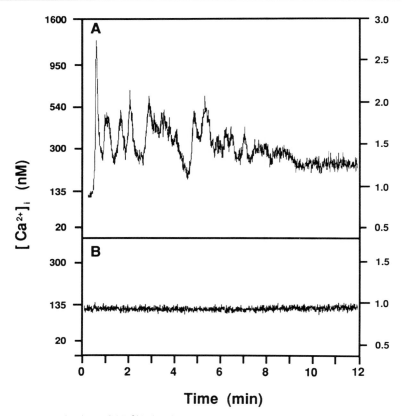

FIG. 9. Monitoring of $[Ca^{2+}]_i$ in single neutrophils adherent to albumin-coated glass surface. (A) Spontaneous $[Ca^{2+}]_i$ oscillations occurring in a single adherent neutrophil in the absence of any added stimulus. (B) $[Ca^{2+}]_i$ level monitored in a representative Ca^{2+}-buffered neutrophil, loaded with 25 μM MAPT/AM in Ca^{2+} medium.

neutrophils were ideal as the model cell because they contain three bio-chemically well-defined storage compartments with different reactivity to exocytotic stimuli.[25–27] We found that the $[Ca^{2+}]_i$ requirement for half-maximal exocytosis (EC_{50}) of specific granules (secondary granules) and secretory vesicles was significantly lower than that for primary (azurophil) granules.

[25] B. Dewald, U. Bretz, and M. Baggiolini, *J. Clin. Invest.* **70,** 518 (1982).
[26] D. G. Wright and J. I. Gallin, *J. Immunol.* **123,** 285 (1979).
[27] D. G. Wright, D. A. Bralove, and J. I. Gallin, *Am. J. Pathol.* **87,** 273 (1977).

Buffering $[Ca^{2+}]_i$ Transients and Assessment of Exocytosis at the Single-Cell Level

The new highly fluorescent Ca^{2+} indicators that have become available after 1985 (such as Fura-2 or Indo I) allow the measurement of $[Ca^{2+}]_i$ at the single-cell level.[28-30] Monitoring of $[Ca^{2+}]_i$ with Fura-2, using rapid dual-excitation microfluorimetry, has allowed us to discover the presence of $[Ca^{2+}]_i$ oscillations in single adherent neutrophils (Fig. 9A).[31] In addition, nonfluorescent Ca^{2+} chelators such as MAPT/AM have also become available, allowing the combined use of Fura-2 and MAPT/AM, both in cell populations as well as in single cells. As shown in Fig. 9B, buffering of $[Ca^{2+}]_i$ with MAPT/AM abolishes the spontaneous $[Ca^{2+}]_i$ oscillations. This approach has allowed us to monitor at the single-cell level $[Ca^{2+}]_i$ transients during phagocytosis followed by assessment of lactoferrin exocytosis (a marker for secondary granules) into the phagocytic vesicle by immunofluorescent staining. We could in parallel deplete the cells of Ca^{2+} or buffer the $[Ca^{2+}]_i$ transients occurring during surface phagocytosis of serum-opsonized yeast particles (MAPT/AM loading in the absence or presence of Ca^{2+}, respectively). Discharge of granule contents into the phagosome was detected in 80% of the cell as a highly fluorescent ring around the ingested particle. This percentage is reduced to 30% in Ca^{2+}-buffered cells and to less than 10% in Ca^{2+}-depleted cells.[32]

With this technique, we could show at the single-cell level that $[Ca^{2+}]_i$ elevations are not necessary for ingestion of particles but control the phagosome–lysosome fusion step (i.e., exocytosis into an intracellularly located particle) during phagocytosis.[32] These experiments indicate that the approach described above for large populations of cells can also be used at the single-cell level (and even at the subcellular level with an image analysis system) for the assessment of the role of Ca^{2+} in the control of secretion.

Acknowledgment

Supported by a grant from the Swiss Research Foundation (No. 3.829.0.87), CNR Special Projects "Biotechnology and Bioinstrumentation" and "Oncology" and by the Italian Association for Cancer Research (AIRC).

[28] A. Malgaroli, D. Milani, J. Meldolesi, and T. Pozzan, *J. Cell Biol.* **105**, 1245 (1987).

[29] P. H. Cobbold and T. J. Rink, *Biochem. J.* **248**, 313 (1987).

[30] W. Schlegel, B. P. Winiger, F. Wuarin, G. R. Zahnd, and C. B. Wollheim, *J. Recept. Res.* **18**, 493 (1988).

[31] M. E. E. Jaconi, R. W. Rivest, W. Schlegel, C. B. Wollheim, D. Pittet, and P. D. Lew, *J. Biol. Chem.* **263**, 10557 (1988).

[32] M. E. E. Jaconi, P. D. Lew, J. L. Carpentier, K. E. Magnusson, and O. Stendahl, *J. Cell Biol.* **109**, 299a (abst.) (1989).

[14] Use of *Tetrahymena* and *Paramecium* in Studies of Exocytosis

By Birgit H. Satir and Lea K. Bleyman

Introduction

Exocytosis is the last step in secretion, in which the cell membrane fuses with the membrane of the secretory vesicle, thereby releasing the secretory product(s) outside the cell. The ciliated protozoa *Tetrahymena* and *Paramecium* have proved to be excellent models for studying this process for the following reasons: (1) the general ultrastructure of these cells is thoroughly characterized, with special emphasis on the membranes involved in the fusion process[1-3]; (2) synchronous release can be induced in both cell types, leading to release events per cell of between 100 *(Tetrahymena)* to 3000 *(Paramecium)*. This results in a high signal-to-noise ratio, irrespective of the parameter being measured[4]; and (3) secretory mutants have been isolated in both species and are available for exocytosis studies.[5-9]

Three cellular compartments are involved in the process of exocytosis: the cell membrane, the secretory vesicles, and the cytosol between them. Each must be primed and ready for the membrane fusion process before exocytosis can occur, and failure in any one will prevent a normal release event. Therefore it is imperative that each of the three compartments be studied not only independently, but in concert with the other two.

This chapter describes the techniques used for quantitative studies on all three compartments. Emphasis is placed on biochemical and ultrastructural techniques. Procedures for *Tetrahymena* and *Paramecium* are described in separate sections; where overlap exists it will be noted.

[1] B. Satir, C. Schooley, and P. Satir, *J. Cell Biol.* **56,** 153 (1973).

[2] B. H. Satir, G. Busch, A. Vuoso, and T. J. Murtaugh, *J. Cell. Biochem.* **36,** 429 (1988).

[3] A. Adoutte, *in* "Paramecium" (H.-D. Gortz, ed.), p. 325. Springer-Verlag, Berlin, 1988.

[4] B. H. Satir, *J. Protozool.* **36,** 382 (1989).

[5] S. Pollack, *J. Protozool.* **21,** 352 (1974).

[6] J. Beisson and M. Rossignol, *in* "Molecular Biology of Nucleocytoplasmic Relationships" (S. Puiseux-Dao, ed.), p. 291. Elsevier, Amsterdam, 1975.

[7] J. Beisson, M. Lefort-Tran, M. Pouphile, M. Rossignol, and B. Satir, *J. Cell Biol.* **69,** 126 (1976).

[8] E. Orias, M. Flacks, and B. H. Satir, *J. Cell Sci.* **64,** 49 (1983).

[9] L. K. Bleyman and B. H. Satir, *J. Protozool.* **37,** 471 (1990).

I. *Tetrahymena*

A. *Organism*

The species of choice is *Tetrahymena thermophila.* It is genetically well defined[10] and easily cultured and manipulated for both biochemical and breeding analyses.[11] It can be grown in monoxenic (one bacterial species), axenic (no live food organism), or defined media. In addition, strains of secretory mutants are available.[8] Strains are available through the American Type Culture Collection (ATCC, Rockville, MD).

B. *Media*

A defined medium is available,[12] but axenic media are sufficient for most experimental approaches. Enriched *Tetrahymena* medium (ETM) is useful for stock maintenance.

1. *Axenic*

PP (2%) 20 g Difco (Detroit, MI) proteose peptone (PP)/liter nanopore filtered H_2O. (This is used as the basic medium to which all other reagents are added. These are given in their final concentrations)

ETM: 2% PP, 0.2 (w/v) glucose (dextrose), 0.1% (w/v) yeast extract (Difco)

PP 210: 2% PP, $FeCl_3$ (10 μM), penicillin G (25 μg) (stock solution, 250 mg/ml), and streptomycin sulfate (250 μg) (stock solution, 250 mg/ml)

PP 210 Ca: PP 210 and $CaCl_2$ (0.5 mM)

PP 105 Ca: 1% (w/v) PP, $FeCl_3$ (5 μM), $CaCl_2$ (0.5 mM), penicillin G (250 μg) (stock solution 250 mg/ml), and streptomycin sulfate (250 μg) (stock solution, 250 mg/ml)

PP 025 Ca: 0.25% w/v PP and $CaCl_2$ (0.5 mM)

2. *Monoxenic*

BP (bacterized peptone) (2%): 2% (w/v) inoculated with *Enterobacter aerogenes* (see Section II,B,2,ii) and cultured for 24 hr at 37°

BP Ca (2%): 2% (w/v) BP and $CaCl_2$ (5 mM)

[10] P. J. Bruns, *in* "The Molecular Biology of Ciliated Protozoa" (J. G. Gall, ed.), p. 27. Academic Press, Orlando, FL, 1986.

[11] E. Orias and P. J. Bruns, *Methods Cell. Biol.* **13,** 247 (1976).

[12] L. P. Everhart, Jr., *Methods Cell. Physiol.* **5,** 220 (1972).

3. Starvation Buffers

Dryl's solution: Na_2HPO_4 (1 mM), NaH_2PO_4 (1 mM), $CaCl_2$ (1.5 mM), and sodium citrate (2 mM) (pH 7.1)[13]

DMC buffer: Dryl's solution at 10% (v/v) normal concentration, $MgCl_2$ (100 μM/flask); label and sterilize for 20 min

C. Stock Maintenance

Stock cultures are kept in test tubes (5 ml ETM/tube) either at room temperature and/or at 15°. Fresh stock tubes are established every 2 weeks at room temperature or once a month at 15°, by transferring 0.5 ml from the old culture into fresh medium. Faster generation times can be obtained at 30°.

Long-term test tube storage,[14,15] as well as preservation of *Tetrahymena* stock, is possible by freezing the cells and storing them in liquid nitrogen.[16]

D. Exocytosis (Capsule Formation)

Quantitative release of the secretory product, mucus, can be induced by specific agents that mimic secretagogues. Induction of synchronous release of mucocyst contents results in formation of a capsule entrapping the live cells (Fig. 1). This provides a method to screen for mutants blocked in exocytosis and for quantitative release of secretory products per cell.[8,17]

1. Induction

a. Cell Preparation. Experimental cultures are established by inoculating 0.1 ml of stock culture into 10 ml of PP 105 Ca, and incubating for 2 days at room temperature. The resulting stationary-phase cells are washed and resuspended in 10 ml (equal volume) of DMC buffer, then left at room temperature for at least 2 hr. Attempting capsule induction within a few minutes of transfer into DMC is lethal to most of the cells.

b. Alcian Blue Induction. Ten microliters of 0.1% (v/v) alcian blue (8GS) is mixed with 40 μl of prepared cells in a polypropylene test tube. Thirty seconds later, cells are refed with a large excess (2.5 ml) of PP 0.25 Ca.

[13] S. Dryl, *J. Protozool.* **6,** Suppl., 25 (1959).

[14] N. E. Williams, J. Wolfe, and L. K. Bleyman, *J. Protozool.* **27,** 327 (1980).

[15] L. Szablewski, P. H. Andreasen, A. Tiedtke, J. Florin Christensen, M. Florin Christensen, and L. Rasmussen, *J. Protozool.* **38,** 62 (1991).

[16] E. M. Simon and M. Flacks, *in* "Round Table Conference on the Cryogenic Preservation of Cell Cultures" (A. P. Rinfret and B. LaSalle, eds.), p. 37. National Academy of Sciences, Washington, DC, 1975.

[17] A. Tiedtke, *Naturwissenschaften* **63,** 93 (1976).

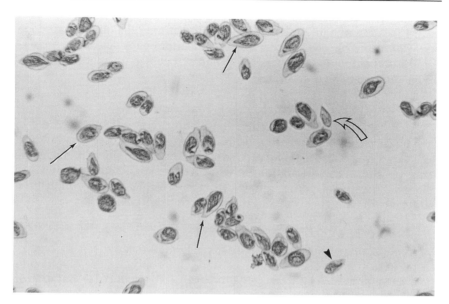

FIG. 1. Population of wild-type *Tetrahymena* after stimulation with alcian blue. The released mucus results in capsules (straight arrows) surrounding each cell. A cell that has escaped from its capsule (arrowhead) and an empty capsule (curved arrow) can be observed. Magnification: ×200.

This treatment allows survival of over 99% of the cells, which rotate or tumble within a blue capsule, until a hole large enough to escape from develops. More than 95% of the cells are still encapsulated 3 min after induction. The next day, 10 to 20% of the cells are still trapped, but active; some have divided once. Alcian blue is an especially desirable "secretagogue" because it stains the capsules blue, thus making them easily detectable under a microscope.

c. Dibucaine Induction. Add 20 μl of prepared cells to a mixture of 10 μl 0.2% (v/v) alcian blue and 20 μl of 10 m*M* dibucaine (pH 6.2). After 30 sec add 2.5 ml of 2% PP. More than 90% of the wild-type cells survive. Unlike in the previous procedure, the cells remain encapsulated, presumably because they are rendered immotile due to dibucaine-induced deciliation.[18]

2. Quantitation

Three minutes after induction, fix the cells by adding formaldehyde to a final concentration of 4%. Concentrate the cells by spinning (30 sec, setting

[18] G. A. Thompson, L. C. Baugh, and L. F. Walker, *J. Cell Biol.* **61**, 253 (1974).

5) in an IEC clinical centrifuge (Needham Heights, MA). Decant and discard the supernatant, and spread a sample of the cells on a microscope slide. Count the number of cells with and without capsules, using a compound microscope under low power.[8] The percentage of cells within capsules is then calculated.

E. Conjugation Rescue of Exocytosis

An essay for cytosolic, as well as membrane, components involved in exocytosis can be performed by using the conjugation rescue method. In these experiments conjugating cell pairs are tested for release by using the alcian blue test. Pair formation between wild-type and mutant cells can result in different categories, ranging from nonconjugation to conjugation that results in complete repair of the mutant phenotype.

1. Conjugation or Pair Formation

Conjugation or pair formation by sexually mature strains of *T. thermophila* is a process that can be induced under appropriate starvation conditions, such as transfer into Tris (Bio-Rad, Richmond, CA) buffer[19] or Dryl's solution.

Early stationary-phase cultures (48 hr, at 30°) of wild-type (*exo⁺*) cells and mutant (*exo⁻*) cells of complementary mating types[11] are transferred to Dryl's solution and starved for 16 hr prior to mixing of the two mating strains. Mix strains at a concentration of 2×10^5 cells/ml each. Observations can be made throughout the period of pairing.[20]

2. Determination of Rescue

Score the following categories by percentage: (1) naked cell pairs (no release), (2) cell pairs with one cell within a capsule, (3) cell pairs within double capsules, (4) single cells naked or within capsules, and (5) empty capsules (single or double, indicating cells have escaped). Category 3, conjugant pairs within double capsules, represents rescue of mutant to wild-type phenotype.

F. Quantitation of Exocytosis in Freeze-Fracture Electron Microscopy

1. Exocytotic Membrane Microdomain

The cell membranes of both *Paramecium* and *Tetrahymena* contain specific exocytic membrane microdomains that have been well character-

[19] P. J. Bruns and T. B. Brussard, *J. Exp. Zool.* **188**, 337 (1974).
[20] B. H. Satir, M. Reichman, and E. Orias, *Proc. Natl. Acad. Sci. U.S.A.* **83**, 8221 (1986).

ized by freeze-fracture techniques. In both species the crucial element involved in the membrane fusion process is an intramembrane particle (IMP) array called the rosette (or fusion rosette), consisting of 11 P and E intramembrane particles (P, proteins associated with the monolayer of the cell membrane facing the protoplasm; E, proteins associated with the exterior monolayer). As the name indicates, this particle array is arranged in a rosette with a single particle at its center.[21]

The rosette and its characteristic changes during exocytosis were first described in *Tetrahymena*.[21] When exocytosis begins, the membrane of the mucocyst fuses with that of the plasma membrane. The initial site of fusion is in the exact center of the rosette.[1] A pore forms and widens, and the content of the mucocyst is released through the pore. The particles forming the circular rim of the rosette disperse radiantly in the plasma membrane; the fate of the central particle is unknown. Exocytosis of one mucocyst leads to dispersal of one assembled rosette and it is this characteristic that can be use as a diagnostic tool for quantitation of exocytosis at the membrane level.[22]

2. Freeze-Fracture Procedure

a. Cell Preparation. Experimental cultures are established by inoculating 0.1 ml of stock culture into 10 ml of ETM, and incubating for 1 to 2 days at 30°. The resulting cells are collected, rinsed, and resuspended in Sorensen's phosphate buffer (20 mM, pH 7.0).[23]

b. Freeze Fracture. Centrifuge prepared cells and fix the resulting pellet in 1.5% (v/v) glutaraldehyde in N-2-hydroxyethylpiperazine-N'-2-ethanesulfonic acid (HEPES; Sigma, St. Louis, MO) buffer (30 mM, pH 7.0) for 2 hr. Transfer the cells into 30% (v/v) glycerol in HEPES buffer and leave them at 0 to 4° overnight. The samples are then frozen in liquid nitrogen[24] and freeze fractured at -100 to $-120°$ in a double-replica device (Warner-Lambert, Morrisplains, NJ) in a Balzer's apparatus model BA301. The replicas are cleaned in sodium hypochlorite (undiluted Clorox bleach) for 10 min and washed in distilled water before mounting on

[21] B. Satir, C. Schooly, and P. Satir, *Nature (London)* **235**, 53 (1972).
[22] S. L. Wissig and B. H. Satir, *J. Submicrosc. Cytol.* **12**, 1 (1980).
[23] A. M. Glauert, "Fixation, Dehydration and Embedding of Biological Specimens." North-Holland Publ., Amsterdam, 1987.
[24] J. E. Rash and C. S. Hudson, eds., "Freeze-Fracture: Methods, Artifacts and Interpretations." Raven Press, New York, 1979.

Formvar-coated grids. The mounted replicas are now ready for electron microscopy. Other methods are available for rapid freeze fracturing of unfixed material.[25]

3. Morphometric Analysis

Morphometry is based on the principles outlined by Weibel[26] and developed further in work done in our laboratory.[22,27,28]

Micrographs of both P and E fracture faces are taken randomly (magnification: $\times 30,000$) and negatives or prints can be used to obtain IMP counts and sizes. The negative or the micrograph is partitioned into areas of 1 cm² using an acetate overlay. The following standardized rules are used as guidelines: (1) IMPs are defined as any resolvable raised element possessing a shadow, regardless of shape; (2) IMP size is determined by the diameter of the particle at its interface with its shadow; (3) only IMPs falling within the range of 3 to 17 nm in diameter are counted; (4) a partitioned 1-cm² area is excluded from the count if more than 6% of the area is composed of a blemish (e.g., a rip) or organelle (e.g., a cilium). The frequency of IMP and other particle arrays (see Section II) can be determined from the beginning to the end of an exocytosis event.

II. Paramecium

A. Organism

The two species generally used for the study of exocytosis of trichocysts (the secretory product) are *Paramecium caudatum* and *Paramecium tetraurelia*. Both species are amenable to genetic analysis. A large number of trichocyst mutants has been isolated in *P. tetraurelia*[3] and we will describe their maintenance. In addition, conjugation rescue experiments can be performed to explore the role of the cytosol.[29]

Axenic[30] and monoxenic media are available. Biochemical data can be obtained from monoxenic cultures by using cells that are grown only to stationary phase (at which point the medium is cleared of bacteria) and that are washed extensively after harvesting. In general, axenic cultures should be used whenever possible. Strains are available from the ATCC.

[25] S. W. Hui, ed., "Freeze-Fracture Studies of Membranes." CRC Press, Boca Raton, FL, 1989.
[26] E. R. Weibel, *Int. Rev. Cytol.* **26**, 325 (1969).
[27] N. J. Maihle and B. H. Satir, *J. Cell Sci.* **78**, 49 (1985).
[28] D. M. Pesciotta and B. H. Satir, *J. Cell Sci.* **78**, 23 (1985).
[29] J. Beisson, J. Cohen, M. Lefort-Tran, M. Pouphile, and M. Rossignol. *J. Cell Biol.* **85**, 213 (1980).
[30] A. T. Soldo, G. A. Godoy, and W. J. Van Wagtendonk, *J. Protozool.* **13**, 494 (1966).

B. Media

1. Axenic: Semidefined Medium

i. Components. For 4 liters:

Proteose peptone	40 g
Trypticase soy broth	20 g
RNA (*Torula* yeast)	4 g
$MgSO_4 \cdot (H_2O)_7$	2 g
Vitamin stock solution	10 ml
Fatty acid stock solution	10 ml
Cephalin stock solution	10 ml
Thioctic acid stock solution	2 ml
Stigmasterol stock solution	10 ml

Adjust the pH to 7.0.

ii. Stock solutions

a. Vitamins:

Folic acid	2 g
Thiamin hydrochloride	1.2 g
Calcium-DL pantothenate	0.8 g
Nicotinic acid amide	0.4 g
Pyridoxamine dihydrochloride	0.2 g
Pyridoxal hydrochloride	0.4 g
Riboflavin	0.4 g
Biotin solution (stock solution: 6.25 mg biotin in 500 ml of distilled water)	8 ml

Add the vitamins to the water to a final volume of 200 ml while stirring. Distribute into screw-cap tubes, 10 ml/tube, and store frozen −20°).

b. Cephalin: Dissolve 19 g of cephalin (L-α-phosphatidylethanolamine) in 190 ml nanopore filtered water while heating. Distribute in 15-ml screw-cap tubes, 10 ml in each. Keep frozen at −20° as above.

c. Fatty acids:

Palmitic acid	1.2 g
Stearic acid	0.8 g
Oleic acid	1.8 ml
Linolenic acid	22 μl
Linoleic acid	88.8 μl

Dissolve fatty acids in water by slowly adjusting the pH to 11 with triethylamine, then bring the volume up to 200 ml. Distribute in 15-ml screw-cap tubes, 10 ml in each. Keep frozen at $-20°$ as above.

d. Stigmasterol: Dissolve 0.4 g stigmasterol in 100 ml of hot ethanol. Stir for about 10 min, let the solution cool, and bring the volume back to 100 ml with ethanol. Store at room temperature.

e. Thioctic acid: Dissolve 10 mg of thioctic acid in 100 ml of 70% ethanol. Keep the solution at $-20°$.

iii. Preparation. Pour about 3600 ml of nanopure water into a 4-liter beaker. Add all reagents and mix thoroughly, using low heat while stirring. Adjust the pH to 7.0 with NaOH (1 or 2 N) and bring the volume to 4 liters. Distribute to tubes (5 ml/15-ml tube), Erlenmeyer flasks (100 ml/500-ml flask), and culture flasks (500 ml or 1 liter/flask). Label, sterilize, and keep in the dark.

2. Monoxenic: Bacterized Cerophyl

i. Cerophyl (Agri-Tech. (Kansas City, MO)). For 1 liter:

Cerophyl	2.5 g
Na_2HPO_4	0.5 g

Boil Cerophyl in 800 ml of nanopore filtered water for 5 min and filter through cotton and then through Whatman (Clifton, NJ) filter No. 1 in a Büchner funnel. Cool the medium to room temperature, add Na_2HPO_4, and bring the total volume up to 1 liter. Distribute into tubes (5 ml/tube) or 500-ml Erlenmeyer flasks (100 ml each). Label and sterilize.

ii. Bacteria. The most commonly used species of bacteria are *Enterobacter aerogenes* and *Klebsiella pneumonia.* These are grown on agar slants at 37° [2 g Bacto-peptone (Difco) and 2 g of agar per 100 ml of water].

Agar is liquefied by heating, the pH is brought to 7 with NaOH, and the agar is distributed into glass screw-cap tubes (5 ml/tube). Autoclave the tubes, keeping the caps loose. While the medium is still liquid, place all tubes in a rack at a 45° angle and tighten the caps when the agar is solidified. Tubes can last for months in a cold room.

iii. Preparation. When inoculating agar slants streak a platinum loopful of bacteria several times on the surface of the slants. Incubate the tubes overnight at 37°. The next day fill the tubes with the Cerophyl medium (5 ml), mix, and distribute 1 ml of the bacterial suspension into 100 ml of Cerophyl medium. Cultures are kept overnight at 37° and the next day *Paramecium* cells are inoculated into this medium, using one tube (5 ml) per 100 ml of bacterized medium.

3. Starvation Buffers

Many buffers can be used. The buffer given here is one the cells tolerate well.

PIPES [piperazine-N',N'-bis(2'-ethanesulfonic acid]	5 mM
MgCl$_2$	20 mM
CaCl$_2$	0.5 mM
KCl	1 mM

Adjust the pH of the buffer to pH 7.0 with NaOH.

C. Stock Maintenance

1. Axenic Stock Cultures

Cultures are kept in test tubes (15 ml) containing 5 ml axenic medium at 27° (transfer once a week) and/or 15° (transfer every 2 weeks) in the dark (vitamins are light sensitive). Use about 10 drops of the previous culture as inoculum.

2. Monoxenic Stock Cultures

Cultures are kept in test tubes (15 ml) containing 5 ml bacterized Cerophyl medium and one to two sterilized wheat grains and refed with 10 drops of bacterized medium once a week. At room temperature the cultures should be transferred every 3 weeks and at 15° once a month. When inoculating the fresh tubes use about 10 drops of the previous culture as inoculum.

D. Exocytosis (Trichocyst Matrix Release)

With the appropriate stimulation, the membrane-bound *in vivo* secretory organelle, the trichocyst, releases proteins called the trichocyst matrix (TMX). The release proteins form a semicrystalline product (40 μm long and 0.3 μm wide) easily seen at the light microscope level (Fig. 2). This protein transformation passes through three stages, from stage I (highly condensed) to stage II (a short-lived intermediate) to stage III the needle-like secreted form).[31] Synchronous release is induced by using secretagogues. Amino dextran beads can be used for *in vivo* experiments.[32] An-

---

[31] R. S. Garofalo and B. H. Satir, *J. Cell Biol.* **99**, 2193 (1984).
[32] H. Plattner, R. Sturzl, and H. Matt, *Eur. J. Cell Biol.* **36**, 32 (1985).

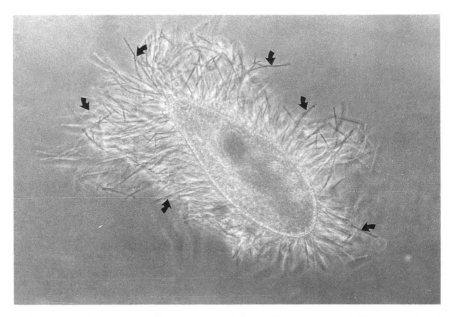

FIG. 2. Wild-type *Paramecium* after stimulation with trinitrophenol. Note the released trichocyst matrices (arrows) surrounding the entire cell body. Magnification: ×650. (Courtesy of A. Vuoso, A. E. C. O. M.).

other secretagogue, trinitrophenol (TNP), fixes the cell as well as the released TMXs which appear as a halo around the cell (Fig. 2). This is useful for quantifying the amount of release per cell and for terminating biochemical reactions.[7,33,34]

1. Cell Preparation and Induction

Grow cells in axenic medium to mid- or late logarithmic phase (3 to 4 days at 27° in the dark) to a cell density of 20,000 to 30,000 cells/ml. Harvest the cells by centrifugation at 800 rpm for 30 sec in pear-shaped flasks in an ICN/HN-SII (Needham Heights, MA) centrifuge and wash twice in PIPES buffer. Place a 15-μl drop of cells in the center of a circle of petroleum jelly on a *clean* microscope slide. Place 15 μl of secretagogue (50 mM TNP) on a coverslip and invert over the cell sample. The petroleum jelly prevents desiccation as well as damage to the cells due to the pressure of the coverslip.

[33] D. M. Gilligan and B. H. Satir, *J. Biol. Chem.* **257**, 13903 (1982).
[34] D. M. Gilligan and B. H. Satir, *J. Cell Biol.* **97**, 224 (1983).

2. Quantitation

The number of TMXs released can be quantitated in two ways in phase-contrast micrographs as described below.

1. Micrographs are taken at magnification ×400, focusing at the cell perimeter. For each experimental point count the visible TMXs per cell perimeter from 100 to 200 cells. The amount of release is classified into five categories: (a) 0 TMX, (b) 1 to 10 TMXs, (c) 11 to 50 TMXs, (d) 51 to 100 TMXs, and (d) > 100 TMXs per cell.

2. Phase-contrast micrographs (×400) of the releasing cells are taken at several different focal levels for each cell. The expanded TMXs are traced on an overlaid transparency for each of three different focal levels. The tracings are superimposed to prevent repeat counting of the same TMX. The amount of release is assigned to one of the five categories given above.

E. Quantitation of Exocytosis in Freeze-Fracture Electron Microscopy

1. Exocytotic Membrane Microdomain

Besides the rosette, the exocytic membrane microdomain in the *Paramecium* plasma membrane is characterized by an additional particle array. In *Paramecium* the diameter of the rosette is 75 nm and the individual rosette IMPs are about 150 Å in diameter. This array is surrounded by a double outer ring of IMPs (300 nm in diameter) and the diameter of the individual IMPs of the rings is about 80 Å. A modified configuration of the outer ring, a parenthesis, can be seen. The spatial sequence and formation of these various arrays are elucidated by using mutant cells.[7,20,35] Formation of a parenthesis is a primary differentiation of the plasma membrane, which is independent of the presence of an underlying secretory organelle, whereas the secondary transformation of parentheses into circular rings and the formation of the central fusion rosette are both triggered by interactions between the secretory vesicle and cell membranes (Fig. 3). Formation of the central fusion rosette is a prerequisite for exocytosis.[7] Knowledge of this sequence of events at the membrane level allows identification of unoccupied sites (parentheses), occupied sites (circular double rings), and competent exocytic sites (double outer circular rings with a central fusion rosette (see Fig. 3).

Electron microscopy is crucial for confirming that exocytosis and membrane fusion are occurring. In pseudoexocytosis expansion of the

[35] H. Plattner, K. Reichel, H. Matt, J. Beisson, M. Lefort-Tran, and M. Pouphile, *J. Cell Sci.* **46,** 17 (1980).

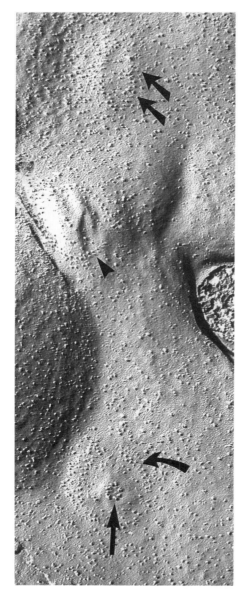

FIG. 3. An example of the P fracture face of the cell membrane of *Paramecium*. Three types of intramembrane particle arrays (IMPs) associated with the exocytic membrane microdomain can be seen. The functional domain is characterized by an assembled rosette of IMPs (arrow, left) in the center of a surrounding ring consisting of a double row of smaller IMPs (curved arrow). Two types of incompetent sites are also present: one has assembled the outer ring (arrowhead) but is missing the assembled central rosette. The second type has a parenthesis (double arrows) that has not transformed into the ring structure or assembled the central rosette. Magnification: ×60,000.

TMXs to stage III takes place without membrane fusion occurring.[34,36] This phenomenon is helpful in separating events involved in TMX expansion from membrane fusion.[33]

2. Freeze-Fracture Procedure

Centrifuge cells (see Section II,D,1) and fix the resulting pellet in 1.5% (v/v) glutaraldehyde in PIPES buffer for 2 hr. Aspirate the supernatant and gently transfer the cells into 30% (v/v) glycerol in HEPES buffer and leave at 0 to 4° overnight. The next day freeze the samples in liquid nitrogen[24] and freeze-fracture at −100 to −120° in a double-replica device in a Balzer's apparatus. The replicas are cleaned in undiluted Clorox bleach for 30 min (or overnight if monoxenic medium is used) and washed in distilled water before mounting on Formvar-coated grids. The mounted replicas are now ready for viewing in the electron microscope. Other methods are available for rapid freeze-fracturing of unfixed material.[25] For morphometric analysis, see Section I,F,3.

F. Isolation of Trichocysts

The secretory organelle and its content can best be isolated by using transport mutants *(tam-6, tam-8)*.[37,38] These mutants contain nonattached trichocysts.[6]

1. Reagents

Trichocyst isolation buffer (TIB-1): 0.3 M sucrose, 10 mM Tris-HCl, 1 mM ethylene glycol-bis(β-aminoethyl ether)-N,N,N',N'-tetraacetic acid (EGTA), 5 mM KCl, 10 mM MgCl$_2$; adjust the pH to 7.0 with NaOH

Trichocyst isolation buffer (TIB-2): 0.3 M sucrose, 10 mM Tris-HCl, 1 mM EGTA, 5 mM KCl; adjust the pH to 7.0 with NaOH and add protease inhibitors:

Leupeptin	10 μg/ml
Aprotinin	0.15 TIU/ml (1 : 200 of stock)
Phenylmethylsulfonyl fluoride (PMSF)	0.5 mM
Isotonic 60% Percoll:	
Percoll solution	30.0 ml

[36] H. Matt, H. Plattner, K. Reichel, M. Lefort-Tran, and J. Beisson, *J. Cell Sci.* **46**, 41 (1980).
[37] O. Lima, T. Gulik-Krzywicki, and L. Sperling, *J. Cell Sci.* **93**, 557 (1989).
[38] G. R. Busch and B. H. Satir, *J. Cell Biol.* **109**, 293a (1989).

TIB-1 (5×) (without magnesium and with 750 nM NaCl; final concentration, 150 mM)	10.0 ml
Distilled H$_2$O	10.0 ml

2. Cell Fractionation

Grow 2 liters of *tam-8* secretory mutant cells to stationary phase (5 days) at 27°. Harvest the cells by centrifugation at 1500 rpm for 35 sec in pear-shaped flasks (125 ml/flask) in an ICN/HN-SII centrifuge. Always keep the cells on ice and perform all subsequent steps at 4°. Wash twice in TIB-1. Concentrate cells to a final volume of 5 to 6 ml. Add protease inhibitors (see TIB-2 for inhibitor concentrations) and homogenize cells with a Dounce (Wheaton, Millville, NJ) homogenizer. At least 80 and up to 250 strokes will break open the cells without damaging the trichocysts. Homogenization should be monitored under a light microscope. Additional homogenization is required until fewer than 10 intact cells per slide are seen. Pipette the resulting homogenate into a single pear-shaped flask containing 35 ml of cold TIB-2 and spin at 1000 rpm for 4.0 min. Set aside the supernatant (the trichocysts) and place on ice. Resuspend the pellet and spin again for a higher yield. Combine both supernatants into a 50-ml polycarbonate centrifuge tube and spin in a RC-5B refrigerated Sorvall (Norwalk, CT) centrifuge with an HB-4 rotor at 1450 *g* for 6 min, bringing down intact trichocysts. Resuspend the pellet in 6 ml of isotonic 60% Percoll solution, using gentle swirling or a large-bore pipette to avoid bubbles.

3. Purification Using Percoll Gradients

Layer 0.5 ml of the above homogenate onto each of four polycarbonate Du Pont (Wilmington, DL) SE-12 centrifuge tubes containing 8 ml of isotonic 60% Percoll solution. Spin the tubes in an RC-5B refrigerated centrifuge fitted with an SE-12 fixed-angle rotor at 37,500 *g* for 60 min. Remove and save the band formed 14 mm from the bottoms of the tubes. Combine the identical bands from the four gradients into a fresh SE-12 centrifuge tube and repeat the Percoll gradient procedure to obtain pure trichocysts (with their membranes). The preparation can be stored at 4° for up to 4 days and is suitable for biochemical, physiological, and ultrastructural analyses.

4. Physiological Analysis

i. Reagents

Acridine orange (AO) solution: 1 mM AO/ml distilled H$_2$O

Carbonyl cyanide *m*-chlorphenylhydrazone (CCCP) stock in dimethyl sulfoxide (DMSO) (2 mM)

Monensin stock in 100% ethanol (10 mM)
Chloroquine stock in distilled H$_2$O (50 mM)

ii. Ionophore treatment. Combine 100 μl of trichocysts with 400 μl of TIB-2 and 5 μl of stock AO solution in a 1.5-ml microfuge tube. Incubate for 1.5 min at room temperature, then spin for 22 sec (15,000 g at room temperature) in an Eppendorf 5414 microfuge fitted with a fixed-angle rotor. Aspirate the supernatant; the light orange pellet contains the trichocysts. Resuspend the pellet in 500 μl of TIB-2 with the ionophore to be studied (20 μM CCCP, 1 to 50 μM monensin, or 500 μM chloroquine). Controls should contain the appropriate solvents (see above). Incubate at room temperature for 15 min, then spin the mixtures in a microfuge for 22 sec, and aspirate the supernatant. Resuspend the pellet in 50 μl of TIB-2. Place 15 μl of the treated trichocysts on a slide and top with a coverslip. The samples are viewed with a 405 ICM Zeiss (Thornwood, NY) epifluorescent microscope fitted with an AO filter (440 to 490-nm excitation). Phase-contrast and fluorescent photomicrographs of the same fields are taken with a Nikon (Garden City, NY) M35 camera and Kodak (Rochester, NY) Tri-X film enhanced by Kodak Diafine developer to 800 ASA. The percentage of fluorescent trichocysts is determined by a comparison of the phase-contrast and fluorescent micrographs, and the effect of drug treatment can be determined.[39]

G. Phosphorylation Studies

The advantage of using these cells for biochemical studies is that, in addition to homogenate-based experiments, live cells can be labeled and the exocytosis process can be analyzed *in vivo*. This enhances the chances of discovering cytosolic and/or membrane components that are integral to the sequence of events leading to exocytosis.

1. Reagents

PIPES buffer: See Section II,B,3
Tris-1 buffer: 5 mM Tris-HCl (pH 7) containing 10 mM MgCl$_2$ and 1 mM KCl
Tris-2 buffer: 50 mM Tris (pH 7) containing 10 mM ethylenediaminetetraacetic acid (EDTA), and proteolytic inhibitors [leupeptin (1 μg/ml), antipain (1 μg/ml), PMSF (1.15 mM), and aprotinin (0.03 TIU/ml)]

2. In Vivo $^{32}P_i$ Labeling

Wild-type cells are grown to early stationary phase (approximately 20,000 cells/ml) in axenic medium and harvested by centrifugation in an IEC clinical centrifuge (500 g, 5 min). Wash the cells four times with

[39] G. R. Busch and B. H. Satir, *J. Cell Sci.* **92**, 197 (1989).

PIPES and concentrate 100 ml of cells (10,000 to 20,000 cells/ml) to approximately 0.3 ml. Add 200 μl of this live cell suspension to 0.1 mC_i of [^{32}P]phosphoric acid (New England Nuclear, Boston, MA) in a plastic cuvette. Incubate for 1 to 2 hr at room temperature. Cell aliquots (20 to 30 μl) are removed from the cuvette to be used for exocytosis stimulation[33] under different conditions.

3. Homogenate $^{32}P_i$ Labeling

Wash early stationary-phase cells twice with 50 ml of Tris-1. Resuspend the resulting pellet in 3 ml of Tris-1 buffer in the homogenizer and add an equal volume of Tris-2. Cells are homogenized on ice; the process is followed under the light microscope until little cell debris is seen. Add the label to the homogenate and incubate as above. Aliquots (2 μl) can be removed during the period of labeling, thoroughly washed, and counted in the scintillation counter to monitor uptake. Spin the final homogenate in an SS34 rotor, Sorvall RC-5B Du Pont centrifuige at 15,000 rpm (270 g) for 10 min at 4°. Collect the resulting low-speed supernatant and spin at 100,000 g in a Beckman (Fullerton, CA) L5-50 at 4° for 1 hr. Concentrate the supernatant in a Centricon (holding about 1 ml each) (M_r 30,000 cutoff units) using the Sorvall SS34 rotor at 5500 rpm (3,650 g) for about 30 to 45 min at 4°. In this manner the supernatant is concentrated to about 500 μl.

This supernatant can be further processed for sodium dodecyl sulfate-polyacrylamide gel electrophoresis (SDS-PAGE), autoradiography, Western blotting, immunoprecipitation, and/or frozen at −80°.

[15] Calcium-Dependent Membrane-Binding Proteins in Cell-Free Models for Exocytotic Membrane Fusion

By CARL E. CREUTZ

Introduction

In the process of exocytosis, secretory vesicle membranes fuse with the plasma membrane, and, in some systems, with other secretory vesicle membranes. Exocytosis is initiated in many cells in response to increases in the cytoplasmic free calcium concentration. Therefore much effort has gone into characterizing proteins that attach to the cytoplasmic surface of secretory vesicle membranes in the presence of calcium.[1-4] Such proteins may alter the character of the membrane surface in the stimulated cell, and act as points of mediation for contacts between the vesicle and the cytoskeleton to promote vesicle movement, between the vesicle membrane and the plasma membrane to promote fusion and externalization of vesicle contents, and between the expended vesicle membrane and clathrin to promote membrane recycling.

This chapter describes methods for isolating calcium-dependent membrane-binding proteins and for characterizing their interactions with membranes. A subset of these proteins is *bivalent* in the sense that they can interact with two membranes and promote their contact. Methods for analyzing this membrane-aggregating activity are described. Membrane contacts formed by the aggregating proteins may undergo fusion, depending on the nature of the membrane, or the addition of cofactors such as cis-unsaturated fatty acids. Methods for monitoring these membrane fusion events are also described.

Isolation of Calcium-Dependent Chromaffin Granule Membrane-Binding Proteins by Calcium-Dependent Affinity Chromatography

Chromaffin granule membranes can be used as an affinity matrix to isolate calcium-dependent membrane-binding proteins.[1,3] The techniques used should be applicable to other biological membranes. Large quantities of chromaffin granules (~ 100 mg protein) are isolated from ~ 40 bovine adrenal glands (~ 75 g of medullary tissue) by centrifugation through a 1.6 or 1.7 M sucrose step gradient.[5] The initial homogenate should contain ethylene glycol-bis(β-amino ethyl ether)-N,N,N',N'-tetraacetic acid (EGTA) (at least 5 mM for a 33% homogenate) to ensure that the membranes are isolated free of calcium-dependent membrane-binding proteins. The granules are ruptured by hypotonic lysis and the membranes washed twice by centrifugation in hypotonic buffer containing EGTA. Thorough washing is important to remove the high content of soluble secretory proteins that could quench reactive sites on the CNBr-activated Sepharose used to form the affinity matrix.

The coupling of membranes to CNBr-activated Sepharose (Pharmacia, Piscataway, NJ) is performed after the resin is hydrated and washed in

[1] C. E. Creutz, *Biochem. Biophys. Res. Commun.* **103,** 1395 (1981).
[2] M. J. Geisow and R. D. Burgoyne, *J. Neurochem.* **38,** 1735 (1982).
[3] C. E. Creutz, L. G. Dowling, J. J. Sando, C. Villar-Palasi, J. H. Whipple, and W. J. Zaks, *J. Biol. Chem.* **258,** 14664 (1983).
[4] C. E. Creutz, W. J. Zaks, H. C. Hamman, S. Crane, W. H. Martin, K. L. Gould, K. M. Oddie, and S. J. Parsons, *J. Biol. Chem.* **262,** 1860 (1987).
[5] S. F. Bartlett and A. D. Smith, this series, Vol. 31, p. 379.

1 mM HCl. Chromaffin granule membranes are resuspended in 10 to 20 ml of coupling buffer (0.5 M NaCl, 0.1 M NaHCO$_3$) and mixed with a slurry of CNBr-activated Sepharose 4B in 1 mM HCl. Coupling is allowed to proceed for 1 to 2 hr at room temperature with continuous shaking. The gel is then washed once with coupling buffer and then the coupling is terminated by the addition of 1 M ethanolamine, pH 8.0, for 1 hr at room temperature with continuous shaking. The resin is then washed in column buffer (see below) by gentle resuspension and settling at 1 g. The resin is not subjected to harsh treatments, such as extremes of pH (usually used at this step to wash away uncoupled protein), if complex membrane activities are to be retained. With 20 mg of chromaffin granule membrane and 7.5 g of dried Sepharose 4B one-half of the membrane protein is coupled. Extraction of the resin with sodium dodecyl sulfate (SDS) and subsequent examination by SDS gels confirms that typical membrane protein components are on the resin, suggesting the coupling procedure has not selectively removed certain membrane components. Electron microscopy indicates that the membranes occupy sites on the outer surface of the beads and do not intercalate into the bead matrix because of their large size. To isolate the full complement of chromaffin granule-binding proteins the resin is usually used within 24 hr. However, some lipid-binding proteins, such as the annexins, may be isolated after storage of the resin at 4° [with ~0.01% (w/v) sodium azide] for at lease 1 month.

The resin with coupled membranes is packed by gravity into a low-pressure column. For 7.5 g of initial dried resin, a 1.6 × 10 cm column with ~20-ml bed volume is appropriate. A flow adaptor is used to provide sharp elution zones. The resin is washed in the column with calcium-containing buffer [e.g., 240 mM sucrose, 30 mM KCl, 1 mM MgCl$_2$, 25 mM N-2-hydroxy ethylpiperazine-N'-2-ethanesulfonic acid (HEPES)–NaOH (adjusted to pH 7.3 at 37°) 4 mM CaCl$_2$, 2 mM EGTA], then with the same buffer containing 0 mM CaCl$_2$, 2 mM EGTA. The column is reequilibrated with calcium buffer and a cytosolic extract containing the same free calcium concentration is applied. At all steps consideration must be given to the gel-filtration properties of the resin. For example, proteins may be advanced to zones without calcium if there is insufficient calcium-containing buffer applied initially to the column. When 75 g of tissue is processed in 2 vol of EGTA-containing extracting solution, a postmicrosomal supernatant volume of ~140 ml is obtained. Typically only one-quarter of this is applied to the column at a time in order to avoid saturating binding sites on the membranes. Repetitive runs of the remaining supernatant have yielded identical results, indicating no evident degradation of the column. After application of the supernatant the column must be washed with at least two bed volumes of calcium buffer. Then the calcium concentration is reduced to elute specifically bound proteins. The calcium may be reduced in steps (or through the use of a gradient of decreasing calcium) to resolve the membrane-binding proteins into classes according to the apparent affinity of calcium-binding sites involved in promoting the membrane interaction.

The temperature of the chromatographic column may be maintained with a thermal jacket at 37° in order to simulate intracellular conditions. This is not essential for isolation of annexins, but surprisingly it does not seem to result in any proteolytic degradation of the major proteins isolated when compared to column runs at 4°. There do, however, appear to be proteolytic changes in the membrane proteins after use of the column, which can be detected on polyacrylamide gels of SDS extracts of the column resin.

The column procedure lends itself to many possible variations used to determine parameters that regulate protein–membrane interactions. The presence of ATP in the elution buffer is essential to elute the major chromobindin, chromobindin A, which otherwise becomes locked in a rigor-like complex on the vesicle membrane that is insensitive to reductions in calcium.[2,6] Conversely, the yield of protein kinase C is reduced greatly if the column buffer contains ATP.[2] Interesting additional variations that have not been carefully explored would include reversing the role of calcium, that is, loading cytosolic proteins in EGTA and eluting with calcium to determine whether certain cytosolic proteins may be removed in a calcium-dependent fashion from secretory vesicles prior to exocytosis, or manipulation of GTP, IP$_3$, or sulfhydryl potential levels.

Alternate Procedure: Affinity Isolation by Centrifugation

The affinity column approach for isolating membrane-binding proteins has the advantage that it is particularly gentle, so that even weak interactions or those of labile proteins may be detected. In addition, through extensive washing the background of noninteracting proteins in the column eluate can be reduced to negligible levels. However, in some cases, repeated centrifugation in calcium-containing buffers followed by extraction in EGTA can provide a convenient alternative procedure. The centrifugation steps must be carried out with sufficient force to pellet the membrane fraction. With intact chromaffin granules this can be accomplished at modest force, for example, 20 min at 20,000 *g*. However, intact organelles can be extremely leaky and thus can increase the background of irrelevant proteins. Washed membranes, or lipid vesicles, give cleaner

[6] W. H. Martin and C. E. Creutz, *J. Biol. Chem.* **262**, 2803 (1987).

backgrounds but require higher centrifugal forces. However, at high forces (~ 100,000 g) certain cytosolic components are precipitated in the presence of calcium. In particular, because typical postmicrosomal supernatant preparations still contain some membranes or lipid particles, the annexins are precipitated from cytosol without the addition of an exogenous membrane substrate. This can confuse the issue if interactions with specific membranes are being examined.

The centrifugation protocol is most useful for preliminary studies, large-scale preparative work, or studies in which a large number of conditions are to be examined that would otherwise require an extensive series of runs on the affinity column. The procedure may also be more convenient for containment of radioactive ligands. In the case of preparative isolation of proteins, the affinity matrix can consist simply of the particulate fraction of a cell or tissue homogenate.

Alternate Procedure: Calcium-Dependent Affinity Chromatography on Phenyl-Sepharose

Annexins have also been isolated by calcium-dependent affinity chromatography on phenyl-Sepharose.[7] It is not clear whether the annexins directly interact with the phenyl groups on this resin. Because the typical isolation protocol involves application of a crude fraction that may contain lipids to the column, interaction of lipid-binding proteins with the column may be through lipids adsorbed to the phenyl-Sepharose. In fact, it is possible to prime the phenyl-Sepharose with lipids to enhance binding of annexins or similar proteins. This can be accomplished simply by applying a sonicated suspension of lipid vesicles (containing acidic phospholipids, e.g., phosphatidylserine) to the column prior to chromatography of cell extracts. It may be helpful to apply the lipids in the presence of EGTA to ensure that the lipids will not be removed during the subsequent reductions in calcium during the chromatography. Alternatively, the column may be prewashed with EGTA-containing buffer.

Conversely, passage of crude annexin preparations over a phenyl-Sepharose column in the continuous presence of EGTA has been found to be helpful in removing lipids from the annexin preparation and hydrophobic proteins that may copurify with annexins in initial membrane-binding steps.[4]

[7] T. C. Südhof and D. K. Stone, this series, Vol. 139, p. 30.

Assay of Calcium-Dependent Interaction of Proteins with Membranes

Affinity chromatography can provide a rough indication of the dependence on calcium or other solution parameters of the interaction of one or several proteins with a given membrane or organelle. However, more detailed analysis requires other methods. The protein of interest must be purified to enable clear interpretation of binding studies, because some membrane binding proteins can inhibit membrane binding by other proteins. This caution may be extended to studies of the effect of protein phosphorylation on the calcium dependence of membrane binding: the phosphorylated and unphosphorylated species should ideally be separated and studied independently. Two examples of binding assays used with purified proteins will be given here: the binding of ^{125}I-labeled annexins to chromaffin granule membranes[8] and the monitoring of fluorescence energy transfer from tryptophan in endonexin to synthetic lipid vesicles containing dansylated phosphatidylethanolamine.[9]

Binding of ^{125}I-Labeled Annexins to Chromaffin Granule Membranes or Lipid Vesicles

Annexins can be labeled with ^{125}I by the Iodogen method to a specific activity of at least 4×10^6 disintegrations per minute (dpm)/μg without apparent alteration of their membrane-binding properties, as judged by comparison of the binding of labeled and unlabeled species.[4] The binding of ^{125}I-labeled protein to membranes is measured by a modification of the method of Burgoyne and Geisow.[10] Intact chromaffin granules, chromaffin granule membranes, or multilamellar phosphatidylserine–phosphatidylcholine (1 : 1) liposomes are preincubated for 15 min at 22° in buffer containing 0.24 M sucrose, 30 mM KCl, 40 mM HEPES–NaOH (pH 7.0). An aliquot of this mixture is then added to a 1.5-ml microfuge tube so as to contain (in final concentration) 0.24 M sucrose, 30 mM KCl, 40 mM HEPES–NaOH (pH 7.0), 52 μg chromaffin granule membrane protein per milliliter (or 50 μg total chromaffin granule protein per milliliter), 2.5 mM calcium chelator (e.g., EGTA) and varying concentrations of CaCl$_2$, ^{125}I-labeled protein, and unlabeled protein (for competition studies). The reactions are started by addition of ^{125}I-labeled protein. The mixtures are allowed to incubate for 15 min at 22° and then centrifuged at

[8] W. J. Zaks and C. E. Creutz, *Biochim. Biophys. Acta* **1029**, 149 (1990).
[9] M. Junker and C. E. Creutz, unpublished observations (1989).
[10] R. D. Burgoyne and M. J. Geisow, *FEBS Lett.* **131**, 127 (1981).

15,600 g for 15 min in an Eppendorf microcentrifuge. The supernatants are aspirated and the pellet and tube wall washed twice with binding buffer containing the same Ca^{2+} concentration as used in the binding step. The radioactivity is measured in both the pellets and the supernatants by scintillation counting. Specific binding is defined as the difference in membrane-bound radioactivity in the absence and presence of excess EGTA or excess unlabeled protein. In the absence of granule membranes or carrier proteins (1 mg ovalbumin or bovine serum albumin per milliliter), ^{125}I-labeled proteins are found to bind to the polypropylene microfuge tube in both a Ca^{2+}-independent and Ca^{2+}-dependent manner. Hence, experiments involving binding to liposomes are best done in the presence of 1 mg bovine serum albumin (BSA) per milliliter. The presence of carrier protein has no effect on the Ca^{2+} dependence of protein binding to chromaffin granule membranes.

Fluorescence Energy Transfer Assay for Annexin–Liposome Binding

A fluorescence energy transfer assay can be used to monitor the binding of annexins to artificial lipid membranes.[9,11,12] The assay depends on the measurement of energy transfer from the tryptophan in the annexin to the chromophore of dansylphosphatidylethanolamine incorporated in lipid vesicles in small amounts (< 5 mol%).

Large unilamellar vesicles are prepared by extrusion, using a Lipex (Vancouver, B.C.) biomembrane extruder according to the method of Mayer et al.[13] The size distribution of the vesicles formed can be determined by dynamic light scattering. Using a polycarbonate filter (0.1-μm pore diameter), vesicles of 0.1 ± 0.03 μm (SD) diameter can be formed. The size distribution is fairly independent of lipid composition, when preparing vesicles with acidic phospholipids mixed in phosphatidylcholine. Although less well defined, sonicated or multilamellar vesicle preparations can be used in this assay. The unilamellar vesicles have advantages in that they have a large radius of curvature that resembles that of biological membranes, and at least one-half of the dansyl label is available on the vesicle surface, thus increasing the signal-to-noise ratio in the energy transfer assay.

Steady state fluorescence is monitored with excitation at 280 nm and emission at 510 nm, employing a 400-nm cutoff filter in the emission beam. In a 300-μl sample containing 100 mM NaCl, 25 mM morpholinepropanesulfonic acid (MOPS) buffer, pH 7.0, the vesicles (4 μM total lipid,

[11] W. L. C. Vaz, K. Kaufmann, and A. Nicksch, *Anal. Biochem.* **83**, 385 (1977).
[12] M. D. Bazzi and G. L. Nelsestuen, *Biochemistry* **26**, 115 (1987).
[13] L. D. Mayer, M. J. Hope, and P. R. Cullis, *Biochim. Biophys. Acta* **858**, 161 (1986).

3 to 5% dansylphosphatidylethanolamine) are mixed with calcium, and the binding reaction initiated by the addition of protein. Calcium alone does not affect the dansyl fluorescence, and the final fluorescence change appears to be independent of the order in which calcium, vesicles, and protein are added. When initiating the reaction by the addition of protein, the maximal fluorescence change occurs too rapidly to measure with hand-mixing procedures, reflecting the rapid kinetics of the membrane binding by annexins. Subsequently, a slow decline in fluorescence may be seen, which correlates with aggregation of the vesicles. Therefore, to characterize the initial binding event, data must be recorded within 1 min. The fluorescence increase associated with the binding event is completely reversible on the addition of excess EGTA, reflecting the reversibility of the binding event.

Examination of the complete excitation spectrum of the dansyl chromophore during the binding reaction reveals that the intrinsic fluorescence of the dye is increased in parallel with the increase in energy transfer. The greatest increase in the excitation spectrum of the dansyl group occurs at wavelengths associated with tryptophan absorbance (~283 nm). Tryptophan emission (350 nm) is also reduced in parallel with membrane binding. Therefore, the annexin–membrane-binding event appears to promote energy transfer from the tryptophan to the dansyl group as predicted, but protein binding alters the membrane surface such that intrinsic dansyl fluorescence is also enhanced. By monitoring fluorescence emission at 510 nm with excitation at 280 nm, the combination of these effects is recorded. Because the two effects appear to reflect the binding event, this is not normally a problem. If it is important to measure energy transfer exclusively, a correction should be made by subtracting the intrinsic dansyl fluorescence measured at 510 nm with excitation at 255 or 315 nm.

Calcium-Dependent Aggregation of Chromaffin Granules

All proteins identified to date that promote a calcium-dependent aggregation of chromaffin granules belong to the annexin family. This suggests the annexins are specifically designed to carry out this "bivalent" activity, and therefore may also be responsible for initiating intermembrane contacts in cells. The chromaffin granule aggregation assay is a convenient assay for monitoring the isolation of annexins, and may also be useful in characterizing the parameters that influence the activity of annexins, such as calcium, phosphorylation, subunit association, or proteolysis. The assay is performed by continuously monitoring the turbidity of a granule suspension.[14] Chromaffin granules prepared by differential centrifugation are

[14] C. E. Creutz, C. J. Pazoles, and H. B. Pollard, *J. Biol. Chem.* **253**, 2858 (1978).

adequate for this assay; however, a more robust response is obtained with granules purified in isotonic gradients (e.g., of sucrose and metrizamide),[15] because contaminating organelles such as mitochondria can inhibit granule–granule interactions.

A modern recording spectrophotometer capable of monitoring multiple reaction cells is most useful for this assay. Absorbance is normally measured at 540 nm, although 350 nm may be used to increase sensitivity. A temperature-controlled cuvette holder is used to maintain temperature at 25 or 37°. Below 25° the aggregation reaction slows rapidly; no aggregation occurs at 0°. Chromaffin granules should be used within 3 to 4 days after preparation if stored at 4°, although the response is greatest on the day the granules are isolated. The standard initial concentration corresponds to an absorbance of 0.3 at 540 nm (approximately 100 μg of granule protein per milliliter). However, by reducing the granule concentration, the sensitivity of the assay can be enhanced, because the amount of annexin necessary to aggregate granules is in constant proportion to the amount of granule protein. The reaction mixture must normally be of low ionic strength, because intact granules are more labile in high salt concentration, particularly chloride salts. The medium must also be isotonic because of the osmotic sensitivity of the intact organelles. A standard composition is 240 mM sucrose, 30 mM KCl, and 28 mM histidine–HCl (pH 6.0) or 28 mM HEPES–NaOH (pH 7.0 to 7.3). In addition, a calcium buffer of 2.5 mM EGTA and various amounts of CaCl$_2$ is used to maintain a constant free calcium concentration. The final calcium concentration should be checked with a calcium-sensitive electrode, as calculations based on literature values of EGTA/Ca^{2+} binding constants have been found to be in error by as much as an order of magnitude, due to the properties of other components of the buffer, pipetting errors, and uncertainties regarding which constants to use (see [12] in this volume). A calcium buffer system is essential when working with intact chromaffin granules as they will leak significant amounts of calcium during the assay.

The reaction mixtures are prepared directly in the spectrophotometer cuvettes by making additions of prewarmed solutions in the following order: 200 μl of 150 mM KCl, 400 μl of 300 mM sucrose/40 mM pH buffer [e.g., histidine (pH 6.0) or HEPES (pH 7.0–7.3)] containing the annexin, and 400 μl of a chromaffin granule suspension prepared immediately before the reaction by mixing one part chromaffin granules (A_{540} of 3.0, in 300 mM sucrose) with three parts 300 mM sucrose/40 mM pH buffer. Baseline absorbances are recorded for all cuvettes for at least 2 min,

[15] H. B. Pollard, H. Shindo, C. E. Creutz, C. J. Pazoles, and J. S. Cohen, *J. Biol. Chem.* **254**, 1170 (1979).

then the reactions are started by the addition of 20 μl of calcium–EGTA buffer and gentle mixing with a Pasteur pipette. The reactions are monitored for approximately 10 min. In cases of high activity, maximum aggregation, corresponding to about a 75% increase in absorbance, occurs within 5 min. However, if activity is quite low it may be fruitful to monitor the reaction for up to a half-hour to detect changes of only a few percent. A baseline trace of granules without annexin or with EGTA must be frequently checked because the granules do undergo some lysis and loss in turbidity during the recording period. This baseline decline should be corrected in calculating a percentage increase in turbidity at any time point.

The aggregation reaction can also be carried out with granule membranes, other membrane fragments, or with liposomes prepared by any of several methods.[16] Measurement of absorbance at 350 nm is useful for enhancing the smaller signals obtained with these particles.

Pitfalls of Aggregation Assay

This assay is nonlinear and subject to threshold and saturation effects: low amounts of annexin, for example, less than 0.5 μg of synexin per milliliter, cannot be detected with the standard chromaffin granule assay described above; the assay is completely saturated by 4 μg of synexin per milliliter.

When assaying crude fractions, contaminating lipases can promote granule lysis and mask the turbidity increase associated with membrane aggregation. Chromophores in crude cytosolic fractions may undergo spectral changes that can be misinterpreted as turbidity changes. When dealing with crude material a wise course is to check for aggregation visually in a phase-contrast microscope.

If more than one annexin, or possibly other membrane-binding proteins, are present in a fraction, they may compete with one another for membrane-binding sites or otherwise mask membrane-aggregating activity. This is particularly true when working with annexin VI, which was previously characterized as the synexin inhibitor "synhibin." [17]

Maintenance of constant pH is essential. Alterations in pH will alter the binding of calcium by EGTA and change the free calcium concentration. Conversely, if the EGTA is challenged with excess calcium, for example, from a column fraction, it will release protons and acidify the medium. If

[16] N. Düzgüneş, J. Wilschut, K. Hong, R. Fraley, C. Perry, D. S. Friend, T. L. James, and D. Papahadjopoulos, *Biochim. Biophys. Acta* **732**, 289 (1983).
[17] H. B. Pollard and J. H. Scott, *FEBS Lett.* **150**, 201 (1982).

the pH drops below ~5.5, granules will spontaneously aggregate and a false-positive result will be obtained.

Fusion of Isolated Chromaffin Granules

After aggregation by an annexin, chromaffin granules will undergo fusion if exposed to low concentrations of free, cis-unsaturated fatty acids.[18,19] The fusion rate and extent can be monitored by turbidity measurement of the granule suspension, as described above, for the aggregation reaction. During fusion the dense core materials of the granules are diluted as the fused granule membranes expand osmotically to encompass the maximum volume. This change in structure is reflected in a dramatic decrease in turbidity back to, or below, the baseline seen with the unaggregated granules. Although this method can give semiquantitative comparisons of the effectiveness of various fatty acids as fusogens, the turbidity change is not specific for fusion. Granule lysis will also result in an indistinguishable drop in turbidity, as would disaggregation of the granules or membranes. (Aggregation of chromaffin granules by an annexin is not reversible by chelation of calcium; however, liposomes and other membranes may dissociate to some extent.) Therefore fusion is best confirmed by direct observation of granule and membrane morphology. In the case of chromaffin granules this can be done at the light microscope level, because fused granules swell to form large vacuole-like structures.[18,19]

Direct Observation of Chromaffin Granule Fusion in Phase-Contrast Microscope

For observation of chromaffin granule fusion in the phase-contrast microscope it is essential to have highly purified organelles. Creation of a vacuole-like structure, 10 μm in diameter, requires the fusion of 1600 chromaffin granules of diameter 2000 Å. Mitochondria can be attached to chromaffin granules *in vitro* by annexin action, but the mitochondria do not subsequently fuse with the chromaffin granules.[18] Therefore contaminating mitochondria act as inhibitors of the fusion reaction and prohibit the formation of large vacuoles that can be seen in the phase-contrast microscope. Purification of chromaffin granules on sucrose–metrizamide isotonic density gradients is ideal.[15] Granules prepared by differential centrifigation alone are generally too impure, although some preliminary work can be done with such material.

Granules are aggregated in the presence of 5 to 10 μg of annexin

[18] C. E. Creutz, *J. Cell Biol.* **91**, 247 (1981).
[19] D. S. Drust and C. E. Creutz, *Nature (London)* **331**, 88 (1988).

(synexin[18] or calpactin[19] have been demonstrated to work well) and calcium, as described above, for the aggregation reaction. A calcium concentration adequate to promote vigorous aggregation is necessary, but after aggregation the calcium is not necessary to promote fusion.[18] The fusion is initiated by the addition of free fatty acid as follows: stock solutions of fatty acids are prepared at 5 mg/ml in ethanol and diluted into buffered sucrose at a concentration of 100 μg/ml before each series of experiments. Arachidonic acid gives the best fusion response and seems to be the most relevant physiologically, but oleic acid is 50% as effective and is more stable. Twenty to 100 μl of the resulting emulsion is added to the 1-ml aggregation reaction, giving final fatty acid concentrations of 2 to 10 μg/ml in the presence of ~100 μg of granule lipid. Above 10 μg/ml, the fatty acids produce too much granule lysis. Fusion begins immediately and is complete in ~10 min at 37° at pH 6.0. Fusion appears not to occur at 0°. The rate of fusion is strongly pH dependent, occurring almost an order of magnitude more slowly at pH 7.2.[18] The fusion event is exciting to observe in real time, as the hundreds of fusing granules form larger and larger "bubbles" that continue to fuse, looking somewhat like a clump of soap bubbles fusing with one another.

Commercial wetting agents used in the preparation of microscope slides will also cause chromaffin granule fusion. Therefore it is important to wash microscope slides in distilled water, or acetone, prior to examining granule aggregates.

The fused granules can be sedimented at 20,000 g to determine the amount of leakage of granule contents that has occurred during fusion. The resulting vacuoles can also be examined by conventional transmission electron microscopy after sectioning of granule pellets. Fixation in glutaraldehyde should be done in suspension before centrifugation, as the delicate vacuoles can be broken during pelleting.

The fused granules should be large enough (10-μm diameter) to permit the study of secretory vesicle ionic conductances by the patch-clamp technique, but this interesting possibility has not yet been explored.

Fluorescence-Based Fusion Assays

A real-time, quantitative assay of chromaffin granule fusion can address certain questions that cannot be resolved by the direct observation method described above. For example, the effects of osmotic pressure on chromaffin granule fusion cannot be determined by phase-contrast microscopy, because the swelling of the vacuoles that follows fusion of granules is also suppressed by an external osmotic pressure.[18] A fluorescence assay of lipid mixing during fusion has been used to study the phenomenology of annexin-mediated chromaffin granule fusion.[20] A population of intact granules is labeled with octadecylrhodamine at self-quenching concentrations, and an increase in fluorescence is seen when these granules fuse with unlabeled granules, causing a reduction in the effective surface concentration of the lipid-bound probe.[21,22]

Chromaffin granules, freshly isolated by differential centrifugation, are labeled with octadecylrhodamine B (R18). A concentrated ethanolic solution of R18 (1 mg/ml) is added dropwise to a vortexed suspension of chromaffin granules in 0.3 M sucrose. The amount of R18 is calculated to be 3 to 5 mol% of the total granule phospholipid (450 nmol/mg protein), and care is taken to keep the total ethanol concentration below 1%. The mixture is incubated in the dark for 1 hr at room temperature and free R18 is removed by two cycles of centrifugation (12,000 g for 10 min at 4°) and the labeled granules are resuspended in 2.0 ml of ice-cold 0.3 M sucrose. Chromaffin granule fusion is measured without stirring at room temperature in a 0.5-cm cuvette containing 0.24 M sucrose, 30 mM KCl, 40 mM HEPES–NaOH (pH 7.0), and a 2.5 mM concentration of a calcium–EGTA buffer system. The sample is excited at 560 nm and fluorescence recorded at 590 nm. The labeled and unlabeled granules are mixed in a 1:4 proportion. Dilute suspensions of granules (total A_{540} is ∼0.03) are used to reduce light scattering. Baseline fluorescence is allowed to stabilize before initiating fusion by the addition of calcium buffer. An immediate drop in fluorescence is seen due to the calcium-dependent interaction of the annexin with the membrane surface. The fusion process is represented by a slower increase of fluorescence that is complete within 20 min.

Under standard conditions this fusion signal appears to report true fusion events correctly as verified by direct microscopic examination. However, on careful examination this system does not behave as theoretically predicted. In particular, if the assay is conducted with labeled granules only there should be no dequenching of the dye on fusion and hence no increase in fluorescence. In fact, the fluorescence increase is found to be linearly proportional to the percentage of labeled granules.[20] In other words, the strongest signal is seen when there are no unlabeled granules. This unorthodox behavior may result from dilution of the probe, as it crosses the bilayer from the outside of the granule to the inside of the granule, during membrane reorganization coincident with fusion. When

[20] W. J. Zaks and C. E. Creutz, *in* "Molecular Mechanisms of Membrane Fusion" (S. Ohki, D. Doyle, T. D. Flanagan, S. W. Hui, and E. Mayhew, eds.), p. 325. Plenum, New York, 1988.

[21] D. Hoekstra, T. de Boer, K. Klappe, and J. Wilschut, *Biochemistry* **23**, 5675 (1984).

[22] D. Hoekstra and K. Klappe, this series, Vol. 220 [20].

comapred side by side, the fusion signal obtained with this assay correlates well with the fusion signal obtained using the transfer of energy from 7-nitrobenz-2-oxa-1,3-diazo-4-yl (NBD)-labeled phosphatidylethanolamine to rhodamine-labeled phosphatidylethanolamine to measure chromaffin granule fusion.[23] However, because the behavior of the R18 assay, when applied to chromaffin granules, is not completely understood at present, it remains advisable to confirm important results obtained with this method by direct morphological observation. This assay has been used to monitor the fusion of R18-labeled neutrophil secretory granules with unlabeled liposomes in the presence of synexin and fatty acids, and correlates well with results obtained with NBD- and rhodamine-labeled liposomes and unlabeled granules.[24]

Caveat Emptor

The methods described in this chapter permit the isolation and characterization of calcium-dependent secretory vesicle-binding proteins. The *in vitro* behavior of some of these proteins suggests that they are involved in critical steps in the secretory pathway. However, the true functional roles of proteins obtained by these methods will be determined only by combining these biochemical approaches with pharmacological, physiological, and genetic analyses performed on cells.

[23] S. J. Morris and D. Bradley, *Biochemistry* **23**, 4642 (1984).
[24] P. Meers, J. D. Ernst, N. Düzgüneş, K. Hong, J. Fedor, I. M. Goldstein, and D. Papahadjopoulos, *J. Biol. Chem.* **262**, 7850 (1987).

Section IV

Intracellular Membrane Fusion

[16] *In Vitro* Studies of Endocytic Vesicle Fusion

By Ruben Diaz, Luis S. Mayorga, Maria I. Colombo, James M. Lenhard, and Philip D. Stahl

Endocytosis is a process carried out by most eukaryotic cells.[1] The endocytosis of fluid and receptor-bound macromolecules results in the formation of intracellular vesicles that are derived from the plasma membrane. These vesicles mediate the transport of macromolecules to different intracellular compartments. Receptor–ligand complexes accumulate in clathrin-coated regions of the plasma membrane that pinch off to form coated vesicles of small diameter. These vesicles appear to fuse to each other, or to other preexisting intracellular vesicles, to form large endocytic vesicles or *endosomes*. The contents of endosomes can be delivered to various intracellular compartments.[2] Ligands are commonly delivered to lysosomes for storage or degradation. Endocytic receptors can either be recycled to the plasma membrane, delivered to lysosomes in conjunction with their ligands, or transported to other intracellular organelles (i.e., trans-Golgi reticulum). Newly synthesized proteins (i.e., endosomal proteases) can also be transported to endosomes from the Golgi apparatus.[3] The incorporation of these proteins within endosomes may constitute a step in the maturation of some endocytic vesicles into lysosomes. Both the delivery and selective retrieval of proteins from endosomes appears to be mediated by vesicle fusion steps.[4] The regulation of these vesicle fusion steps may determine the fate of the contents of endosomes.

Vesicle Fusion in Cell-Free Assay

Although the occurrence of vesicle fusion has been widely reported in studies with intact cells, the mechanism(s) of the steps that lead to vesicle fusion remains poorly understood. The intracellular environment in which vesicles fuse is inaccessible to experimental manipulation in intact cells. Broken cell preparations and semiintact cells have proved useful in reconstitution studies of vesicle-mediated transport and vesicle fusion.[5,6] These preparations have made possible the manipulation of the cytoplasm without disrupting intracellular organelles.

[1] J. L. Goldstein, G. W. Anderson, and M. S. Brown, *Nature (London)* **279**, 679 (1979).
[2] T. Wileman, C. Harding, and P. Stahl, *Biochem. J.* **232**, 1 (1985).
[3] S. Kornfeld and I. Mellman, *Annu. Rev. Cell Biol.* **5**, 483 (1989).
[4] J. Gruenberg and K. E. Howell, *Annu. Rev. Cell Biol.* **5**, 453 (1989).
[5] W. E. Balch and J. E. Rothman, *Arch. Biochem. Biophys.* **240**, 413 (1985).
[6] C. J. M. Beckers, D. S. Keller, and W. E. Balch, *Cell (Cambridge, Mass.)* **50**, 405 (1987).

Vesicle fusion can be determined by monitoring the degree of mixing of either membrane or intravesicular contents. For example, the mixing of an enzyme in one compartment with a substrate in another compartment can be monitored by the enzyme-mediated conversion of the substrate to a product that can be assayed. The change of some measurable physical property of a marker when it comes into close proximity with another marker (i.e., resonance energy transfer of two fluorescent molecules) can also be used to assess vesicle fusion. Alternatively, if two markers with an affinity for each other are present in different vesicles, the formation of a measurable complex (i.e., precipitable immune complex) within the lumen of fused vesicles can serve as a measure of vesicle fusion.

The methodology designed in our laboratory has been used to study vesicle fusion during early stages of endocytosis in cell-free systems. By using broken-cell preparations, the reconstitution of (1) plasma membrane-derived vesicle fusion to endosomes, (2) endosome–endosome fusion, and (3) endocytic vesicle fusion to a protease-containing compartment has been characterized. For these purposes, ligands that are recognized by endocytic receptors (i.e., the macrophage mannose and Fc receptors) serve as biochemical markers for vesicle fusion. These ligands either have a high affinity for each other, or serve as substrates for proteolytic enzymes present in endocytic vesicles. Thus, vesicle fusion can be assessed by the formation of a measurable complex when these ligands are present in the same compartment or, alternatively, by the appearance of proteolytic products, as ligands present in endocytic vesicles fuse with vesicles containing proteases. Because these ligands have a high affinity for each other or are substrates for intrinsic enzymes, the volume of the endocytic compartment does not substantially affect their interaction when they are present in the same compartment.

Vesicle fusion can also be determined by using morphological criteria. For example, fusion of two sets of endosomes containing different sizes of gold particles results in the colocalization of these gold particles in the same compartment. Fusion can thus be monitored by electron microscopic analysis.

Basic Properties and Preparation of Fusion Probes

The ligands used to study vesicle fusion events have an affinity for the macrophage cell surface mannose and Fc receptors.[7,8] The mannose receptor recognizes high-mannose glycoproteins, whereas the Fc receptor has a

---

[7] R. Diaz, L. Mayorga, and P. Stahl, *J. Biol. Chem.* **263**, 6093 (1988).
[8] L. Mayorga, R. Diaz, and P. Stahl, *J. Biol. Chem.* **263**, 17213 (1988).

high affinity for the Fc domains of aggregated IgG. The four ligands used are (1) mannosylated monoclonal anti-dinitrophenyl (DNP) IgG (anti-DNP Man-IgG), (2) monoclonal anti-DNP IgG aggregated with rabbit anti-mouse IgG (aggregated anti-DNP IgG), (3) dinitrophenol-derivatized rat preputial β-glucuronidase (DNP–β-glucuronidase), and (4) monoclonal anti-DNP IgG aggregated with dinitrophenol-derivatized radiolabeled bovine serum albumin (IgG–[^{125}I]DNP–BSA). The antibody used to prepare the first two ligands has an affinity for DNP-derivatized proteins and avidly binds DNP–β-glucuronidase. Thus vesicle fusion can be monitored by measuring the degree of mixing of antibody present in one set of vesicles with the derivatized enzyme present in a complementary population of vesicles. The amount of immune complex formed by antibody and DNP–β-glucuronidase can be measured by immunoprecipitating the complex and assaying the pellet for β-glucuronidase activity. The sensitivity of this assay is high, because it can detect small quantities of enzyme.

Anti-DNP Man-IgG is prepared by derivatizing anti-DNP monoclonal antibody (IgG$_1$) with methylate-activated cyanomethyl-1-thioglycoside-D-mannopyranoside to yield a mannosylated conjugate, as described by Diaz *et al.*[7] The conjugation reaction is performed in the presence of a low-affinity hapten, 7-nitrobenzene-2-oxa-1,3-diazole-ε-aminocaproic acid, to protect the antibody-binding site from being modified. This hapten can be removed by dialysis, and free antibody can be recovered for these studies. The derivatized antibody binds mannose receptors with high affinity and is internalized by receptor-mediated endocytosis. DNP–β-glucuronidase is prepared by incubating dinitrophenyl fluoride with purified rat preputial β-glucuronidase.[7] Optimum derivatization is obtained under conditions in which the inactivation of the enzyme by the coupling procedure is minimized, and the immunoprecipitablity of the conjugated form of the enzyme by the anti-DNP IgG is maximized. Six dinitrophenol molecules can be conjugated to the protein without substantial loss of enzymatic activity. Anti-DNP Man-IgG and DNP–β-glucuronidase form an immune complex when mixed together. Although macrophages contain endogenous β-glucuronidase activity, the macrophage-derived enzyme does not interfere with the formation of immune complex, because the antibody recognizes only DNP-derivatized molecules. Because the formation of the immune complex is relatively pH independent, these probes are likely to form a complex in vesicles that maintain an acidic lumen (e.g., endosomes). Both probes are also fairly resistant to proteolytic degradation, which makes them suitable candidates for transporting studies along the endocytic pathway. The presence of hydrolytic enzymes in different compartments of this pathway should not substantially interfere with the activity of either probe.

Aggregated anti-DNP IgG binds to Fc receptors and has properties similar to those of the mannosylated form of the antibody.[9] Unlike the mannose receptor, binding to the Fc receptor is not affected by sucrose, a component present in the preparation of broken cell homogenates. Aggregated anti-DNP IgG bound to the cell surface Fc receptor is used as a marker for plasma membrane-derived vesicles, because this ligand remains bound to cell surface receptors during the steps of vesicle preparation. Aggregated anti-DNP IgG associated with the cell surface is commonly prepared by adding monoclonal anti-DNP IgG and rabbit anti-mouse IgG (4:1 molar ratio) to cells at 4°. Only the immune aggregate that forms during this incubation has an affinity for the Fc receptor, and nonaggregated antibody can be removed by resuspending cells in ligand-free medium.

Intravesicular proteolysis of IgG–[^{125}I]DNP–BSA is another enzymatic reaction used for measuring vesicle fusion. As endocytic vesicles containing IgG–[^{125}I]DNP–BSA fuse with protease-containing compartments, the degradation of ligand provides a measure of vesicle fusion. IgG–[^{125}I]DNP–BSA is prepared by derivatizing BSA with dinitrophenol fluoride and mixing the product with anti-DNP IgG [5:1.25 (w/w) ratio].[7] Prior to immune complex formation, DNP–BSA is labeled with ^{125}I by using standard radiolabeling techniques.[10] Mixing is carried out in the presence of cells at 4°. As immune complexes form, they bind to cell surface Fc receptors. Unbound ligand can be removed by resuspending cells in ligand-free buffer. Mayorga et al.[11] have shown that DNP–BSA, as opposed to the other ligands, is readily degraded by endosomal proteases. The products of degradation can be separated from intact protein by trichloroacetic acid precipitation of undegraded protein. The appearance of acid-soluble radioactivity can serve as a measure of fusion of endocytic vesicles containing IgG–DNP–BSA to a protease-containing compartment.

Vesicle and Cytosol Preparation

In principle, any cell type that expresses the mannose receptor and/or the Fc receptor can be used in studies of endocytic vesicle fusion. Cells derived from the reticuloendothelial system, including Kupffer's cells, bone marrow, and alveolar macrophages, are good candidates because they express both receptors. The cells chosen for these studies are the murine

[9] I. Mellman and H. Plutner, J. Cell Biol. 98, 1170 (1984).
[10] P. D. Stahl, P. H. Schlesinger, E. Sigardson, J. S. Rodman, and Y. C. Lee, Cell (Cambridge, Mass.) 19, 207 (1980).
[11] L. S. Mayorga, R. Diaz, and P. D. Stahl, J. Biol. Chem. 264, 5392 (1989).

macrophage-like cell line J774-E clone characterized by Diment *et al.*[12] This line is grown as a monolayer in minimum essential medium containing Earle's salts supplemented with 10% (v/v) fetal calf serum.

Uptake of Ligands

Cells are washed and resuspended in uptake medium (2×10^7 cells/ml) containing Hanks' balanced salt solution buffered with 10 mM 4-(2-hydroxyethyl)-1-piperazineethanesulfonic acid (HEPES) and 10 mM 2-[2-hydroxy-1,1-bis(hydroxymethyl)ethyl]aminoethanesulfonic acid (TES) and supplemented with 10 mg/ml BSA. Ligands in uptake medium are added to the cells for 5 min at 37° at concentrations that saturate cell surface receptors [Man-IgG (10 μg/ml) and DNP–β-glucuronidase (20 μg/ml)] to allow internalization into endosomes. Aggregated anti-DNP–BSA (5 μg/ml) is incubated for 60–90 min at 4° to label plasma membrane-derived vesicles. IgG–[^{125}I]DNP–BSA (5 μg/ml) is internalized for 2 min at 37° to label earlier endosomes. The shorter internalization time prevents this ligand from entering an endocytic compartment with proteolytic activity. After the internalization step, cells are kept at 4° in all subsequent steps. Uptake and binding steps are followed by at least two washes in uptake buffer. The cells are next washed with phosphate-buffered saline (PBS) supplemented with 5 mM ethylenediaminetetraacetic acid (EDTA). EDTA is included to remove ligands that are associated with the cell surface mannose receptor. Binding to this receptor is Ca^{2+} dependent. Cells are then washed in homogenization buffer [250 mM sucrose, 0.5 mM ethyleneglycol-bis(β-aminoethyl ether)-N,N,N',N'-tetraacetic acid (EGTA), 20 mM HEPES–KOH (pH 7)] prior to the homogenization step.

Homogenization of Cells

After cell surface binding or uptake of fusion probes into endosomes, cells are homogenized using conditions that result in the maximal degree of cell disruption, with minimal release of intravesicular components caused by organelle damage. Macrophages are quite resistant to disruption by standard methods of homogenization (e.g., glass–glass homogenizer). For these studies, a homogenization method has been developed that consists of passing the cells suspended in homogenization buffer (5×10^7 cells/ml) through two 27-gauge needles, connected in series with plastic tubing.[13] At either end, there is a syringe that serves as a reservoir for the cell suspension. Cells are transferred from one syringe to the other as forcibly as

[12] S. Diment, M. S. Leech, and P. D. Stahl, *J. Leukocyte Biol.* **42**, 485 (1987).
[13] L. S. Mayorga, R. Diaz, and P. D. Stahl, *Methods Cell Biol.* **23**, 179 (1989).

possible at 4°. The required number of passes is determined by the percentage of cells that remain intact (i.e., exclude trypan blue), and depends on the ability of the operator to apply the appropriate pressure. Under these conditions, when less than 10% of the cells remains intact, that is, exclude trypan blue, less than 30% of the internalized ligands is released into the extravesicular compartment. The homogenate is centrifuged at low speed (800 g, 5 min) to eliminate unbroken cells and nuclei. The postnuclear supernatant can be used immediately for fusion studies, or further purified to obtain an enriched endosomal population. Alternatively, the vesicles may be frozen rapidly in liquid nitrogen and stored at −80°. When ready for use, the vesicles are thawed rapidly at 37°.

Enrichment for Endocytic Vesicles

Vesicle populations can be separated using different fractionation techniques. The fate of endocytic vesicles can be followed easily by using a radiolabeled derivative of the ligand that is internalized into endosomes. Unfortunately, pure endosomal populations that remain fusion competent are not obtained easily. A simple and rapid differential sedimentation step, however, can help remove large vesicles devoid of fusion markers (e.g., mitochondria) without significant sedimentation of endocytic vesicles.[14] This centrifugation step is performed at 37,000 g for 1 min at 4°. A fusion-competent endocytic vesicle fraction can be obtained by a second centrifugation step at 50,000 g for 5 min at 4°. This endosomal pellet can be resuspended in the appropriate medium and used for fusion assays. To remove large nonendocytic vesicles from plasma membrane-derived vesicle preparations, the first centrifugation step is at a much lower speed (15,000 g, 1 min at 4°). Higher speeds would remove a significant fraction of vesicles with cell surface fusion markers within them, suggesting that a portion of these vesicles are larger in diameter than early endocytic vesicles.

Other fractionation techniques can be used with various degrees of success. For example, Percoll density gradients can separate vesicles according to density.[15] These self-forming gradients are useful in separating plasma membrane-derived vesicles from denser organelles (starting Percoll density, 1.04 g/ml in 250 mM sucrose), or lysosomes from lower density vesicles (starting Percoll density, 1.05 g/ml in 250 mM sucrose). A disadvantage of this technique is that Percoll particles must be removed prior to performing fusion assays. This is done by centrifugation of fractions from Percoll gradients at high speed (100,000 g, 2 hr at 4°). The vesicles sedi-

[14] R. Diaz, L. S. Mayorga, L. E. Mayorga, and P. D. Stahl, *J. Biol. Chem.* **264**, 13171 (1989).
[15] T. Wileman, R. L. Boshans, P. Schlesinger, and P. D. Stahl, *Biochem. J.* **220**, 665 (1984).

ment over a layer of compact Percoll. Although they are competent for fusion studies, these vesicles show a much lower efficiency of fusion. In general, any manipulation that alters the integrity of endocytic vesicles tends to decrease their fusion activity.

Preparation of Cytosol

Cytosol from J774-E clone macrophages is obtained by resuspending the cells in 10 vol of homogenization buffer and disrupting the cells by nitrogen cavitation at 350 psi for 10 min, or by passing them through 27-gauge needles (see above). A postnuclear supernatant is prepared as described above. Membranes are then removed by centrifugation at 35,000 g for 15 min followed by two 300,000 g spins for 60 min, all at 4°. The protein content of these preparations ranges from 3 to 8 mg protein per milliliter. Supernatants can either be immediately used, or frozen in liquid nitrogen and stored at −80°. Cytosol fractions are gel filtered through 1-ml Sephadex G-25 spin columns prior to use in fusion assays. This step removes low molecular weight components (i.e., ATP or calcium) without altering the protein content of cytosol. Cytosol prepared from other sources (e.g., rabbit alveolar macrophages or L-929 mouse fibroblasts) also supports fusion.

Biochemical Fusion Assays

Because these fusion assays rely on the mixing of endocytic markers, they detect only the final step in what may be a complex number of biochemical events. Other events, such as vesicle recognition, vesicle aggregation, and vesicle membrane realignment, may precede vesicle fusion. The fusion assays do not measure those steps directly. However, if any of the events that precede fusion is altered, the fusion reaction may be affected. The morphological approach (see Morphological Characterization of Vesicle Fusion, below) makes it possible to distinguish at least one additional step (i.e., endosome aggregation) that appears to precede fusion. Finally, crude preparations cannot overrule other vesicle fusion events, because the fusion probes report only on fusion events between vesicles containing endocytic markers. Vesicles devoid of marker may also fuse to endocytic vesicles. These vesicles may or may not have even emerged from endocytosis.

In vitro reconstitution of endocytic vesicle fusion requires the incubation of endosomes under conditions that support fusion. These conditions include the presence of cytosol, ATP, KCl, and incubation at 37°. Both postnuclear supernatants and vesicle fractions containing endosomes supplemented with cytosol are used. There are several advantages in using vesicle fractions. First, the concentration of vesicles can be increased, relative to the concentration of vesicles present in postnuclear supernatants. The efficiency of vesicle fusion improves with increasing vesicle concentration. Second, cytosol concentration can be varied easily without altering vesicle concentration. Third, postnuclear supernatants contain a substantial amount of fusion marker released from endocytic vesicles during the homogenization step. This fraction of the ligand should not be allowed to contribute to the measurement of fusion activity. By obtaining a vesicle fraction from postnuclear supernatants, the extravesicular fusion marker is removed. Finally, vesicle and cytosol fractions can be manipulated separately (e.g., incubation with trypsin or other inhibitors) prior to mixing for the fusion reaction.

Endosome–Endosome Fusion

In the standard assay for reconstitution of endosome–endosome fusion, vesicles containing anti-DNP Man-IgG are mixed with equal amounts of vesicles containing DNP–β-glucuronidase to a final volume of 10 to 20 μl. The vesicles are incubated at 37° for 30 min in fusion buffer [homogenization buffer supplemented with 1.5 mM MgCl$_2$, 50 mM KCl, and 1 mM dithiothreitol (DTT)] containing DNP–BSA (50 μg/ml). Because fusion markers may be released from endocytic vesicles during the incubation, DNP–BSA is added as a scavenger to block the binding of anti-DNP Man-IgG to DNP–β-glucuronidase in the extravesicular compartment. The medium also contains an ATP-regenerating system consisting of 1 mM ATP, 8 mM creatine phosphate, and 31 units/ml creatine phosphokinase. The regenerating system maintains a constant level of ATP during the incubation. When vesicle fractions are used in the fusion assay, cytosol is included at saturating concentrations required for fusion activity (1.5–2 mg protein/ml). Fusion reactions can be stopped by cooling at 4°. To precipitate the fusion-dependent immune complex formed, vesicles are solubilized by addition of 150 μl of solubilization buffer [1% (w/w) Triton X-100, 0.2% (v/v) methylbenzethonium chloride, 1 mM EDTA, 0.1% (w/v) BSA, 150 mM NaCl, 10 mM Tris-HCl, pH 7.4] containing 50 μg of DNP–BSA per milliliter and 2 μl of *Staphylococcus* A protein [10% (v/v) suspension], coated with rabbit anti-mouse IgG. *Staphylococcus* A protein (IgGsorb; The Enzyme Center, Inc.) is coated by incubation of 2 μl of a 10% (v/v) suspension of bacteria with 1 μl of rabbit anti-mouse IgG (rabbit IgG fraction, 4 mg/ml; Organon Teknika Corporation) for 30 min at room temperature, followed by three washes with solubilization buffer. The samples are incubated at 4° for 30 min, diluted with 1 ml of solubilization buffer, and pelleted at 1500 g for 5 min. The *Staphylococcus* A protein-bound immunoprecipitates are then washed twice with 1 ml of solubilization buffer. To quantify the immune complex formed, pellets are resuspended in 100 μl of solubilization buffer and an equal volume of β-glucuronidase substrate (4 mM 4-methylumbelliferyl-β-D-glucuronide in 0.1 M acetate buffer, pH 4.5) is added. Samples are then incubated at 37° for 1 to 2 hr and the reaction is stopped with 1 ml of glycine buffer (133 mM glycine, 67 mM NaCl, and 83 mM Na$_2$CO$_3$, adjusted to pH 9.6 with NaOH). The fluorescence of umbelliferone is measured in a spectrofluorometer at 366-nm exitation, 450-nm emission, or in a fluorometer with appropriate filters.

When the reaction is carried out under incomplete fusogenic conditions, fusion should not be observed. Little fusion is observed when certain required components are removed from the assay (e.g., cytosol or KCl). Inhibition of fusion is achieved by active depletion of a component from the fusion mixture. ATP, for instance, is depleted by addition of a mixture of hexokinase (25 units/ml) and 5 mM mannose.

Several controls can be included in the assay. Reactions containing detergent, but lacking the scavenger DNP–BSA during the fusion reaction, measure the total amount of immune complex that can be formed. Frequently, fusion efficiency is expressed as a percentage of this amount, because it represents the activity that should result from the complete mixing of the compartments containing the probes. This amount may be overestimated because some of the ligands are not in sealed vesicles. This is significant when postnuclear supernatants are used, because ~ 30% of endocytic ligand is present in the extravesicular compartment due to endosome disruption during the homogenization step. The proportion of a given probe in closed vesicles can be estimated by measuring the proportion of immune complex that is obtained when vesicles containing the ligand are mixed with a solution of the complementary probe in the presence and absence of detergent.

The rate of fusion increases with the total amount of vesicles present in the system. However, at high vesicle concentrations, some components of the reaction (e.g., ATP) can be depleted. For postnuclear supernatants, good fusion is obtained by mixing 30 to 50 μg of total proteins from each preparation to a final volume of 20 μl. For endosomal fractions, 10 to 15 μg of vesicle protein from each preparation is required for optimum fusion. Under these conditions, the efficiency of fusion ranges between 30 and 40%. It is possible that a fraction of the ligand is present in fusion-incompetent vesicles. These vesicles may have lost their ability to fuse within intact cells. Alternatively, endosomes may become fusion incompetent during vesicle preparation or *in vitro* incubation.

This assay permits the study of endosome fusion requirements.[7] Fusion is time dependent, reaching a plateau after a 30-min incubation at 37°. Fusion is not observed at temperatures below 18° and reaches a maximal rate at 37°. By altering the conditions, it is possible to determine the energy, ionic, and cytosol requirements for optimal fusion (Table I). The requirement for cytosol is saturable at about 1.5 mg/ml of J774 cytosol. Furthermore, the assay provides information about specific cytosol and membrane-associated protein requirements.

Only early endosomes appear to be fusion competent. As fusion probes

TABLE I

EFFECT OF IONS, IONOPHORES, NUCLEOTIDES, AND ATPASE INHIBITORS ON FUSION REACTION[a]

Experimental condition[b]	Relative fusion[c]	Experimental condition	Relative fusion
Nucleotide		Ions	
Regenerating system[d]	1.00	KCl (50 mM)	1.00
Depleting system[e]	0.00	NaCl (50 mM)	0.98
ADP (1 mM)	0.00	Potassium gluconate (50 mM)	0.97
ATP (1 mM)	0.62	Sucrose	0.00
GTP (1 mM)	0.34		
PNP–AMP[f] (1 mM)	0.00		
ATPase inhibitors		Ionophores and amines	
Sodium vanadate (1 mM)	0.95	NH_4Cl (10 mM)	1.10
Sodium azide (1 mM)	0.93	Chloroquine (1 mM)	0.92
Oligomycin (10 μg/ml)	0.89	Nigericin (10 μM)	0.88
N-Ethylmaleimide[g] (1 mM)	0.05	CCCP[i] (10 μM)	0.84
DCCD[g,h] (100 μM)	0.93	Valinomycin (10 μM)	0.83

[a] Values presented are averages for a minimum of three experiments.

[b] Fusion reactions were performed in the presence of complete fusion conditions for 30 min at 37° except when one of the components is omitted or substituted for another component. When the effects of drugs were tested, they were included during the incubation at the indicated concentration. An ATP-regenerating system was always present, except when the effect of nucleotides was tested. To assess the ionic requirement of fusion, other salts or sucrose were substituted for KCl in the fusion buffer.

[c] Fusion activity for each condition in a given experiment was compared to the fusion observed in the presence of a standard fusion buffer and an ATP-regenerating system, which was assigned a value of 1.

[d] ATP (1 mM), 8 mM creatine phosphate, and 31 units/ml creatine phosphokinase.

[e] Hexokinase (25 units/ml) and 5 mM mannose.

[f] PNP–AMP, Adenylimidodiphosphate.

[g] Treatments with N-ethylmaleimide and DCCD were carried out at 4° for 30 min prior to the fusion reaction. These treatments were followed by either the addition of 2 mM DTT to quench unreacted NEM or 2 mM glycine to quench unreacted DCCD.

[h] DCCD, Dicyclohexylcarbodiimide.

[i] CCCP, Carbonyl cyanide m-chlorophenylhydrazone.

are internalized for longer periods of time, probes are enriched in late endocytic compartments. These compartments are not able to fuse to each other, or to early endosomes *in vitro,* even though they are likely to participate in vesicle fusion events in intact cells. Either the assay conditions may not be appropriate for fusion of late endosomes, or the fusion properties of late endosomes are lost in the preparations. Mullock *et al.*[16] have shown that late endosomes, isolated from liver, associate with lysosomes in their preparations. The experimental conditions that favor late endosome–lysosome association are similar to these conditions. It is possible that the choice of tissue and subtle differences in methodology result in changes of vesicle properties that, in turn, can affect the results of the fusion assay.

Plasma Membrane-Derived Vesicle–Endosome Fusion

The fusion of plasma membrane-derived vesicles to endosomes is carried out by using the same protocol described for endosome–endosome fusion, except that vesicles are obtained from cells that have aggregated anti-DNP IgG bound to their surfaces.[8] The fusion-dependent formation of aggregated anti-DNP IgG and DNP–β-glucuronidase is measured. The efficiency of fusion of plasma membrane-derived vesicles with endosomes appears to be lower than that of fusion among endosomes. Several explanations can account for this observation. First, postnuclear supernatants contain a larger fraction of the aggregated anti-DNP IgG in the extravesicular compartment. Second, morphological studies have shown that a large fraction of ligand is trapped within vesicles that are surrounded by larger vesicles. Therefore some plasma membrane-derived vesicles may not be accessible to fuse with other vesicles. Finally, it is possible that plasma membrane-derived vesicles are less fusion competent than endosomes. The lower efficiency of fusion, nevertheless, can be overcome by adding a larger fraction of plasma membrane-derived vesicles to the assay.

This assay measures the fusion of ligands in vesicles derived from the cell surface to an intracellular compartment, and has been found to have the same requirements observed for endosome–endosome fusion. Prior to homogenization, the aggregated anti-DNP IgG is localized in coated pit regions of the cell surface. On homogenization, these pits close to form coated vesicles that contain the fusion markers. This suggests that both coated vesicles and pits, probably after uncoating, are capable of fusing with endosomes.

[16] B. M. Mullock, W. J. Branch, M. van Schaik, L. K. Gilbert, and J. P. Luzio, *J. Cell Biol.* **108,** 2093 (1989).

Proteolysis

The proteolysis assay measures the transport of an endocytic marker (e.g., IgG–[^{125}I]DNP–BSA) to a protease-containing compartment. As ligand enters a protease-containing compartment the rate of hydrolysis increases, which in turn constitutes a measure of fusion.[11] Cells are allowed to internalize surface-bound immune complexes. Vesicle fractions are obtained, incubated under fusion conditions, and dispensed in 5- to 10-μl aliquots, each containing approximately 20,000 counts per min (cpm) of radioactivity. To measure the time course of fusion, aliquots are incubated for different periods of time at 37°, followed by dilution with 200 μl of hydrolysis buffer [250 mM sucrose, 40 mM acetate buffer (pH 4.5), 1 mg BSA per milliliter], prewarmed to 37°. Dilution is used to prevent further fusion during this step. To measure the time course of degradation the samples are incubated again at 37° for different periods of time, and the reaction is stopped with trichloroacetic acid (TCA) [10% (w/v) final concentration]. The samples are cooled on ice for 10 min and centrifuged at 12,000 g for 15 min at 4°. Radioactivity is measured in the pellet and in the supernatant. The time-dependent appearance of TCA-soluble products can then be plotted for every incubation time under fusion conditions. This value is usually expressed as a percentage of the total radioactivity of the sample.

The percentage of ligand degraded after incubation under fusion conditions increases two to three times over that observed after incubation at acidic pH alone (i.e., without fusion). An increase in ligand degradation is observed for vesicles generated at all times of internalization, suggesting that fusion with intracellular protease-containing compartments could be reconstituted, even with ligand-containing vesicles derived from the plasma membrane. Homogenates from cells that have internalized IgG–[^{125}I]DNP–BSA for 2 min or less are commonly used. The short time of internalization impedes the access of ligand to proteases in intact cells and lowers the initial rate of degradation observed in the assay. Several factors influence the sensitivity of the assay. Low background activity increases the sensitivity. Additionally, as the initial radioactivity present in each sample increases, so does the sensitivity.

An increase in ligand degradation can be interpreted as a consequence of the fusion of endosomes with a protease-containing compartment. Alternate explanations can be proposed to explain the increase of ligand degradation on *in vitro* incubation. For example, the stability of the vesicles during the incubation may be preserved better under complete fusion conditions, thereby allowing more efficient intravesicular degradation. Alternatively, the protease(s) responsible for the observed degradation may be activated under fusion conditions (namely, the presence of ATP, cytosol, ions). Vesicle dilution would not be expected to affect the increase in degradation, if vesicle interaction were not necessary for the onset of degradation. Endosome–endosome fusion shows a significant dependence on vesicle dilution. Fusion-dependent degradation in this assay is not dilution dependent, which suggests that the interaction of different vesicles with each other is required for this process.

Morphological Characterization of Vesicle Fusion

To visualize endocytic vesicles that are fusion competent in a cell-free system, different populations of endosomes can be marked with colloidal gold particles of different sizes.[14] Fusion is assessed by colocalization of two or more different markers in the same vesicle, following the incubation of endosomes under fusion conditions.

Ligands

Colloidal gold of 5-, 10-, and 20-nm diameter is used. Mannose-derivatized BSA (Man-BSA), a neoglycoprotein with high affinity for the mannose receptor,[10] is used to coat colloidal gold (6–15 μg/ml of colloid) according to the technique described by Roth.[17] Alternatively, the colloidal gold can be coated with aggregated anti-DNP IgG to make it a ligand for the macrophage Fc receptor. This ligand is used in morphological studies with plasma membrane-derived vesicles.

Vesicle Preparation and Fusion Reaction

The internalization of gold particles coated with Man-BSA is carried out under the same conditions described for soluble ligands, except that, to improve the efficiency of internalization, the particles are first bound to cell surface receptors by incubation with J774-E clone macrophages for 1 hr at 4°. Each probe (i.e., different sizes of gold) is bound to a separate population of macrophages, and the ligand is internalized by warming the cells to 37° for 5 min. Postnuclear supernatants from each population are prepared and fusion reactions are carried out by following the protocols described above.

Preparation of Samples for Electron Microscopy

After the fusion reaction, vesicles are fixed in suspension with 2% (v/v) glutaraldehyde prepared in homogenization buffer. After fixation, vesicles

---

[17] J. Roth, *in* "Techniques in Immunocytochemistry" (G. R. Bullock and P. Petrusz, eds.), Vol. 2, p. 207. Academic Press, London, 1983.

are pelleted by centrifugation (15,000 g, 15 min), washed in 0.1 M sodium cacodylate, pH 7.2, and postfixed in 1% (v/v) OsO_4 in cacodylate buffer for 45 min at room temperature. Vesicles are then rinsed, dehydrated, and embedded in plastic according to standard techniques.[18] Thin sections are cut and analyzed by transmission electron microscopy.

Glutaraldehyde is added to the sample at 37° immediately following the fusion reaction. Subsequently, the samples are incubated at room temperature for 5 min and are then transferred to ice for an additional 15 min. Under these conditions the presence of endosomal aggregates is observed. After the vesicles are collected by centrifugation, these aggregates are preferentially found at the bottom of the pellet. To have a full view of the sample the pellet is cut transversally, prior to embedding in plastic. When quantitation is required, the whole transversal section is screened, and fusion or aggregation is expressed as number of events per square micrometer.

Morphological Analysis of Endosome–Endosome Fusion

When postnuclear supernatants containing endosomes labeled with either 5-, 10-, or 20-nm diameter colloidal gold particles are mixed under fusion conditions, clustering of vesicles is commonly observed. These clusters contain between 5 and 20 vesicles. The majority of these vesicles are loaded with gold particles; however, some empty vesicles are also present. The latter could be endosomes that form in the absence of endocytic markers or, alternatively, Golgi-derived vesicles destined to fuse with endosomes. Aggregates are often composed of vesicles containing mixtures of different sizes of gold, indicating that some endosomes have already fused (Fig. 1a). These aggregates are not observed in the absence of ATP or cytosol, whereas the presence of KCl, a requirement for fusion, does not seem to be strictly necessary for aggregation. Endosomal aggregation may, therefore, constitute an intermediate step in the fusion process, which can be assessed by this morphological approach.

Prior to incubation under fusion conditions, colloidal gold is found in a heterogeneous population of structures, consisting mainly of small-diameter tubules and vesicles (100 to 200 nm). Incubation at 37° results in the appearance of large vesicles (600 to 1000 nm). These large endocytic vesicles are likely to be the products of multiple fusion events. In addition, the colocalization of three different sizes of colloidal gold particles in endosomes provides further evidence that endosomes fuse multiple times (Fig. 1b).

[18] M. A. Hayat, "Basic Techniques for Transmission Electron Microscopy." Academic Press, Orlando, FL, 1986.

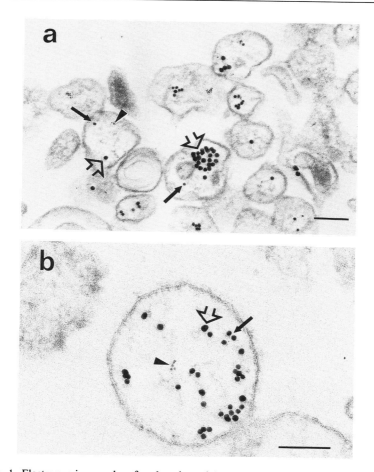

Fig. 1. Electron micrographs of endocytic vesicle aggregation and fusion. Endosomes loaded with Man–BSA-coated colloidal gold particles of different sizes (▶, 5 nm; ➡, 10 nm; ⇨, 20 nm) are incubated under fusion conditions for 30 min at 37°. The preparations are fixed in suspension and analyzed by transmission electron microscopy. (a) *In vitro* clustering of endosomes loaded with colloidal gold particles. Colocalization of different sized gold within the same vesicles indicates endosome fusion. (b) Large endosomal vesicle containing 5-, 10-, and 20-nm colloidal gold particles. Bars: 100 nm.

Requirements for Endosome Fusion

The assays described above demonstrate similar requirements for vesicle fusion, suggesting that fusion between vesicles in the early steps of the endocytic pathway share comparable mechanisms. Fusion is ATP, KCl, time, and temperature dependent. Fusion requires both cytosolic and membrane-associated factors. *N*-Ethylmaleimide (NEM) treatment of cytosol and vesicle fractions inhibits fusion.[19] A NEM-sensitive factor, active in vesicle-mediated transport in other intracellular pathways, appears to be the target of this specific inhibition. Furthermore, GTP-binding proteins may play an important role in the regulation of vesicle fusion, because GTPγS has an inhibitory effect at high cytosol concentrations, whereas it activates fusion at low cytosol concentrations.[20,21] Understanding the specific role of the described requirements may be amenable to further study, using these fusion assays.

[19] R. Diaz, L. S. Mayorga, P. J. Weidman, J. E. Rothman, and P. D. Stahl, *Nature (London)* **339,** 398 (1989).
[20] L. S. Mayorga, R. Diaz, and P. D. Stahl, *Science* **244,** 1475 (1989).
[21] L. S. Mayorga, R. Diaz, M. I. Colombo, and P. D. Stahl, *Cell Regul.* **1,** 113 (1989).

[17] Preparation of Semiintact Cells for Study of Vesicular Trafficking *in Vitro*

By S. PIND, H. DAVIDSON, R. SCHWANINGER, C. J. M. BECKERS, H. PLUTNER, S. L. SCHMID, and W. E. BALCH

Introduction

In eukaryotic cells the transport of proteins and lipids between the subcellular compartments generally involves the successive budding and fusion of carrier vesicles.[1] Considerable understanding of both the exocytic and endocytic pathways has been obtained through a variety of morphological and biochemical approaches in intact cells. However, in order to identify and characterize the essential components driving vesicular transport, it has been necessary to develop cell-free systems that efficiently reconstitute events observed *in vivo*. In several instances significant success has been achieved by the use of cell homogenates or partially purified membrane fractions.[2,3] Unfortunately, such techniques are not always applicable, especially when one or more of the compartments under study is sensitive to homogenization. In such cases a more selective approach is required. This has led to the development of several methods of cell perforation that result in the removal of discrete fragments of the plasma

[1] G. Palade, *Science* **189,** 347 (1975).
[2] J. E. Rothman and L. Orci, *FASEB J.* **4,** 1460 (1990).
[3] J. Gruenberg and K. E. Howell, *Annu. Rev. Cell Biol.* **5,** 453 (1989).

membrane to yield cells that have lost their soluble content, but which retain the majority of their subcellular organelles in an intact form within the cytoskeletal matrix.[4-6]

This chapter focuses on techniques used in our laboratory to reconstitute endoplasmic reticulum (ER)-to-Golgi transport *in vitro*, using adherent tissue culture cells that have been rendered semiintact by gently scraping them from the dish after hypotonic swelling. This technique is discussed in the context of other protocols that allow the investigator to perforate both adherent and nonadherent cells.

Principle of Endoplasmic Reticulum-to-Golgi Transport Assay

The ER-to-Golgi transport assay is based on the observation that individual N-linked oligosaccharide-processing enzymes reside in distinct subcellular compartments. Thus delivery of glycoproteins to the cis Golgi can be detected by the action of α-mannosidase I, a resident protein of this compartment, which converts high mannose (Man_9) asparagine-linked oligosaccharides to the Man_5 form.[7] *In vivo* this event may be masked by the subsequent conversion of the sugar chains to more complex forms resulting from the action of other mannosidase and glycosyltransferase activities located in more distal compartments. However, in semiintact cells it is possible to prevent these additional reactions from occurring simply by omitting uridine diphosphate N-acetylglucosamine (UDPGlc-NAc) from the assay. This sugar nucleotide is the substrate of transferase I (TrI), the glycosyltransferase that catalyzes the addition of GlcNAc to the Man_5 oligosaccharide core, and which is essential for the formation of "complex" oligosaccharides containing galactose and sialic acid.[7] In practice, the need to eliminate UDPGlcNAc from the assay is accomplished by using a Chinese hamster ovary (CHO) cell mutant, clone 15B, deficient in TrI activity.[8] This cell line also allows us to measure ER-to-cis Golgi transport directly *in vivo*.

Any protein that transverses the secretory pathway can be used to mark successive transport steps between exocytic compartments. However, to alleviate the need for an immunoprecipitation step, we use cells infected with vesicular stomatitis virus (VSV). In this case we can follow the movement of the VSV-G, protein, the surface glycoprotein of the mature virus that behaves in all respects like a normal plasma membrane protein

---

[4] C. J. M. Beckers, D. S. Keller, and W. E. Balch, *Cell (Cambridge, Mass.)* **50**, 523 (1987).
[5] K. Simons and H. Virta, *EMBO J.* **6**, 2241 (1987).
[6] R. Brands and C. A. Feltkamp, *Exp. Cell Res.* **176**, 309 (1988).
[7] R. Kornfeld and S. Kornfeld, *Annu. Rev. Biochem.* **54**, 631 (1985).
[8] C. Gottlieb, J. Baenziger, and S. Kornfeld, *J. Biol. Chem.* **250**, 3303 (1975).

in transit through the exocytic pathway. The distinct advantage is that 4 hr postinfection VSV-G is the predominant glycoprotein being synthesized and can be observed directly by sodium dodecylsulfate (SDS) gel electrophoresis after radiolabeling cells with [^{35}S]methionine. Moreover, the use of a temperature-sensitive mutant of VSV, strain tsO45, which is incompetent in export from the ER at the restrictive temperature (40°) but is synchronously exported on shift to the permissive temperature (32°),[9] allows us to radiolabel VSV-G protein prior to the preparation of semiintact cells, with confidence that subsequent manipulations *in vitro* will focus exclusively on transport of VSV-G between the ER and the Golgi compartments. Processing from the high mannose to the Man$_5$ form results in an increase electrophoretic mobility. To accentuate this difference, we use a postincubation processing reaction with endoglycosidase D (Endo D), an enzyme that specifically cleaves the chitobiose core of N-linked oliogsaccharides of the Man$_5$ type of certain glycoproteins such as VSV-G protein, giving a more distinctive shift in electrophoretic mobility.[10] Use of Endo D allows direct quantitative analysis of transport by densitometry without the need for carbohydrate-specific lectins, or high-performance liquid chromatography (HPLC) analysis of carbohydrate structure, which might otherwise be needed to quantitate the appearance of the Man$_5$ structure.

Experimental Procedures

Growth of Cells and Virus

The 15B clone of CHO cells (originally obtained from S. Kornfeld, Washington University, St. Louis, MO) is maintained in standard monolayer culture (10-cm dishes) in α-minimal essential medium (α-MEM; Earle's salts, with glutamine and nucleosides) supplemented with 8% (v/v) fetal bovine serum (FBS), 100 IU penicillin per milliliter, and 100 μg streptomycin per milliliter. The cells are routinely passaged such that on the morning of use they form a complete monolayer while still maintaining a well-spread morphology. For 15B cells this represents approximately 2000 cells/mm^2 (1.5×10^7 cells/dish). Cells that have become tightly packed generally give poor preparations of semiintact cells. A tsO45 VSV stock [originally obtained from N. Kong (Massachusetts Institute of Technology, Boston, MA)] is propagated in baby hamster kidney (BHK) cells by infecting cells at 0.1 plaque-forming unit (pfu) in Dulbecco's minimal essential medium (DMEM) (without fetal bovine serum) supplemented

[9] H. F. Lodish and N. Kong, *Virology* **125**, 335 (1983).
[10] T. Mizuochi, J. Amano, and A. Kobata, *J. Biochem. (Tokyo)* **95**, 1209 (1984).

with 10% (v/v) tryptose phosphate broth. After 36–48 h, the virus is harvested by pelleting the cells at 5000 g for 10 min in 250-ml conical tubes, the cell-free supernatants are combined, and aliquots (400 μl) are rapidly frozen by immersion in liquid N_2. Virus can be stored indefinitely at $-80°$. For use, each portion can be thawed and refrozen ($-80°$) at least twice without significant reduction in apparent infectivity.

Infecting 15B Cells with VSV tsO45

Cells grown in confluency as described above are prepared for virus infection by aspiration of the medium and washing of the monolayer with 5 ml phosphate-buffered saline (PBS).[11] Virus is thawed by incubation in a 30° water bath (higher temperatures will result in virus aggregation and inhibition of infectivity), and 100 μl of the stock (approximately 2×10^9 pfu/ml, to provide 10–20 pfu/cell) is mixed with 1 ml of α-MEM containing 25 mM N-2-hydroxyethylpiperazine-N'-2-ethanesulfonic acid (HEPES; pH 7.2) and 5 μg of actinomycin D (freshly added from a 1-mg/ml stock in ethanol). The mixture is transferred to a plate, which is rocked gently in all directions to ensure uniform distribution of virus. It is then placed on a rocking platform (Labquake shaker, Labindustries, Berkeley, CA) inside a 32° CO_2 incubator and gently rocked for 45 min (16 oscillations/min). After virus binding, 5 ml of a postinfection medium containing α-MEM supplemented with 8% FBS is added, and the incubation continued without rocking for an additional 4 hr at 32°. Generally, cells postinfected for 3.5 to 4.5 hr express high levels of VSV-G protein and actively transport protein between the ER and Golgi compartments in semiintact cells.

Labeling of Vesicular Stomatitis Virus tsO45-Infected 15B Cells

To follow the transport of tsO45 VSV-G protein between the ER and the Golgi compartment, infected cells are labeled with [^{35}S]methionine at the restrictive temperature (40°) prior to the preparation of semiintact cells. For this purpose, cells are transferred to a 40° water bath arranged such that the tissue culture dish rests on a level, perforated stainless steel platform situated just below the surface of the water. The depth of the water (a few millimeters above the platform) is just sufficient to immerse the base of the tissue culture dish without the dish floating in the water bath when the lid is removed. The medium is aspirated and the monolayer washed twice with 3-ml portions of prewarmed labeling medium [methionine-deficient labeling medium [Cat. No. 7270, Sigma, St. Louis, MO;

---

[11] R. Dulbecco and M. Vogt, *J. Exp. Med.* **99,** 167 (1954).

supplemented with leucine, lysine, and HEPES–KOH (pH 7.2)]. To deplete endogenous tRNA^{Met} pools the cells are then incubated for 5–10 min at 40° in 4 ml of labeling medium. Finally the medium is replaced with a further 1.5 ml of labeling medium containing 10 μl (100 μCi) of [^{35}S]methionine (Trans-label; ICN Biomedicals, Costa Mesa, CA), the plate rocked gently to ensure even coverage, and then incubated for 10 min at 40°. During the labeling period the plate is briefly rocked at 1- to 2-min intervals by lifting and lowering opposite sides while maintaining contact with the water bath. This ensures the uniform distribution of label and prevents the cells from drying out. It is important that handling of the plate is done in such a way as to ensure that the temperature of the labeling medium does not drop below 40° (as VSV-G protein will exit the ER). After 10 min of labeling, the medium is supplemented with 30 μl of 0.25 M methionine and a further 2 ml of prewarmed labeling medium and incubated for an additional 2 min at 40°. After this chase, to complete synthesis of radiolabeled VSV-G protein, the cells are ready to be perforated.

Preparation of Semiintact Cells

To prepare semiintact cells, the chase medium is rapidly removed by aspiration, and the dish immediately transferred to an ice–water bath arranged in a similar fashion to the 40° bath. Immediately on transfer to the ice–water bath, 3–4 ml of 50/90 H/KOAc [ice-cold 50 mM HEPES (adjusted to pH 7.2 with KOH) containing 90 mM potassium acetate (KOAc)] is added to the cells. Plates are washed a total of three times with 50/90 H/KOAc. After the final wash, the buffer is replaced with 5 ml of a fivefold dilution of the wash buffer [10/18 H/KOAc: 10 mM HEPES (pH 7.2)/18 mM KOAc] and incubated for 10 min on ice to swell the cells and render them susceptible to perforation. The 10/18 H/KOAc is then replaced with a further 3 ml of 50/90 H/KOAc and the cells immediately scraped from the plate with a rubber policeman (Macalester-Bicknell, New Haven, CT; Cat. No. 36300-0014), using smooth, firm strokes. The suspension is transferred by Pasteur pipette to a 15-ml polystyrene centrifuge tube and centrifuged at 800 g for 3 min (3°). The supernatant is aspirated and the cells resuspended in 3 ml of 50/90 H/KOAc, using a 1-ml Gilson Pipetman (Rainin Instrument Co., Emeryville, CA). After subsequent centrifugation (800 g, 3 min), the cells are resuspended in 4 vol of 50/90 H/KOAc (generally 200–250 μl buffer per 1.5 × 10^7 15B cells). This yields a final concentration of approximately 6 mg cell protein/ml of cell suspension. Resuspended cells are stored on ice until further use. Cells generally retain efficient transport activity after incubation on ice for at least 4 hr.

The "semiintact" cell index of the suspension can be determined with trypan blue by phase-contrast microscopy. Semiintact cells will bind the dye in the nucleus, yielding a dark staining morphology, whereas intact cells retain their usual highly refractile appearance (Fig. 1). At high magnification ($\times 100$) perforation of cells is often readily apparent. Routinely the above procedure results in perforation of 95–99% of 15B cells.

In Vitro Reconstitution of Endoplasmic Reticulum to Golgi Transport

Reagents. The following buffer and salt solutions are prepared and stored in 1- or 10-ml aliquots at $-20°$: 1 M HEPES, pH 7.2 (adjusted with KOH); 1 M KOAc; 100 mM magnesium acetate (MgOAc) (neutralized); a calcium/ethylene glycol-bis(β-aminoethyl ether)-N,N,N',N'-tetraacetic acid (EGTA) buffer containing 50 mM EGTA (pH 7.0), 18 mM CaCl$_2$, and 20 mM HEPES adjusted to a final pH of 7.2 with KOH.

ATP-Regenerating System. An ATP source is prepared by mixing 5 parts of 40 mM ATP (sodium form, neutralized) with 5 parts of 200 mM

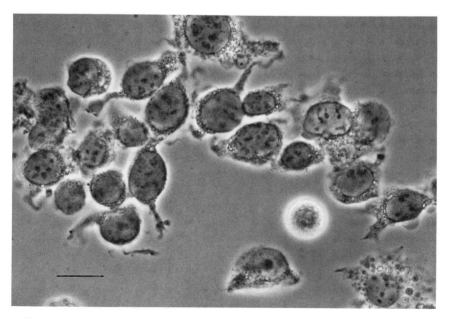

FIG. 1. Morphology of semiintact cells. Semiintact cells were prepared by the swelling–scraping procedure as described in text. A 10-μl aliquot was stained with trypan blue [final concentration, 0.01% (v/v)] and examined by phase-contrast microscopy. Magnification: $\times 800$. Bar: 5 μm.

creatine phosphate and 1 part of rabbit muscle creatine phosphokinase (1000 U/ml) (Cat. No. C3755; Sigma): This mixture is stored at −80° in 110-μl portions.

Gel-Filtered Cytosol. Cytosol is prepared from 15B cells (or other sources such as wild-type CHO cells, HeLa, rat liver, rat brain, etc.) by homogenizing a cell (or tissue) pellet with 4 vol of 25/125 H/KOAc and centrifuging at 100,000–150,000 g for 60 min at 4°. After discarding the lipidic surface layer the resulting clear supernatant is gel filtered on a Sephadex G-25 (medium) column in 25/125 H/KOAc. The void fraction is pooled and 125-μl portions flash-frozen in liquid N_2 and stored at −80°. The final protein concentration of the cytosol is generally 5–10 mg/ml.

Procedure. In vitro transport assays are conducted in 1.5-ml microcentrifuge tubes (Cat. No. 72.690; Sarstedt, Princeton, NJ). (*Note:* Some brands may contain chemical residues or surfactants that inhibit transport and should be compared to Sarstedt tubes before use.) The tubes are placed in anodized aluminum blocks (18 × 10 × 4 cm) that have been machined with 5 rows of ten 1.2 × 3 cm holes into which the tubes can be easily inserted or removed. Placing these blocks on ice effectively keeps the tubes and the contents at 1–3° during preparation of an assay mixture.

The following components are added *in the indicated order* to each tube held on ice: water, 20 μl (or to make a final volume of 40 μl), 1 μl of 1 M HEPES (pH 7.2), 1 μl of 100 mM MgOAc, 2 μl of KOAc, 4 μl of calcium/EGTA buffer, 2 μl of the ATP regenerating mix, 5 μl of cytosol, and 5 μl of semiintact cells. Assay tubes prepared in this way will contain 36.5 mM HEPES (pH 7.2), 2.5 mM Mg^{2+}, 77 mM K$^+$, 82 mM acetate, 1.8 mM Ca^{2+}/5 mM EGTA (100 nM free Ca^{2+}), 1 mM ATP, 5 mM creatine phosphate, and 0.2 IU creatine phosphokinase. These conditions have been optimized for our system (15B cells and 15B cytosol) and may vary for different cell or cytosol types.

When adding inhibitors or other factors (antibodies, purified proteins, etc.) it is important not to alter the ionic balance grossly. Thus, components that we add in aqueous solutions (e.g., N-ethylmaleimide) are compensated for in the amount of water added to the individual tubes, whereas it is often necessary to reduce the amount of the various salt solutions when adding proteins (typically dialyzed into 25/125 H/KOAc). In all cases control tubes are included to exclude nonspecific effects.

When using tsO45, solutions that contain dithiothreitol (DTT) should be avoided because less than 10 μM DTT changes the folding of tsO45 G protein and prevents its subsequent processing by Endo D. Glutathione (reduced form, up to 5 mM) can be substituted for DTT without adverse effect on the transport assay. Similarly, sucrose and related polyols, and amines (Tris, triethylamine, etc.), which also inhibit ER-to-Golgi transport in semiintact cells, should be avoided or controlled whenever they are added to the assay mixture.

Balance sheets are constructed to keep track of the exact experimental conditions for each tube. In practice, many of the components (water, salts/buffers, ATP-regenerating system) can be added to the individual tubes as a mixture and held on ice while semiintact cells are being prepared. Cytosol is added to the assay tubes after the semiintact cells have been prepared, and the cells are always added last, just prior to initiation of transport. Generally incubations are performed at 30°, and the transport reaction is usually complete after 90 min. Frequent shaking or rocking of the tubes during the incubation reduces the observed levels of transport, so tubes are left standing still in the water bath. Transport is terminated after the appropriate time by transferring the tube back to the ice-cold blocks.

Postincubation with Endoglycosidase D. After incubations are complete, the cooled cells are pelleted by a 30-sec spin at 10,000 to 15,000 g in a microcentrifuge. The supernatant is aspirated and 35 μl of endoglycosidase D (Endo D) digestion buffer [50 mM sodium phosphate (pH 6.5), 5 mM ethylenediaminetetraacetic acid (EDTA), 0.2% (w/v) Triton X-100] and 5 μl of Endo D are added. Endo D (Boehringer Mannheim, Indianapolis, IN) is reconstituted to 0.1 IU/ml in 10 mM Tris-HCl (pH 7.4), 200 mM NaCl, and stored at $-80°$. We also prepare Endo D from *Diplococcus pneumoniae* by published procedures.[12] The tubes are either vortexed or triturated to resuspend and solubilize the cell pellet, and the solubilized cell pellets incubated in capped tubes overnight at 37°.

Sample Analysis. The fraction of VSV-G protein transported to the cis Golgi compartment and processed to the Man$_5$ Endo D-sensitive form is determined using SDS-gel electrophoresis, fluorography, and densitometry. For this purpose, each tube is supplemented with 10 μl of a of a stock 5$\times$ concentrate gel sample buffer prepared by mixing 12.5 ml of 1 M Tris-HCl (pH 6.8), 20 ml glycerol, 0.5 g DTT, 4.0 g SDS, and 4.0 mg bromphenol blue in a final volume of 40 ml.[13] After briefly vortexing, samples are heated at 95° for 5 min, and centrifuged for 5 sec to pellet any condensate. Samples are loaded onto 7.5% (w/v) SDS-acrylamide gels (16-cm wide, 20-well sample combs), and electrophoresis carried out at 30–45 mA, constant current. Gels are subsequently soaked for 20 min in 100 ml of a fluorographic enhancement solution prepared by dissolving salicylic acid (Na$^+$ form, pH 7.0) in water, prior to the addition of methanol to bring to a final concentration of 0.125 M salicyclic acid, 30% methanol.[14] Dried gels are exposed to Kodak (Rochester, NY) XAR-5 film

[12] L. R. Glasgow, J. C. Paulson, and R. L. Hill, *J. Biol. Chem.* **252,** 8615 (1977).
[13] U. K. Laemmli, *Nature (London)* **227,** 680 (1970).
[14] J. P. Chamberlain, *Anal. Biochem.* **98,** 132 (1979).

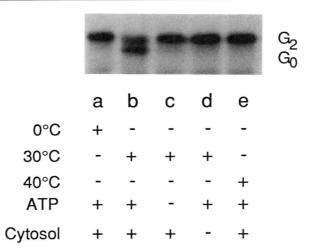

	a	b	c	d	e
0°C	+	-	-	-	-
30°C	-	+	+	+	-
40°C	-	-	-	-	+
ATP	+	+	-	+	+
Cytosol	+	+	+	-	+

FIG. 2. Transport from ER to Golgi in semiintact cells. Semiintact cells containing [^{35}S]methionine-labeled VSV-G protein were prepared by the swelling–scraping procedure as described in text. Aliquots (25 μg protein) were incubated for 90 min in complete assay mix at 0, 30, 40° (lanes a, b, and e), or at 30° in mixes without ATP (lane c) or cytosol (lane d). Samples were then processed with Endo D and analyzed by SDS-PAGE and fluorography. G_2 is the undigested (Man$_{8-9}$) ER form of VSV-G protein, and G_0 is the transported (deglycosylated) form.

at $-80°$ (typically an overnight exposure is required). The fraction of VSV-G protein processed to the Endo D-sensitive form, which has greater electrophoretic mobility than the high mannose (ER-associated, Man$_{8-9}$ form) (Fig. 2), is determined by densitometry of the exposed autoradiogram. In a standard reaction, transport is generally 50–80% efficient. Details of the properties of the transport assay can be found in earlier publications.[4,15]

Transport in Vitro to Medial Golgi Compartment Using Wild-Type Vesicular Stomatitis Virus-Infected Cells

In addition to the reconstitution of transport to the cis Golgi compartment by using 15B cells, transport to additional Golgi compartments can be readily reconstituted in vitro by using wild-type cell lines. This is a more typical situation than the specialized case for 15B in that most cell lines contain a full complement of processing enzymes modifying oligosaccharides to the complex form containing galactose and sialic acid. Thus transport to the medial Golgi compartment can be assessed by the acquisition of

[15] C. J. M. Beckers and W. E. Balch, *J. Cell Biol.* **108**, 1245 (1989).

resistance of VSV-G protein to the enzyme endoglycosidase H (Endo H), following modification of the N-linked oligosaccharides by the sequential activity of TrI and mannosidase II present in the medial Golgi.[7] Although this assay can be conducted with either tsO45 or wild-type virus, we generally use the latter.

Using wild-type virus, the basic procedure is as described above with the exception that the labeling time with [35S]methionine is reduced to 3 min and the chase time to 30 sec (because newly synthesized VSV-G protein exits the ER with a half-time of less than 10 min[16]). In addition, the assay is supplemented with 1 mM UDPGlcNAc (by the addition of 1 μl of a 40 mM stock solution of UDPGlcNAc in water), and carried out at 37° for 2–3 hr. To perform Endo H digestions, the assays are terminated by transfer to ice and the cells pelleted as described above. Each cell pellet is suspended in 20 μl of 100 mM sodium acetate (pH 5.5), containing 0.3% (w/v) SDS and 30 μl of 2-mercaptoethanol/ml (added just prior to use), boiled for 5 min, and centrifuged for 5 sec. Subsequently, 40 μl of 100 mM sodium acetate (pH 5.5) is added to each tube. When the samples are cool, 3 μl of recombinant Endo H (1 U/ml; Boehringer Mannheim) is added and the samples are mixed and incubated for 12–16 hr at 37° in capped tubes. Endo H digestions are terminated by the addition of 16 μl of the 5× gel sample solution and processed as described above.

Use of Nitrocellulose to Perforate Cells

A different technique, pioneered by Simons and Virta for the study of polarized secretion[5] but also applicable to the study of ER-to-Golgi transport,[4] involves the preparation, using nitrocellulose, of semiintact cells. For this technique all operations are carried out in the cold room (3°) to prevent export of VSV-G protein from the ER during manipulations. A confluent dish of cells is washed three times with 50/90 H/KOAc. A nitrocellulose filter (Cat. No. 162-0115; Bio-Rad Laboratories, Richmond CA) is trimmed to be equal in diameter to the inner dimensions of a 10-cm dish, and is saturated with 50/90 H/KOAc. Before use, the filter is blotted between two pieces of Whatman (Clifton, NJ) 3MM filter paper under a heavy weight (10 kg) for 1 min to remove excess buffer, and an edge of the filter bent at a 90° angle to facilitate handling with forceps. To attach the filter to cells, the tissue culture dish is drained for 45 sec in a vertical position. The filter is subsequently lowered gently onto the plate, starting at one edge to avoid trapping air. After 10 min the filter is removed by lifting one edge with the forceps while a second pair of forceps holds the opposite edge of the filter in place. The filter is transferred to the tissue culture lid

---

[16] J. K. Rose and R. W. Doms, *Annu. Rev. Cell Biol.* **4**, 257 (1988).

(with the side having attached cells facing up), and overlaid with 3 ml of 50/90 H/KOAc. Under the correct conditions of confluency, the majority of the cells remain attached to the plate and are perforated when the filter is removed. Either the dish or the filter can be gently scraped with a rubber policeman to release the attached cells. The suspended semiintact cells are washed as described above. Alternative procedures using nitrocellulose filters have been described for BHK and Madin–Darby canine kidney (MDCK) cell lines.[5,6]

Use of Poly(L-lysine) to Attach Cells for Perforation

Semiintact cells can also be prepared by scraping cells that have been plated on dishes pretreated with poly(L-lysine). This procedure has proved useful for poorly adherent cell lines or for reactions sensitive to hypotonic swelling because cells plated on poly(L-lysine) need only be scraped from the plate to be rendered semiintact. Additionally, in some cases, it may be applicable to cells that normally grow in suspension.

The procedure for preparing semiintact cells is as follows: Tissue culture dishes (10 cm) are washed with 10 ml of sterilized, ultrapure water and then incubated for 60 min at 37° with 2.5 ml of a stock solution of poly(L-lysine) (5–50 μg/ml) (Cat. No. P1524; Sigma) in ultrapure water. Dishes are subsequently washed three or four times with phosphate-buffered saline (PBS) and seeded with cells for overnight culture. The number of cells for seeding is sufficient so that the cells will be 70–80% confluent after 16–20 hr of culture. The concentration of poly(L-lysine) used in pretreating plates needs to be optimized for each cell type by scraping cells after overnight incubation and assessing "semiintactness" with trypan blue. We use the lowest concentration of poly(L-lysine) that consistently results in >90% trypan blue-positive cells after scraping. It is possible to have the cells so tenaciously attached that scraping results in preparation of a cell homogenate. For scraping, plates are washed four times in an appropriate assay buffer, and cells scraped with a rubber policeman into 3 ml of assay buffer as described above. We have found that other reagents such as Cell-Tak (Biopolymers, Inc., Farmington, CT) can also be used, in a manner similar to poly(L-lysine), to increase cell adherency for perforation.

Comments on Experimental Procedures

We have described an approach for the preparation of semiintact cells and their use in an assay of transport of proteins between the ER and the Golgi. We and others have successfully followed transport, using acquisition of Endo D susceptibility or Endo H resistance, in a large variety of cell lines including wild-type CHO, NRK, HeLa, Vero, COS, A431, BHK, and MDCK.[4-6] It is apparent that these and other cell lines are readily amenable to perforation and retention of functional organelles. An essential requirement for the application of this methodology is that cells must stick to a dish with sufficient adherence to be sheared during scraping with a rubber policeman or during "rip-off" with nitrocellulose. As indicated above, the degree of adherence can be dramatically increased through the use of poly(L-lysine), which, in some cases, may allow this methodology to be applied to suspension cells. Conversely, cells that adhere tenaciously to the plate may not need to be swollen if they readily perforate during scraping. It should be noted that the degree of cytosol-dependent transport occurring when using the swelling or nitrocellulose techniques reflects the efficiency of perforation, and that this may vary between preparations and cell lines. In cases in which cytosol dependence is weak, indicating a low efficiency of perforation, semiintact cell suspensions can be gently homogenized [10 to 20 strokes with a "loose" Dounce homogenizer (Wheaton, Millville, NJ)] prior to pelleting.

A number of other methods for perforating cells have been described, including those using digitonin,[17] electropermeabilization[18,19] and bacterial toxins.[20-22] In general these reagents create pores or channels in the plasma membrane that make the cells accessible to low molecular weight factors, but do not allow the efficient exchange of cytoplasmic proteins. In some circumstances streptolysin O- and digitonin-induced pores may be sufficiently large to cause loss of cytosolic proteins, and to allow penetration by antibodies.[17,21] However, they must be carefully optimized for any given cell type because low concentrations of these reagents may also disrupt intracellular cholesterol-containing membranes such as those of the ER. In contrast, the swelling and nitrocellulose perforation techniques described above appear more generally applicable, and routinely render cells sufficiently permeable to reconstitute strong cytosol dependence without gross disruption of subcellular organelles.

Although we routinely use these techniques to assay the transport of VSV-G protein between the ER and various Golgi compartments, similar conditions can be used to study the transport of other proteins in both the

[17] G. Fiskum, S. W. Craig, G. L. Decker, and A. L. Lehninger, *Proc. Natl. Acad. Sci. U.S.A.* **77**, 3430 (1980).

[18] D. E. Knight and M. C. Scrutton, *Biochem. J.* **234,** 497 (1986).

[19] D. E. Knight and M. C. Scrutton, this volume [10].

[20] R. Fussle, S. Bhakdi, A. Sziegoleit, J. Tranum-Jensen, T. Kranz, and H. J. Wellensiek, *J. Cell Biol.* **91**, 83 (1981).

[21] T. W. Howell and B. D. Gomperts, *Biochim. Biophys. Acta* **927**, 177 (1987).

[22] G. Ahnert-Hilger, B. Stecher, C. Beyer, and M. Gratzl, this volume [11].

exocytic and endocytic pathways. Potential markers include endogenous proteins, other viral glycoproteins, and proteins acquired through transfection. Moreover, because both the interior and exterior membranes of the cell are jointly accessible to a wide range of reagents and macromolecules, semiintact cells may provide a useful model system for the study of a broad range of problems in cell biology, including signal transduction, organization of the cytoskeletal matrix, and gene activation.

Acknowledgments

This work was supported by U.S. Public Health Service Grant GM33301 to W.E.B., and by the Harold G. and Leila Y. Mathers Charitable Foundation, the Medical Research Council of Canada, and the Science and Engineering Research Council of Great Britain. We thank Dr. E. Smythe for critical discussion of the manuscript.

[18] Fluorescence Methods for Monitoring Phagosome– Lysosome Fusion in Human Macrophages

By Nejat Düzgüneş, Sadhana Majumdar, and Mayer B. Goren

Introduction

Macrophages are professional scavenger cells that phagocytose and degrade invading pathogens. The phagocytic vacuole, the phagosome, is believed to fuse with lysosomes, thereby exposing the phagocytosed pathogens to several digestive enzymes present in the lysosomes.[1] The molecular control of the fusion of phagosomes and lysosomes is not understood well. Certain pathogens, such as *Mycobacterium tuberculosis, Toxoplasma gondii,* and *Legionella pneumophila,* survive inside macrophages by inhibiting phagosome–lysosome fusion by unknown mechanisms.[2-4]

Early studies on phagosome–lysosome fusion used the fluorescent probe acridine orange as a convenient marker for lysosomes, because it accumulates readily in the acidic interior of lysosomes.[5,6] Because acridine

[1] P. J. Edelson, *Rev. Infect. Dis.* **4,** 124 (1982).

[2] J. A. Armstrong and P. D. Hart, *J. Exp. Med.* **134,** 713 (1971).

[3] T. C. Jones and J. G. Hirsch, *J. Exp. Med.* **136,** 1173 (1972).

[4] M. A. Horwitz, *J. Exp. Med.* **158,** 2108 (1983).

[5] M. B. Goren, P. D. Hart, M. R. Young, and J. A. Armstrong, *Proc. Natl. Acad. Sci. U.S.A.* **73,** 2510 (1976).

[6] M. C. Kielian and Z. A. Cohn, *J. Cell Biol.* **85,** 754 (1980).

orange traverses biological membranes readily, however, it can redistribute into those compartments where it is bound most efficiently.[7] Here an alternative assay based on the use of sulforhodamine or rhodamine (R)-labeled dextran is described.[8] The assay is based on the initial uptake of sulforhodamine or R–dextran into secondary lysosomes of macrophages, and the subsequent colocalization of the rhodamine label and phagocytosed fluorescein-labeled yeast cells on the fusion of phagosomes with secondary lysosomes.[9]

Reagents. The following stock solutions of fluorescent probes are prepared by dissolving the material in phosphate-buffered saline (PBS):

Rhodamine (R)–dextran of average M_r 70,000 (Sigma Chemical Co., St. Louis, MO): 25 mg/ml
Sulforhodamine (Molecular Probes, Eugene, OR): 20 or 7 mg/ml
Fluorescein isothiocyanate (FITC) (Molecular Probes or Sigma): 10 mg/ml

The solutions are filter sterilized by passing them through 0.22-μm pore size filters (Schleicher & Schuell, Keene, NH), and stored in the dark at 4°.

Labeling of Yeast *(Saccharomyces cerevisiae)* with Fluorescein Isothiocyanate

Baker's yeast cells are suspended in PBS at 10^9 cells/10 ml, and incubated with 10 μg/FITC per milliliter (final concentration) for 30 min at room temperature in the dark. Unconjugated FITC is removed by repeated (three times or more) centrifugation of the yeast cells at 3000 rpm in a Sorvall (Norwalk, CT) RT6000 cell centrifuge for 10 min. The labeled cells are killed by incubation at 90° for 30 min, harvested by centrifugation, resuspended at a cell density of 2×10^8 cells/ml in a 1:1 mixture of fetal bovine serum (FBS) and RPMI 1640 medium, and incubated for 45 min at 37°. The FITC-labeled and opsonized yeast cells are then diluted in RPMI 1640/10% (v/v) heat-inactivated FBS (medium A) at a density of 10^6 cells/ml, and cultured with macrophages, as described below. FITC-labeled yeast cells can be stored in isotonic saline with sodium azide (0.02%, w/v).

[7] M. B. Goren, C. L. Swendsen, J. Fiscus, and C. Miranti, *J. Leukocyte Biol.* **36**, 273 (1984).
[8] Y. Wang and M. B. Goren, *J. Cell Biol.* **104**, 1749 (1987).
[9] M. B. Goren and N. Mor, *in* "Virulence Mechanisms of Bacterial Pathogens" (J. A. Roth, ed.), p. 184. American Society for Microbiology, Washington, DC, 1988.

Fluorescence Labeling of Human Peripheral Blood Monocyte-Derived Macrophages

Monocytes are isolated from fresh human blood or buffy coats from a blood bank. It is preferable to use blood tested at the blood bank for human immunodeficiency virus and hepatitis B virus; only blood found to be seronegative for these pathogens should be used. The blood is layered on Histopaque (Sigma) in 50-ml plastic culture tubes, the mononuclear cell band formed after centrifugation (2000 rpm for 30 min in a Sorvall RT6000 cell centrifuge, at room temperature) is removed, the cells counted in a hemacytometer, and placed in 24-well plastic culture dishes (10^6 cells/well) for 24 hr. The nonadherent cells are removed by washing three times with medium A. The monocytes are cultured for 5 days at 37° in a CO_2 (5%) incubator. The cells are then washed with PBS, and incubated for 24 hr with medium A containing 0.1–0.5 mg/ml R–dextran. The viability of the macrophages is 99% following this labeling. The cells are washed with PBS to remove all unincorporated probe molecules, and again incubated for 2 hr in medium A. Alternatively, the cells can be labeled with 35–200 μg sulforhodamine[8] per milliliter. As with R–dextran, the concentration of sulforhodamine to be used is determined by the tolerance of the cells for the fluorophore, and the level of labeling achieved.

Similar experiments can be performed with murine peritoneal macrophages[8] and with the murine macrophage cell line J774.[10]

Induction of Phagocytosis

The culture medium is aspirated from the macrophage monolayer and replaced with 1 ml of the FITC-labeled yeast suspension (about 10^6 cells/ml of medium A). The yeast-to-macrophage ratio can be altered to achieve optimal visualization of the process. After a 45-min incubation at 37° unattached yeast cells are removed by repeated washing with PBS. Medium A is added, and the cells are observed under an inverted epifluorescence microscope equipped with filter cubes for fluorescein and rhodamine [e.g., a Nikon Diaphot microscope (Garden City, NY)].

Observations with Assay

Following the incubation with R–dextran, punctate red fluorescence is observed in the secondary lysosomes of the macrophages.[11] After the initial 45-min incubation of the macrophages with the yeast particles, yeast cells

---

[10] N. Düzgüneş, J. Goldstein, and M. B. Goren, unpublished data (1988).
[11] S. Majumdar and N. Düzgüneş, unpublished data (1991).

with green fluorescence are observed within macrophage phagosomes. Within the next 4 hr many of the yeast particles appear orange, as a result of the delivery of lysosomal R–dextran into phagosomes. Some particles remain green. The range of colors observed, "almost pure green (no fusion) to very bright orange (heavy fusion), contribute a vivid panorama of the spectrum of activities that should be expected for the varied behavior of a biological system." (See Ref.[9] p. 192.) After 24 hr only red fluorescence is observed, indicating the fusion of all the phagosomes with lysosomes. The assay is described schematically in Fig. 1.

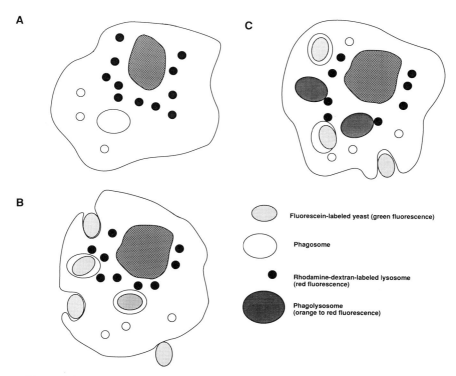

FIG. 1. Schematic representation of phagosome–lysosome fusion monitored by the fluorescence assay described in text. (A) A macrophage has ingested rhodamine–dextran by pinocytosis and delivered it to secondary lysosomes. (B) A rhodamine–dextran-labeled macrophage phagocytosing fluorescein-labeled yeast. (C) Fusion of secondary lysosomes with yeast-containing phagosomes. Phagocytosed yeast particles that have not encountered lysosomes will be bright green in the microscope, whereas phagosomes that have fused with the rhodamine–dextran-labeled lysosomes will appear light or dark orange.[11] The different intensities of orange probably reflect the amount of rhodamine–dextran delivered into the phagosomes, and hence the number of lysosomes that have fused with a phagosome.

This method has been used to examine the effect of various cytokines on phagosome–lysosome fusion.[11] Interleukin 2 (IL-2), preincubated with the cells overnight at a concentration of 1000 U/ml, was found to inhibit phagosome–lysosome fusion. At 4 hr the ingested yeast particles were still green, whereas in the untreated macrophages they appeared orange. After 24 hr the particles appeared slightly red, whereas some green particles were still observable. All the particles were red at this time point in the untreated control cells. Tumor necrosis factor alpha (TNF-α) and γ-interferon (IFN-γ) were, at 100 or 1000 U/ml, even more inhibitory, with many green particles being observed even after 24 hr of incubation. In contrast, macrophage colony-stimulating factor (M-CSF) at 1000 U/ml appeared to enhance phagosome–lysosome fusion.

The macrophage pathogen *Mycobacterium avium* is thought to persist in phagosomes by preventing the fusion of the phagosomes with lysosomes.[12] This hypothesis was examined using a modification of the technique outlined above. *Mycobacterium avium* was labeled with FITC, and shown to be viable following this treatment.[13] When labeled *M. avium* was incubated with R-dextran-pretreated human macrophages, no red fluorescence was noted in areas where the green-fluorescing microorganisms were localized, even after 24 hr. When the macrophages were preincubated overnight with 1000 U M-CSF per milliliter, the green areas colocalized with the red fluorescence within 4 hr.[14] With M-CSF-treated macrophages, the green fluorescence of heat-killed *M. avium* colocalized with red fluorescence within 2 hr, whereas with control macrophages the heat-killed organisms colocalized with red fluorescence more slowly. These observations support the hypothesis that live *M. avium* inhibits phagosome–lysosome fusion, and that M-CSF can activate macrophages to overcome the inhibition of phagosome–lysosome fusion. The molecular mechanisms of the inhibition of fusion, and of the reversal of this process by M-CSF, remain to be investigated.

[12] C. Frehel, C. de Chastellier, T. Lang, and N. Rastogi, *Infect. Immun.* **52,** 252 (1986).
[13] N. Düzgüneş, S. Majumdar, P. Minasi, P. Nassos, D. Yajko, and D. Daleke, unpublished data (1991).
[14] N. Düzgüneş, S. Majumdar, P. Nassos, and M. B. Goren, unpublished data (1991).

[19] *In Situ* Resonance Energy Transfer Microscopy: Monitoring Membrane Fusion in Living Cells

By PAUL S. USTER

Introduction

Background

The resolution limit of fluorescence light microscopy (about 400 nm) can be exceeded indirectly by using resonance energy transfer (RET) microscopy. This technique is able to visualize the spatial location of two different, fluorescently labeled membrane probes and determine if they are in the same membrane, or in physically adjacent but separate bilayers (10 to 400 nm apart.

The use of RET to follow fusion events and the microscopic visualization by RET were originally conceived by Snipes and colleagues.[1] Herpes virus-induced fusion of hamster embryo lung (HEL) cells was photographed in the fluorescence microscope using a band-pass filter that excited both probes, and an emission filter that visualized both probes simultaneously. Cells labeled with a fluorescein probe were green, cells labeled with a rhodamine probe were yellow-orange, and cells of an intermediate color were believed to be fused. However, RET (and fusion) cannot be *unequivocally* demonstrated with such a microscopic configuration because probes in separate but overlying bilayers also have an intermediate color. In this chapter we describe the procedures for using RET microscopy to visualize lipid-phase coalescence (fusion) of fluorescently labeled membranes and how to outfit a stock epifluorescence microscope to record RET on film emulsion.

Resonance energy transfer microscopy was developed to study the spatial and temporal distribution of fluorescent probes in model membranes and in living cells.[2,3] Particular attention was paid to the mechanics of visualizing definitively the colocalization of both membrane probes in the same bilayer. This microscope configuration has been used to study the ATP-dependent liposome fusion with the Golgi apparatus of permeabilized, cultured skin fibroblasts.[4]

---

[1] P. M. Keller, S. Person, and W. Snipes, *J. Cell Sci.* **28**, 167 (1977).
[2] P. S. Uster and R. E. Pagano, *J. Cell Biol.* **103**, 1221 (1986).
[3] P. S. Uster and R. E. Pagano, this series, Vol. 171, p. 850.
[4] T. Kobayashi and R. E. Pagano, *Cell (Cambridge, Mass.)* **55**, 797 (1988).

METHODS IN ENZYMOLOGY, VOL. 221

Principle of Resonance Energy Transfer

Resonance energy transfer is the transfer of photon energy from one fluorescent compound (i.e., the donor probe) to a chemically different acceptor probe molecule, only when both are in close physical proximity. For conventional probes with fluorescence lifetimes on the order of 10 nsec, the physical proximity must be less than 10 nm.[5] It is also critical that the fluorescence emission spectrum of the donor should overlap the absorption spectrum of the acceptor as much as possible. As a result of this transfer of photon energy, the fluorescence intensity of the donor probe is quenched. Although it is not required that the acceptor probe be fluorescent, a fluorescent probe has the useful property of reemitting the transferred photon energy at the characteristic emission wavelengths of the acceptor. This sensitized emission of the acceptor probe is the useful hallmark of RET, which enables it to be distinguished in the microscope from other potential donor quenching mechanisms.[6]

To observe this sensitized emission and colocalize it with the donor, a three-channel system of spectral windows has been developed that is readily installed on existing epifluorescence microscopes. The "donor channel" uses an exciter filter, dichroic beam splitter, and barrier band-pass filter to limit fluorescence observation only to the donor probe. The acceptor probe cannot be visualized in this channel. The "acceptor channel" uses similar components with different spectral windows to limit observation only to acceptor probes directly excited by appropriate light wavelengths. The "transfer channel" is used to observe the sensitized acceptor emission from locations in which the donor and acceptor molecules are within 10 nm of each other. The exciter band-pass filter and dichroic beam splitter from the donor channel are combined with the barrier emission filter of the acceptor channel. As illustrated for model membranes,[2,3] donor probe and acceptor probes in separate but adjacent membranes cannot be readily observed in the transfer channel. Additionally, quenched donor fluorescence is observed in the donor channel only in colabeled membranes, that is, those containing both probes in the same lipid bilayer.

Materials

Lipids

Egg phosphatidylcholine, (95%; iodine value 40) and egg phosphatidylglycerol (95%) are obtained from Asahi Chemical (Tokyo, Japan). The

[5] D. D. Thomas, W. F. Carlsen, and L. Stryer, *Proc. Natl. Acad. Sci. U.S.A.* **75**, 5746 (1978).
[6] R. A. Badley, *in* "Modern Fluorescence Spectroscopy" (E. L. Wehry, ed.), Vol 2, p. 91. Plenum, New York, 1976.

fluorescent donor, N-(7-nitrobenz-2-oxa-1,3-diazol-4-yl)-dipalmitoyl-L-α-phosphatidylethanolamine (NBD-PE), and the fluorescent acceptor, N-(Texas Red-sulfonyl)dipalmitoyl-L-α-phosphatidylethanolamine (SRH-PE), are obtained from Molecular Probes (Eugene, OR).

Liposome Preparation

Egg phosphatidylcholine/egg phosphatidylglycerol (9:1, mol/mol) liposomes containing equimolar quantities of NBD-PE and SRH-PE at different probe densities are prepared by thin film hydration in distilled water containing 2.0 mM sodium edetate. The phospholipid dispersions are sonicated to optical clarity using a Branson model 5200 bath sonicator.

Prior to spectrofluorimetric analysis, all liposome dispersions are diluted as appropriate to give a final probe concentration of 0.6 μM NBD-PE and SRH-PE *each*. All measurements are made with a Shimadzu RF-540 spectrofluorometer, using an excitation wavelength (E_x) at 450 nm.

Procedures

Selection of Membrane Probes

The microscope system described below is useful for donor probes with blue light absorption and green light emission, coupled to an acceptor with green light absorption and red light emission.[7] The absorption and fluorescence emission spectra of potential probes should be evaluated in appropriate solvents, using a spectrofluorometer. The emission spectrum of the donor must overlap the absorption (excitation) spectrum of the acceptor as much as possible. Also, the absorption spectrum of the donor should overlap the absorption spectrum of the acceptor as little as possible. This reduces cross-channel interference, which decreases sensitivity and increases the fraction of acceptor molecules available for RET.

The effect on RET of changing probe density in membranes should be evaluated in liposomes, using a spectrofluorometer. Probe density can be approximated as the mole percent fluorescent probe (mol%) of total lipid. The effect of changing probe density on fluorescence spectra is illustrated in Fig. 1. As the probe density is increased fourfold, the green fluorescence of the donor decreases markedly, while the red fluorescence of the acceptor increases. Figure 2 shows that such donor fluorescence quenching and

---

[7] Another potentially useful combination that would require different filter combinations would require a donor probe with UV absorption and visible blue light emission combined with an acceptor with blue light absorption and green light emission.

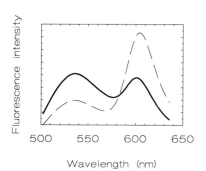

Wavelength (nm)

FIG. 1. Spectroscopic consequences of RET illustrated with liposomes. Increasing individual probe density from 0.13 mol% (—) to 0.50 mol% (– – –) decreases the fluorescence emission (E_m) of NBD-PE at 530 nm while increasing the sensitized emission of SRH-PE at 600 nm.

sensitized acceptor emission are not linearly proportional to mole percent probe.

To monitor membrane coalescence it is also critical that the fluorescent probes be sufficiently lipophilic to prevent significanty transfer by collision or diffusion-mediated processes. A simple method for assessing this is to label liposomes by either of the techniques listed below and monitor changes in fluorescence as a function of incubation time, temperature, and lipid concentration.[8] Significant changes in fluorescence indicate a serious artifact, which must be remedied by a different choice of one or both membrane probes.

Dual-Probe Dilution Labeling Method

Dual-probe dilution labeling is particularly useful for quantitatively monitoring the fusion of liposomes with cells in culture. Direct labeling of the plasma membrane or organelles is avoided, thereby minimizing potential effects on viability.

This labeling method was introduced by Struck *et al.*[9] to monitor liposome–liposome fusion. A liposome dispersion is prepared in which donor and acceptor probes reside in the same bilayer at about 1 to 2 mol% of each probe. This dispersion is subsequently incubated with unlabeled membranes, either liposomal or biological in origin. As a result of fusion, both probes are diluted into the unlabeled membranes. The decrease in

[8] J. W. Nichols and R. E. Pagano, *Biochemistry* **21,** 1721 (1982).
[9] D. K. Struck, D. Hoekstra, and R. E. Pagano, *Biochemistry* **20,** 4093 (1981).

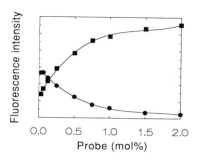

FIG. 2. Fluorescence intensity changes with changing probe density. Liposome dispersions colabeled with NBD-PE and SRH-PE at various densities are illuminated at 450 nm and the fluorescent emission at (●) 530 nm and (■) 605 nm is recorded.

probe density reduces RET efficiency, which is calculated from donor quenching (Fig. 3).

Because it is difficult to determine unquenched donor fluorescence *in sutu,* monitoring fusion by donor quenching efficiency is impractical for RET microscopy. The nonlinear relationship between RET efficiency and probe density also makes determination of kinetics problematic (Fig. 3). A more straightforward approach is to record fluorescence intensity in both the transfer and donor channel spectral windows. For quantitative studies in which a photomultiplier or digital image analyzer is attached to the RET microscope, the ratio of acceptor fluorescence intensity (transfer channel) to donor fluorescence intensity (donor channel) can be used to determine actual probe densities. In Fig. 4 liposome dispersions containing equimolar amounts of donor and acceptor probes at different densities are prepared and examined in a spectrofluorometer. The fluorescence emission from the transfer channel and donor channel is modeled by exciting at 450 nm and

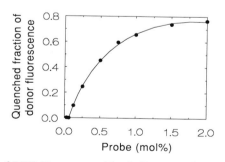

FIG. 3. Efficiency of RET. Donor quenching in liposome dispersions is calculated from $1 - F/F_0$, where F is the donor fluorescence in the presence of RET and F_0 is donor fluorescence in its absence.

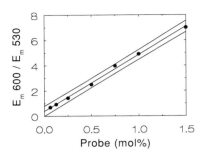

FIG. 4. Fluorescence emission ratios correlate with probe density (mol%). The ratio of fluorescence intensities at 600 and 530 nm are linearly proportional to probe density (mol%). The linear regression ($r^2 = 0.997$) and 95% confidence interval bands are shown.

recording the fluorescence emission (E_m) at 600 and 530 nm, respectively. The ratio of transfer (E_m 600) to donor (E_m 530) fluorescence intensity is inversely proportional to probe density from 0.06 to 1.5 mol%. This ratio method has the added advantage of being independent of actual fluorescence intensities.

Single-Probe Labeling Method

In this method, originally introduced by Keller *et al.*,[1] the two membranes of interest are individually labeled with donor or acceptor probe. It is important to evaluate the effects of labeling on membrane or cell viability. This labeling procedure is especially useful for studying the intracellular distribution of membrane-bound probes. It is important to achieve an acceptor probe labeling of 1 to 2 mol%, because RET efficiency is strictly dependent on acceptor probe density but is independent of donor probe density (see Fig. 3).[2,10] For membrane-bound probes, sensitized acceptor emission is linearly proportional to the mole percentage of donor probe entering the same bilayer.[2] The kinetics of donor probe entry into the acceptor-labeled membrane can be monitored readily in the transfer channel, without additional calculations.

Modification of Epifluorescence Microscope

To date, RET microscopy studies have been conducted with a Zeiss (Thornwood, NY) IM-35 microscope with Planapo ×40/0.9 NA or Planapo ×63/1.3 NA objective lenses. Fluorescence illumination is provided by an HBO 100 mercury lamp connected to electronic shutters, used to

[10] B. K.-K. Fung and L. Stryer, *Biochemistry* **17**, 5241 (1978).

control specimen exposure. Neutral density filters are also used to control light intensity.

Resonance energy transfer can be visualized by modifying the filter packs of an epifluorescence microscope with the following Zeiss components or their optical equivalent. The donor channel uses a BP436/17 exciter filter and FT510 dichroic beam splitter to limit excitation wavelengths from 428 to 444 nm. A BP515–565 barrier filter limits emission wavelengths from 515 to 565 nm. The acceptor channel uses a BP546/12 exciter filter and FT580 dichroic beam splitter to limit excitation wavelengths from 540 to 552 nm. A LP610 barrier filter allows emission wavelengths greater than 610 nm. The transfer channel uses the BP436/17 exciter filter, FT510 dichroic beam splitter, and LP610 barrier filter. These filter combinations are effective in studies using donor fluorescent probes based on fluorescein or 7-nitrobenz-2-oxa-1,3-diazol-4-yl (NBD), and acceptor fluorescent probes based on rhodamine or sulforhodamine (SRH).

Photomicroscopy

Qualitative evidence of changes in fluorescent probe density and distribution can be documented on photographic emulsion. Black-and-white negatives can be recorded on Kodak (Rochester, NY) Tri-X Panatomic film push-processed to ASA 1600, using Diafine developer. Color photomicrographs can be recorded on Kodak Ektachrome film and push-processed to ASA 1200.

In the system described above, it is critical to eliminate potential artifacts (and thereby reduce subjective bias) by observing the following caveats.

Potential photobleaching should be minimized by selecting the microscopic field of interest in low-illumination phase contrast, using a tungsten lamp. As a general rule, the shorter the emission wavelength of a fluorescent molecule, the faster will be the rate of photobleaching. Thus, fine tuning the focal plane depth of field should be done only in low-level fluorescence, using the acceptor channel. This image should be recorded on film emulsion, using an appropriate exposure time. It is critical that all exposure times be controlled electronically by the shutter attached to the mercury lamp; manual shutter control can introduce substantial variation.

Depending on the application, the filter pack should then be switched to the donor or transfer channel, and the image in this channel recorded on film emulsion. Because of donor probe photobleaching, it is usually not possible to record images in both the donor and transfer channels unless the specimen is heavily labeled. For most applications, the best comparisons are between the same field of view in the acceptor and transfer channels.

Negatives and photomicrographs must be exposed and processed as similarly as possible. Appropriate controls must be on the same roll of film as on the experimental treatment, to ensure identical processing of the negatives. Valid comparisons can be made only if the controls are of the same exposure time and intensity of illumination as the experimental treatment. Negatives of control and experimental treatment should be printed at the same magnification. The prints should be made on the same lot of film paper, and the enlarger should have an electronically controlled shutter.

The modifications described above can be achieved with minimal investment. However, combining RET microscopy with low light-level detector technology and digital image analysis will facilitate both image acquisition and kinetic studies. The use of two-stage or three-stage charge-coupled cameras will allow low illumination levels to be used and reduce photobleaching to undetectable levels.[11] Microprocessor-controlled digital imaging will quantitatively determine fluorescence intensity at discrete locations in the microscopic field of interest, and will calculate wavelength ratios.[12]

Acknowledgments

Supported in part by Liposome Technology, Inc., a Carnegie Corporation of New York Fellowship to P.S.U., and U.S. Public Health Service Grant GM-22942 to Richard E. Pagano (Carnegie Institution of Washington).

[11] G. T. Reynolds and D. L. Taylor, *BioScience* **30**, 586 (1980).
[12] J. DiGuiseppi, R. Inman, A. Ishibara, K. Jacobson, and B. Herman, *BioTechniques* **3**, 394 (1985).

Section V

Membrane Fusion in Fertilization

[20] Detection of Sperm–Egg Fusion

By Frank J. Longo and Ryuzo Yanagimachi

The purpose of this chapter is to summarize methods used to detect sperm–egg fusion, a process that follows gamete attachment and is believed to be instrumental in the activation of the egg. Unequivocal determination of sperm–egg fusion is especially important in studies examining specific events of gamete interaction, which may be causal to egg activation, and where its verification is important, so that later developmental events or their absence may be interpreted correctly. Methods that have been employed to detect sperm–egg fusion reliably are reviewed and include light microscopy, dye transfer, electron microscopy, and membrane conductance/capacitance changes. The main principles involved in analyses of interacting gametes leading to their fusion are outlined, as well as advantages and disadvantages of individual methods. For detailed methodologies the reader is encouraged to consult the original literature.

The following terms and their definitions are employed here:

Gamete attachment: Physical contact of the gametes directly observable in living specimens and involving the adherence of the sperm to the egg.

Gamete continuity: A situation in which the sperm and egg are electrically coupled and share diffusible components, but their plasma membranes have not necessarily fused. This is essentially equivalent to the status of functional contact of fusing cells.[1]

Sperm–egg membrane fusion (sperm–egg fusion or gamete fusion): A process involving the fusion of sperm membrane with the egg plasma membrane, such that the continuity of the gamete plasma membranes is established. This follows gamete attachment and results in the adjoining of the sperm and egg cytoplasms. As a consequence of such a process, the contents of the sperm, that is, its nucleus, mitochondria, and axonemal complex (tail components), are topologically within the egg cytoplasm.

Methods

Light Microscopy

In the 1800s, sperm incorporation into the egg cytoplasm and development of the male pronucleus were described in numerous species. It was

---

[1] B. A. Pethica, *in* "Cell Fusion" (D. Evered and J. Whelan, eds.), p. 1. Pitman, London, 1984.

not until the advent of the electron microscope and the pioneering work by Colwin and Colwin[2] that investigators came to realize that the sperm nucleus was incorporated into the egg cytoplasm through a fusion of the gamete membranes. Hence in specimens examined by light microscopy, in which the sperm nucleus is observed within the egg cytoplasm and develops into a male pronucleus, it is a reasonable presumption that sperm–egg membrane fusion has occurred. Because the site of gamete fusion is restricted to a small area involving the sperm and egg membranes, precise determination of when gamete fusion occurs is beyond the resolving power of the light microscope. Additionally, it is important to recognize that merely observing a sperm in an egg by light microscopy may not be sufficient evidence to support the notion that gamete fusion has occurred. For example, in some experimental situations sperm may be phagocytosed.[3-6] In these cases, true gamete membrane fusion is not established. Although sperm nuclear transformations may mimic stages of male pronuclear development, they are degenerative, characteristic of the contents within a phagosome.

The absence of a decondensing sperm nucleus, a male pronucleus, or egg activation does not necessarily imply the failure of sperm–egg fusion. For example, when hamster and starfish oocytes at the germinal vesicle stage are inseminated, sperm–egg fusion occurs but sperm nuclei in the egg cytoplasm remain "unchanged" until the germinal vesicle breaks down.[7,8]

Unless the undispersed sperm nucleus is deep within the egg cytoplasm, determination of whether it is inside or outside the egg is difficult to ascertain by light microscopy. Incorporated sperm, transforming into male pronuclei within the egg cytoplasm, can be observed in living whole-mount preparations with phase-contrast or interference contrast optics (Fig. 1). Exceptions do exist and include porcine and bovine zygotes, which contain dense inclusions, making them ill-suited for such analyses. Living specimens can also be stained with the DNA-intercalating dye Hoechst 33342 to reveal, using fluorescence microscopy,[9] the presence of the incorporated

[2] L. H. Colwin and A. L. Colwin, in "Fertilization" (C. B. Metz and A. Monroy, eds.), Vol. 1, p. 295. Academic Press, New York, 1967.

[3] A. Bendich, E. Borenfreund, S. S. Witkin, D. Beju, and P. J. Higgins, Prog. Nucleic Acid Res. 17, 43 (1976).

[4] R. Pijenborg, S. Gordts, and I. Brosens, in "Hamster Quality and Fertility Regulation" (R. Rolland ed.), p. 313. Elsevier, New York, 1985.

[5] K. Kyozuka and K. Osanai, Gamete Res. 21, 127 (1988).

[6] K. Kyozuka and K. Osanai, Gamete Res. 22, 123 (1989).

[7] N. Usui and R. Yanagimachi, J. Ultrastruct. Res. 57, 276 (1976).

[8] F. J. Longo and A. W. Schuetz, Biol. Bull. (Woods Hole, Mass.) 163, 453 (1982).

[9] S. J. Luttmer and F. J. Longo, Gamete Res. 15, 267 (1986).

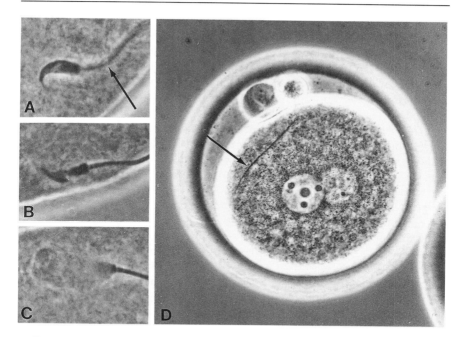

Fig. 1. Phase-contrast micrographs of fertilized hamster eggs. Living zygotes are compressed between a slide and coverslip, prior to examination. (A) Sperm nuclei following sperm–egg fusion, (B) undergoing decondensation, and (C) in an advanced state of decondensation. (D) Fertilized egg containing male and female pronuclei. Portions of incorporated sperm tail are shown at the arrows.

sperm nucleus. Hoechst staining of living sperm, eggs, or both gametes has been employed to reveal the presence of the maternal, paternal, or both genomes, respectively, in fertilized eggs and embryos.[9]

The presence of incorporated sperm nuclei has been demonstrated in fixed eggs, either as whole mounts or in sections. (For methods with which to prepare specimens for light microscopy see Ruthman,[10] Bedford,[11] and Longo and Anderson.[12] In whole mounts of marine specimens, the fertilized eggs are pelleted and resuspended in ethanol–acetic acid fixative (3:1). After fixation for 10–60 min, the eggs are washed in distilled water and resuspended in a small volume of lactoorcein, which is prepared by mixing an equal volume of 2% (w/v) orcein in acetic acid and 85% (v/v)

[10] A. Ruthmann, "Methods in Cell Research." Cornell Univ. Press, Ithaca, NY, 1970.
[11] J. M. Bedford, in "Methods in Mammalian Embryology" (J. C. Daniel, ed.), p. 37. Freeman, San Francisco, 1971.
[12] F. J. Longo and E. Anderson, J. Microsc. 96, 255 (1972).

lactic acid in distilled water. Stained preparations are gently flattened between a coverslip and slide, for observation with bright field optics. For mammalian eggs, the cumulus cells are removed prior to fixation and staining with lactoorcein. Fixed, whole-mount preparations may also be examined with phase and interference contrast optics, or stained with a variety of different fluorochromes, specific for DNA, that reveal the incorporated sperm nucleus as well as the maternal chromatin when viewed with fluorescence microscopy.[13] Such preparations can be stunning in their clarity and have been used to follow the transformations of both the male and female pronuclei during the course of fertilization. Fixed, inseminated eggs may also be embedded and sectioned for light microscopy, according to standard protocols used for examination of other cell types. Sections stained with basophilic dyes are capable of revealing the organization of the male and female pronuclei and chromosomes.

Light microscopy is a relatively quick and inexpensive method to detect sperm–egg fusion, especially with regard to equipment and supplies required for analyses. It is unequivocal when the incorporated sperm nucleus is observed to transform into a male pronucleus, and normal cleavage and embryonic development ensue. A major disadvantage of the method is that it lacks sufficient spatial and temporal resolution to determine when fusion has actually occurred. Therefore it may be equivocal under some experimental conditions. With the light microscope, sperm nuclei taken into the egg cytoplasm by phagocytosis may appear morphologically similar to those incorporated by true membrane fusion. Additionally, if the sperm nucleus fails to decondense (or transform into a male pronucleus) within the egg cytoplasm, or if the egg fails to activate, this may be erroneously taken as evidence of the failure of sperm–egg fusion.

Dye Transfer

Detection of sperm–egg fusion using the DNA-specific fluorochrome Hoechst 33342 involves preloading unfertilized eggs with the dye. When fusion occurs the fluorochrome enters the sperm and stains its DNA, resulting in the appearance of a bright, fluorescent sperm nucleus at the egg surface (Fig. 2). This method was initially used to establish the time of sperm–egg fusion in the sea urchin *Lytechinus variegatus*[14,15]; it has since been employed in similar types of experiments with mammalian gametes.[16] Unfertilized eggs are treated for 5 to 60 min with Hoechst 33342

[13] C. Mori, H. Hashimoto, and K. Hoshino, *Biol. Reprod.* **39,** 737 (1988).
[14] R. E. Hinkley, B. D. Wright, and J. W. Lynn, *Dev. Biol.* **118,** 148 (1986).
[15] R. E. Hinkley, R. N. Edelstein, and P. I. Ivonnet, *Dev. Growth Differ.* **29,** 211 (1987).
[16] J. D. Conover, and R. B. L. Gwatkin, *J. Reprod. Fertil.* **82,** 681 (1988).

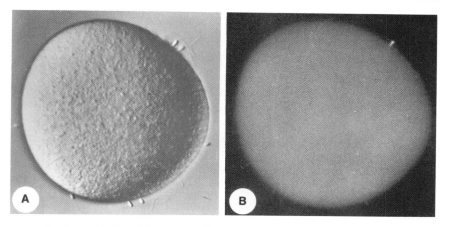

FIG. 2. Nomarski (A) and fluorescence (B) micrographs of a Hoechst 33342-pretreated egg fertilized and then fixed at 10 sec postinsemination. Several sperm have bound to the egg surface, but only one is fluorescent. (Reproduced with permission from Hinkley *et al.*[14])

(10 µg/ml), dissolved in the solution in which the gametes are suspended. At the conclusion of the staining period, the eggs are washed to remove all unincorporated fluorochrome and then mixed with sperm. Aliquots of the interacting gametes are removed and fixed in 5% (w/v) paraformaldehyde, or 1% (v/v) glutaraldehyde, in an appropriate buffer (e.g., for marine specimens, seawater; for mammals, 100 mM phosphate buffer, pH 7.3). Additional samples may be removed and fixed at appropriate intervals throughout fertilization. In the case of sea urchins, in which fertilization events occur rapidly,[17] samples need to be taken and fixed at closely spaced intervals (seconds) in order to determine the time of sperm–egg fusion.

The method of Hinkley *et al.*[14,15] is a relatively simple, rapid technique, requiring fluorescence optics to detect sperm–egg fusion at early time periods after insemination. Video recording of living Hoechst 33342-pretreated eggs demonstrates that fluorochrome transfer to the fertilizing sperm is relatively slow, requiring 45 to 90 sec. The marked delay in dye transfer from the preloaded egg to the fused sperm is a major disadvantage of the dye transfer method and indicates that Hoechst 33342 may be of limited usefulness as an indicator of the exact time of sperm–egg fusion in living specimens. However, this problem can be circumvented with fixed specimens. The use of fixed specimens, however, may induce other changes that can interfere with an accurate assessment of the timing of

[17] F. J. Longo, J. W. Lynn, D. H. McCulloh, and E. L. Chambers, *Dev. Biol.* **118**, 155 (1986).

sperm–egg fusion. Such potential problems are discussed in the following section concerning the detection of gamete fusion at the ultrastructural level of observation.

The technique of preloading of eggs with Hoechst dye for the detection of sperm–egg fusion has been used successfully for various marine invertebrates including sea urchins, mollusks, and starfish.[9,14,15] Although this method allowed detection of sperm–egg fusion in the mouse, it was not satisfactory in hamster.[16] Hoechst dye diffused out of preloaded hamster eggs with time and stained sperm nuclei outside of the egg.[16] If, however, hamster eggs preloaded with Hoechst dye are rinsed thoroughly, inseminated, and then fixed, followed by rinsing, only fused sperm show fluorescence.[18]

It is important to recognize that with this method the correlation of dye transfer and sperm–egg fusion is inferred. That the gametes are actually fused at the time of dye transfer is assumed and has not been established unequivocally. That dye transfer may take place as a result of an association of the sperm and egg plasma membranes not involving their fusion has not been eliminated. Such an association may be akin to gamete continuity or functional contact.[1,19]

Electron Microscopy

The ultrastructural investigations by Colwin and Colwin[2] and Bedford[20] defined the manner in which the sperm was incorporated into the egg, through fusion of the sperm and egg membranes (Figs. 3 and 4).[21,21a] Since then, sperm–egg fusion has been observed in many different species, and in each case appears to involve similar mechanisms (see Yanagimachi[21]). The method of analysis used to determine sperm–egg fusion at the electron microscopic level of observation is relatively straightforward; the gametes are mixed and at different time intervals, depending on how quickly they fuse, samples are fixed and prepared for ultrastructural examination. The method of fixation may be quite variable, depending on the species in question. (The original literature for specific formulas and procedures should be examined.) Once fixed, the specimens may be embedded, sectioned and examined in a transmission electron microscope

[18] J. Stewart-Savage and B. Bavister, *Dev. Biol.* **128**, 150 (1988).
[19] F. J. Longo, S. Cook, D. H. McCulloh, P. I. Ivonnet, and E. L. Chambers, *in* "Mechanisms of Fertilization: Plants to Humans" (B. Dale, ed.), p. 203. Springer-Verlag, New York, 1990.
[20] J. M. Bedford, *Am. J. Anat.* **133**, 213 (1972).
[21] R. Yanagimachi, *in* "Membrane Fusion in Fertilization, Cellular Transport and Viral Infection" (N. Düzgüneş and F. Bronner, eds.), p. 3. Academic Press, San Diego, 1988.
[21a] J. M. Bedford and G. W. Cooper, *Cell Surf. Rev.* **5**, 65 (1978).

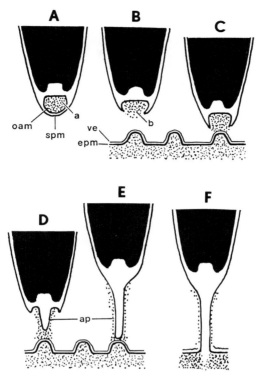

FIG. 3. Successive stages of the acrosome reaction (A–C) and sperm–egg fusion (D–F) in the sea urchin. a, Acrosome; ap, acrosomal process; b, bindin; epm, egg plasma membrane; oam, outer acrosomal membrane; spm, sperm plasma membrane; ve, vitelline envelope. (Reproduced with permission from Yanagimachi.[21])

(TEM), or critical point dried and examined with a scanning electron microscope (SEM) to detect sperm–egg fusion. The latter technique is relatively rapid in comparison to TEM; however, detection of early stages of gamete fusion are often difficult to determine with SEM.

Finding fusing or fused gametes in sections examined with TEM can be tedious, because the areas involved in fusion are small, relative to the size of the gametes. Considerable time can be spent in searching for the site of sperm–egg interaction. Such difficulties can be ameliorated by examining (1) polyspermic preparations and/or (2) embedded specimens, prior to sectioning, to identify interacting sperm on the egg surface. The embedded specimen may then be oriented and trimmed so that the area containing interacting gametes is reduced for sectioning. As an aid in locating the site of gamete fusion, it is important to recognize that the process is polarized,

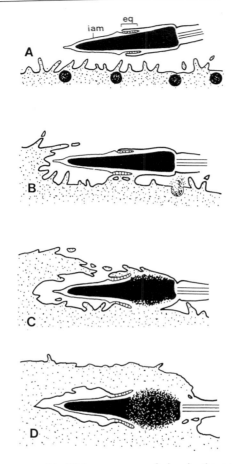

FIG. 4. Successive stages (A–D) in sperm–egg fusion in the eutherian mammal. The sperm depicted in (A) has undergone the acrosomal reaction. iam, Inner acrosomal membrane; eq, equatorial segment of the acrosome. (Redrawn with permission from Bedford and Cooper.[21a])

that is, in many nonmammalian species gamete fusion is mediated where the sperm apex (acrosomal process) comes in contact with the egg surface (Fig. 3). In mammals it is the plasma membrane over the equatorial segment of the acrosome that first fuses with the egg plasma membrane (Fig. 4). The sperm plasma membrane of this specific region of mammalian spermatozoa is fusogenic only after the acrosomal reaction.[21]

Electron microscopy provides sufficient resolution to determine unequivocally whether the sperm and egg have fused. It does not, however, provide sufficient temporal resolution to indicate exactly when fusion occurred. To this end, experiments correlating ultrastructural and electrophysiological studies of sperm–egg interactions have established when sperm–egg fusion takes place.[17] Using the sperm-induced conductance increase as time zero, the chronology and sequence of events leading to gamete fusion and egg activation have been investigated in the sea urchin, *L. variegatus*.[17] Eggs with only one attached sperm are observed and the change in membrane conductance, recorded with a microelectrode inserted into the egg, serves as a well-defined zero time. From 1 to 5 sec after the onset of electrical activity the tip of the sperm acrosomal process is in contact with the vitelline layer, separated by a distance of 50 to 150 Å from the egg plasma membrane. Continuity between the sperm and egg plasma membranes occurs approximately 5 sec after the onset of electrical activity, and results in the adjoining of the sperm and egg cytoplasms. The fusion of the sperm and egg plasma membranes observed at the ultrastructural level occurs, as described by Palade and Bruns[22] for vesicular fusion with the plasma membrane in endothelial cells. The sperm and egg plasma membranes contact one another and the outer leaflets of the two membranes then establish continuity with one another. This is followed by the "fusion" of the inner leaflets and establishment of cytoplasmic continuity of the two cells. These observations are consistent with molecular mechanisms of membrane fusion in other systems, based on ultrastructural, optical, capacitance, and conductance measurements.[23–27]

The actual mechanism of membrane fusion and how ultrastructural preparative techniques affect this process are unknown. Studies examining specimens at the time of sperm–egg fusion need to consider factors that may have a significant bearing on the acquisition of data and their interpretation. For example, almost all preparative techniques for the detection of sperm–egg fusion at the ultrastructural level of investigation rely on aldehyde-fixed specimens. No known studies have been published examining how rapidly aldehyde fixatives stop membrane fusion when added to isolated cells. In addition, it is unclear how well intermediate stages of membrane fusion are preserved during fixation. Miniature end-plate potentials recorded intracellularly at single neuromuscular junctions, and

[22] G. E. Palade and B. R. Bruns, *J. Cell Biol.* **37**, 633 (1968).

[23] L. V. Chernomordik, G. B. Melikyan, and Y. A. Chizmadzhev, *Biochim. Biophys. Acta* **906**, 309 (1987).

[24] A. J. Verkleij, J. Leunissen-Bijvelt, B. de Kruijtt, M. Hope, and P. R. Cullis, in "Cell Fusion" (D. Evered and J. Whelan, eds.), p. 45. Pitman, London, 1984.

[25] D. Shotton, *Nature (London)* **272**, 16 (1978).

[26] E. Neher, *Biochim. Biophys. Acta* **373**, 377 (1974).

[27] C. A. Helm, J. N. Israelachvili, and P. M. McGuiggan, *Science* **246**, 919 (1989).

resulting from exocytotic fusion of synaptic vesicle membranes with the plasma membrane, continue to be recorded for up to 1 min following the addition of glutaraldehyde to the intact muscle.[28-31] This suggests that synaptic vesicle–plasma membrane fusion is not immediately halted by the addition of fixative. No information is available, however, on how rapidly the external medium is replaced by fixative, nor is it possible to assess the role of diffusion barriers in the intact muscle tissue. In addition, conventional fixation techniques have serious shortcomings when applied to lipids, because they may lead to membrane artifacts such as lipid extraction and redistribution, particularly in fusing membrane systems.[32-38] Low-temperature techniques, including the combination of freeze substitution and low-temperature embedding, as well as freeze-fracture replication, are potentially capable of overcoming these limitations.[39-41] However, such protocols are currently difficult to apply to studies of early fertilization events.

In both the ultrastructural observations of electrically recorded eggs[17] and the dye transfer studies of Hinkley et al.,[14,15] it is presumed that the fusing sperm and egg plasma membranes are sufficiently stable during fixation and subsequent processing that their specific morphological relationship to one another is retained. To what degree this is true is not known. Studies have shown that the membrane bilayer can remain in a labile state during fixation with glutaraldehyde,[36,42] because these structures are insufficiently stabilized against secondary changes during subsequent processing of the cells.[10] The possibility exists that, for up to 5 sec, membrane changes associated with gamete fusion are transitional, unstable

[28] J. I. Hubbard and M. B. Laskowski, *Life Sci.* **11**, 781 (1972).
[29] A. W. Clark, *J. Cell Biol.* **69**, 521 (1976).
[30] J. Heuser, in "Motor Innervation of Muscle" (S. Thesleff, ed.), p. 51. Academic Press, New York, 1976.
[31] J. E. Smith and T. S. Reese, *J. Exp. Biol.* **89**, 19 (1980).
[32] J. A. Petersen and H. Rubin, *Exp. Cell Res.* **60**, 383 (1970).
[33] R. E. Scott, *Science* **194**, 743 (1976).
[34] D. L. Hasty and E. D. Hay, *J. Cell Biol.* **78**, 756 (1978).
[35] G. Poste, C. W. Porter, and D. Paphadjopoulos, *Biochim. Biophys. Acta* **510**, 256 (1978).
[36] D. E. Chandler and J. Heuser, *J. Cell Biol.* **69**, 521 (1979).
[37] A. M. Dvorak, H. F. Dvorak, S. P. Peters, E. S. Shulman, D. W. MacGlashan, K. Pyne, V. S. Harvey, S. J. Galli, and L. M. Lichtenstein, *J. Immunol.* **131**, 2965 (1983).
[38] E. J. Neufeld, P. W. Majerus, C. M. Krueger, and J. E. Saffitz, *J. Cell Biol.* **101**, 573 (1984).
[39] C. Weibull, W. Villiger, and E. Carlemalm, *J. Microsc.* **135**, 213 (1984).
[40] A. J. Verkleij, B. Humbel, D. Studer, and M. Muller, *Biochim. Biophys. Acta* **812**, 591 (1985).
[41] G. Knoll, K. N. J. Burger, R. Bron, G. van Meer, and A. Verkleij, *J. Cell Biol.* **107**, 2511 (1988).
[42] E. M. Eddy and B. M. Shapiro, *J. Cell Biol.* **71**, 35 (1976).

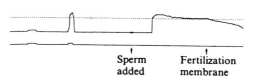

↑   ↑
Sperm   Fertilization
added   membrane

FIG. 5. Activation potential of a monospermic sea urchin egg. The top trace is voltage against time; the bottom trace is current against time. The dotted line indicates 0 mV. The sudden change in egg plasma membrane potential occurs at about the time of sperm–egg membrane fusion. (Reproduced with permission from Jaffe.[44a])

or tenuous, and are not maintained or are reversed when fixative is added. An increase in capacitance, indicative of the establishment of gamete continuity, followed by a loss of capacitance,[43,44] suggest that reversal of gamete continuity can occur in eggs clamped at negative membrane potentials, even in the absence of fixatives. Presently, however, there is no evidence that fusion of the sperm and egg plasma membranes is reversed after the onset of electrical activity.

Although the actual mechanism of membrane fusion has not been determined, the technique of electron microscopy is capable of determining when and where such a process has occurred. This method is capable of providing unequivocal proof in establishing that the sperm and egg plasma membranes and cytoplasms are confluent. At present it is one of the best and most reliable methods available. However, the techniques involved in detecting sperm–egg fusion at the ultrastructural level can be relatively time consuming and tedious. The method requires access to sufficient material for examination, appropriate fixation of specimens, and relatively expensive instrumentation for analysis.

Electrical (Conductance/Capacitance) and Other Methods

It has been claimed that sperm–egg fusion occurs at the same time as the conductance and/or capacitance changes measured in eggs inserted with microelectrodes, and then inseminated (Fig. 5).[44a] (See Refs. 45–48 for procedures detailing the handling of eggs for electrical recording at

[43] D. H. McCulloh and E. L. Chambers, J. Gen. Physiol. **88**, 38 (1986).

[44] D. H. McCulloh and E. L. Chambers, J. Cell Biol. **103**, 236 (1986).

[44a] L. A. Jaffe, Nature (London) **261**, 68 (1976).

[45] E. L. Chambers, in " Mechanisms of Egg Activation" (R. Nuccitelli, G. N. Cherr, and W. H. Clark, Jr., eds.), p. 1. Plenum, New York, 1989.

[46] J. W. Lynn, in "Mechanisms of Egg Activation" (R. Nuccitelli, G. N. Cherr, and W. H. Clark, Jr., eds.), p. 43. Plenum, New York, 1989.

[47] S. Miyazaki, in "Mechanisms in Egg Activation" (R. Nuccitelli, G. N. Cherr, and W. H. Clark, Jr., eds.), p. 231, Plenum, New York, 1989.

[48] R. Kado, this volume [22].

fertilization.) Verification that sperm–egg fusion has occurred at the onset of the conductance or capacitance change of the fertilizing egg is not available and it is possible that the initiation of electrical changes in the egg is brought about by a situation in which the gametes are electrically coupled and share diffusible components, but their plasma membranes have not fused.[19] Correlative electrophysiological and ultrastructural studies with sea urchin gametes have demonstrated that sperm–egg membrane fusion follows by approximately 5 sec the onset of the change in electrical activity of the egg.[17]

A number of studies have associated sperm–egg fusion with the sudden cessation of flagellar activity and the abrupt immobilization of the fertilizing sperm.[49-52] However, investigations by Lynn and Chambers,[53] Hulser and Schatten,[54] Hinkley et al.,[14] and Longo et al.[17] demonstrate that in the sea urchin, sperm–egg fusion occurs after the conductance change, and sperm immobilization and gamete fusion do not occur simultaneously. Sperm–egg fusion precedes the cessation of sperm mobility by approximately 6 sec.

Associated with gamete fusion is the activation of the egg. Changes, such as the cortical granule reaction, are initiated as a consequence of this dynamic membrane interaction and are often cited as evidence that sperm–egg fusion has occurred. This assumption is usually valid; however, there are instances (e.g., in cytochalasin B-treated sea urchin eggs) in which egg activation takes place, but sperm entry fails to occur.[55] At what level gamete interaction leading to the egg activation and sperm entry is affected by this drug has not been determined unequivocally. It is possible that in cytochalasin-treated eggs either sperm–egg fusion is blocked, or fertilization cone formation and sperm entry, which normally follow sperm–egg fusion, are arrested. Determination of which of these processes is affected by cytochalasin is important to gain an understanding of how egg activation is initiated and of mechanisms involving sperm entry into the egg cytoplasm.

Acknowledgments

The investigations cited herein were supported by National Institutes of Health Grants HD-03402 (R.Y.) and HD-15510 and HD-22085 (F.J.L.).

[49] D. Epel, N. L. Cross, and N. Epel, *Dev. Growth Differ.* **19**, 15 (1977).
[50] R. Yanagimachi, *Curr. Top. Dev. Biol.* **12**, 83 (1978).
[51] G. Schatten, *Dev. Biol.* **86**, 426 (1981).
[52] M. J. Whitaker and R. A. Steinhardt, *Dev. Biol.* **95**, 244 (1983).
[53] J. W. Lynn and E. L. Chambers, *Dev. Biol.* **102**, 98 (1984).
[54] D. Hulser and G. Schatten, *Gamete Res.* **5**, 363 (1982).
[55] F. J. Longo, *Dev. Biol.* **64**, 249 (1978).

[21] Identification of Molecules Involved in Sperm–Egg Fusion

By W. J. LENNARZ and N. RUIZ-BRAVO

Introduction

Gamete fusion and egg activation have been the subjects of scientific investigation for decades. Only in the last 20 years have the molecular mechanisms involved in these processes been clarified. Recognition and binding of sperm and egg take place via a cell surface-binding molecule, the sperm receptor, on the surface of the egg and a complementary ligand on the surface of the sperm.[1-3] Using fertilization in the sea urchin *Strongylocentrotus purpuratus* as a model system, the molecular nature of the species recognition and binding of the sperm receptor and its ligand has been investigated.[2] Progress has been made in characterizing the sperm receptor in mammals, in particular the mouse.[1,3] This brief overview emphasizes the investigations carried out with sea urchins and, where appropriate, will allude briefly to the differences and similarities between the sea urchin and mouse sperm receptors. The molecular nature of the complementary ligand on the surface of the sea urchin sperm will also be discussed in the context of its interaction with the sperm receptor.

Fertilization can be thought of as a series of membrane fusion events (Fig. 1). The first membrane fusion event is the acrosome reaction, that is, the fusion of the sperm acrosomal membrane with the sperm plasma membrane. The acrosome reaction is triggered by the fucan sulfate polysaccharide component of the jelly coat, the gelatinous outer covering of the egg.[4] The acrosome reaction releases the contents of the acrosomal granule, and in sea urchins induces the formation of the acrosomal filament. The acrosomal filament is coated with bindin, a 30-kDa protein that is the ligand for the egg cell surface sperm receptor.[5,6]

Bindin species specifically binds to a sperm receptor, part of the vitelline layer of the egg. The involvement of bindin in sperm–egg fusion is well documented. Isolated bindin binds species specifically to dejellied

---

[1] P. Wasserman, *Science* **235**, 553 (1987).

[2] N. Ruiz-Bravo and W. J. Lennarz, *in* "The Molecular Biology of Fertilization" (H. Schatten and G. Schatten, eds.), p. 21. Academic Press, San Diego, 1989.

[3] M. B. Macek and B. D. Shur, *Gamete Res.* **20**, 93 (1988).

[4] G. K. SeGall and W. J. Lennarz, *Dev. Biol.* **86**, 87 (1981).

[5] G. W. Moy and V. D. Vacquier, *Curr. Top. Dev. Biol.* **13**, 31 (1979).

[6] V. D. Vacquier and G. W. Moy, *Proc. Natl. Acad. Sci. U.S.A.* **74**, 2456 (1977).

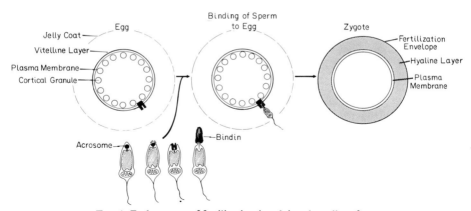

FIG. 1. Early events of fertilization involving the cell surface.

eggs, resulting in their agglutination.[7] Bindin binds to the isolated sperm receptor *in vitro,* although the binding in this case is not species specific.[8] In addition, bindin has been localized to the site of sperm and egg fusion.[5] Bindin has been extensively studied at the biochemical and molecular biological level and, like viral fusogenic proteins, aggregates lipid vesicles *in vitro,* facilitating their fusion.[9] Thus, the role of bindin in fertilization may be a dual one, binding and membrane fusion. Receptor-binding molecules from the sperm have been described in mouse (95 kDa,[10] 42 kDa,[11] 60 kDa,[12] 200/220 kDa,[11] trypsin inhibitor-sensitive site[13]), guinea pig,[14] boar,[15] and rabbit,[16] but none has been so thoroughly characterized as bindin from sea urchin sperm.

In addition to bindin, sperm possess protease activities that may play a role in membrane fusion. A metalloendoprotease activity has been found to be essential for the fusion of the sperm acrosomal membrane with the plasma membrane to release the contents of the acrosomal granule.[17] This activity is analogous to one described in myoblast fusion to form myotu-

[7] C. G. Glabe and W. J. Lennarz, *Nature (London)* **267,** 836 (1981).

[8] D. P. Rossignol, B. J. Earles, G. L. Decker, and W. J. Lennarz, *Dev. Biol.* **104,** 308 (1984).

[9] C. G. Glabe, *J. Cell Biol.* **100,** 800 (1985).

[10] L. Leyton and P. Saling, *Cell (Cambridge, Mass.)* **57,** 1123 (1989).

[11] L. Leyton, A. Robinson, and P. Saling, *Dev. Biol.* **132,** 174 (1989).

[12] B. D. Shur and C. A. Neely, *J. Biol. Chem.* **263,** 17706 (1988).

[13] D. A. Benare and B. T. Storey, *Biol. Reprod.* **39,** 235 (1988).

[14] P. Primakoff, H. Hyatt, and J. Tredick-Kline, *J. Cell Biol.* **104,** 141 (1987).

[15] R. N. Peterson and W. P. Hunt, *Gamete Res.* **23,** 103 (1989).

[16] M. G. O'Rand, *Biol. Reprod.* **25,** 611 (1981).

[17] H. A. Farach, D. I. Mundy, W. J. Strittmatter, and W. J. Lennarz, *J. Biol. Chem.* **262,** 5483 (1987).

bules.[18] A metalloendoprotease activity in the egg has been implicated in the second membrane fusion event of fertilization, the fusion of the sperm plasma membrane with that of the egg.[19] Given the well-documented fusogenic properties of bindin, it is possible that a metalloendoprotease activity is involved in generating fusogenic peptides from bindin, thus facilitating fusion of the sperm and egg plasma membranes. A metalloendoprotease activity has also been detected in a number of mammalian sperm, including human.[20]

In the sea urchin, sperm and egg recognition and binding of sperm via bindin to the sperm receptor results in egg activation, cortical granule exocytosis, and other intracellular events designed to prevent polyspermy and to initiate the process of embryogenesis. How the binding of sperm and egg actually bring about egg activation is still not understood, but these events seem to involve second messengers similar to those found in other systems. Numerous data support a model in which sperm–egg binding activates a guanine nucleotide-binding (G) protein, which in turn triggers the breakdown of phosphatidylinositol. The breakdown products, inositol 1,4,5-trisphosphate (IP_3) and diacylglycerol (DAG), are responsible for the observed increases in intracellular calcium and pH necessary to initiate additional activation events.[21-24] It is possible that occupation of the sperm receptor sites by bindin activates a G protein, which in turn triggers the subsequent events of egg activation. In support of this hypothesis, serotonin and acetylcholine receptors have been expressed in *Xenopus* eggs and, on addition of the appropriate receptor ligands,[25] three of the responses characteristic of fertilization have been observed: namely, egg membrane depolarization, cortical granule exocytosis, and endocytosis. Similar results have also been obtained when serotonin receptors were expressed in starfish eggs.[25a] These results support the idea that sperm–egg receptor interactions occur via a similar system. Indeed, the bindin-like ligand isolated from sperm of the marine worm *Urechis* has been shown to activate eggs *in vitro*.[26] However, the situation may be more complex because, although isolated bindin binds to sea urchin eggs, it does not

[18] C. B. Couch and W. J. Strittmatter, *Cell (Cambridge, Mass.)* **32**, 257 (1983).
[19] J. L. Roe, H. A. Farach, W. J. Strittmatter, and W. J. Lennarz, *J. Cell Biol.* **107**, 539 (1988).
[20] E. Diaz-Perez, P. Thomas, and S. Meizel, *J. Exp. Zool.* **248**, 213 (1988).
[21] P. R. Turner, M. P. Sheetz, and L. A. Jaffe, *Nature (London)* **310**, 414 (1984).
[22] M. Whitaker and R. F. Irvine, *Nature (London)* **312**, 636 (1984).
[23] P. A. Turner, L. A. Jaffe, and P. Primakoff, *Dev. Biol.* **120**, 577 (1987).
[24] D. Kline, *Dev. Biol.* **126**, 346 (1988).
[25] D. Kline, L. Simoncini, G. Mandel, R. A. Maue, R. T. Kado, and L. A. Jaffe, *Science* **241**, 464 (1988).
[25a] F. Schilling, G. Mandel, and L. A. Jaffe, *Cell Regul.* **1**, 465 (1990).
[26] M. Gould, L. Stephano, and Z. Holland, *Dev. Biol.* **117**, 306 (1986).

activate them.[7] This could be due to the loss of the activation function during bindin purification. Alternatively, bindin binding may be only a component of the egg activation trigger in sea urchins. In the mouse, the relationship between ZP3, the sperm receptor protein in the zona pellucida, and egg activation is likewise poorly understood. Although DAG and IP$_3$ induce zona pellucida modifications,[27,28] the mechanism whereby binding of ZP3 to its ligand initiates the release of these second messengers is unknown. Exploration of the molecular link between the sperm receptor and egg activation in sea urchins is currently being investigated.

Biochemical Nature of Sperm Receptor

Because of the species recognition and sperm-binding activities of the sperm receptor, it is important to determine the chemical composition and molecular structure of the receptor. Three partially pure preparations of biologically active sperm receptor have been obtained.[8,29-32] The first, the intact receptor, is derived from egg membranes and is able to inhibit fertilization in a species-specific manner in a competition bioassay.[8,31,32] The isolated receptor is a highly complex proteoglycan-like molecule of high molecular weight ($> 10^7$) that is soluble only in strong chaotropic agents. N-terminal sequencing of the intact receptor reveals three detectable N termini, indicating that it is not a gross mixture of polypeptides. The second receptor preparation, the pronase glycopeptide (PGP), is derived from the first by exhaustive proteolytic digestion with pronase.[8] This soluble carbohydrate-rich receptor fragment contains little to no detectable polypeptide and is also of high molecular weight ($> 10^6$). The carbohydrate of the PGP is glycosaminoglycan-like, and contains fucose, galactosamine, iduronic acid, and sulfate. Although the PGP, like the intact receptor, is able to inhibit fertilization in a competition bioassay, it is not able to do so in a species-specific manner.

The loss of species specificity, but not sperm-binding ability, by the carbohydrate-rich PGP suggests that the carbohydrate chains are responsible for binding and that the polypeptide core is responsible for conferring species specifity to the binding. A somewhat similar situation has been

[27] S. Kurasawa, R. M. Shultz, and G. S. Kopf, *Dev. Biol.* **133**, 295 (1989).
[28] Y. Endo, R. M. Schultz, and G. S. Kopf, *Dev. Biol.* **119**, 199 (1987).
[29] N. Ruiz-Bravo and W. J. Lennarz, *Dev. Biol.* **117**, 204 (1986).
[30] N. Ruiz-Bravo, D. E. Earles, and W. J. Lennarz, *Dev. Biol.* **118**, 202 (1986).
[31] N. Ruiz-Bravo, D. P. Rossignol, G. L. Decker, L. I. Rosenberg, and W. J. Lennarz, *in* "The Molecular and Cellular Biology of Fertilization" (J. L. Hedrick, ed.), p. 293. Plenum, New York, 1986.
[32] Ruiz-Bravo, J. Janak, and W. J. Lennarz, *Biol. Reprod.* **41**, 323 (1989).

found in the mouse: the carbohydrate chains of ZP3, the mouse sperm receptor, are responsible for binding sperm. In this case, however, instead of conferring species specificity to the binding, the polypeptide core induces the acrosome reaction.[33]

A third receptor preparation, derived by limited proteolytic digestion of the egg cell surface, supports the idea that binding ability in sea urchins is due to the carbohydrate, and that species specifity is conferred by the polypeptide.[29] Tryptic fragments of the sperm receptor are able to inhibit fertilization in a species-specific manner as long as they retain a minimum amount of polypeptide (30% protein by weight). If the remaining polypeptide is further digested by pronase, the trypsin-derived fragments are still able to inhibit fertilization, indicating that they are able to bind sperm, but they do not have species specificity.[29] How the polypeptide confers species specificity to binding is not known. It is possible that removal of the polypeptide induces a conformational change, which in turn abolishes species specificity. In support of this hypothesis, binding of bindin to sulfate fucans has been found to be sensitive to changes in the spatial orientation of the sulfate esters.[34] Because of its solubility and its demonstrated correlation between species specificity and the presence of the polypeptide core, this third preparation of receptor, consisting of a soluble, species-specific, proteolytic fragment, may prove extremely useful for further investigation of the molecular nature of sperm–egg binding.

Sperm Receptor Preparation

Preparation of Receptor from Cell Membranes

Materials:

Lysis buffer: 10 mM ethylene glycol-bis(β-aminoethyl ether)-N,N,N',N'-tetraacetic acid (EGTA), 37 mM glycine ethyl ester, 2 mM aminotriazole (ATAZ), 50 mM Tris-HCl, pH 8

Soybean trypsin inhibitor (SBTI)

Phenylmethylsulfonyl fluoride (PMSF)

Artificial sea water (ASW): (10.3 mM KCl, 422.6 mM NaCl, 10.0 mM CaCl$_2$, 48.7 mM MgCl$_2 \cdot$6H$_2$O, 26 mM NaSO$_4$, 2.4 mM NaHCO$_3$, pH 8, filter through a 0.45-μm pore size filter

HCl (0.1–6 N)

Sucrose (30%, w/w) in 0.5 M NaCl, 0.02% (w/v) sodium azide

Sucrose (78%, w/w) in 0.5 M NaCl, 0.02% (w/v) sodium azide

[33] J. D. Bleil and P. M. Wasserman, *Dev. Biol.* **95**, 317 (1983).
[34] P. L. DeAngelis and C. G. Glabe, *J. Biol. Chem.* **262**, 13946 (1987).

Column (2 in × 18 in) containing Sepharose CL-4B
Column buffer: 4 M guanidine hydrochloride, 10 mM
  dithiothreitol (DTT), 10 mM Tris-HCl, pH 8

Procedure

1. Collect eggs into ASW, allow them to settle, and make a 10% suspension in fresh ASW. Approximately 1 mg of highly purified receptor can be obtained from 800 ml of eggs.
2. Remove the jelly coat. Add HCl dropwise (0.1–6 N, to minimize dilution of egg suspension) until the pH reaches 5.0–5.5. Use the pH meter to monitor pH constantly, and stir the eggs gently with a plastic rod. After 2 min neutralize the egg suspension by adding 2 M Tris-HCl, pH 8.0. To remove the excess Tris, gently wash the eggs by allowing them to settle, and resuspend them in at least 10 vol of MFASW. Wash the eggs three times.
3. Add SBTI (10 μg/ml) and PMSF (100 μM) to lysis buffer.
4. Resuspend the dejellied eggs in a 10-fold volume of ice-cold lysis buffer with protease inhibitor and stir vigorously with a stir bar for 10–15 min on ice until the eggs have lysed.
5. Centrifuge at 37,000 g for 20 min at 4°. Discard the supernatant and gently resuspend the pellet containing egg "ghosts" (membranes) in ice-cold lysis buffer with protease inhibitors. Repeat the centrifugation and resuspension twice. The egg membrane pellet can be stored frozen at this stage.
6. Resuspend the egg "ghosts" in lysis buffer with protease inhibitors so that the final volume is approximately equal to 25–50% of the volume of eggs used (no more than 15 mg protein/ml).
7. Add Triton X-100 to a final concentration of 1.5% (w/v) and mix well to solubilize the membranes.
8. Layer the solubilized membranes onto a gradient composed of 10 ml of 78% sucrose and 21 ml of 30% sucrose, and 7–8 ml of detergent-solubilized membranes/gradient tube.
9. Centrifuge in an SW-28 rotor at 76,000 g (average) for 16 hr at 4°. The crude sperm receptor will band at the interface of the 30 and 78% sucrose.
10. Collect the crude receptor and resuspend it in lysis buffer with protease inhibitors. To remove the sucrose, centrifuge the receptor at 37,000 g for 20 min at 4° and resuspend it in lysis buffer with protease inhibitor. Repeat. The crude receptor can be frozen at this stage.
11. Solubilize the receptor in a minimum volume of 4 M guanidine hydrochloride buffer and load it onto a Sepharose CL-4B column.

The receptor will be in the excluded volume of the column (V_0). This preparation of sperm receptor has four detectable N termini and contains a small amount of the fucan sulfate polysaccharide from jelly coat. The receptor can be dialyzed against water, ASW, or buffer for testing of biological activity or for biochemical analysis.

12. Further purification of the sperm receptor (three detectable N termini, no fucan sulfate polysaccharide) can be obtained by density gradient centrifugation. Add solid CsCl so the Sepharose CL-4B fractions containing the receptor to a final density of 1.4 g/ml, add fresh dithiothreitol to 10 mM, and centrifuge at 150,000 g (average) for 48 hr at 20°. The receptor can be found at the top of the gradient.

Preparation of Sperm Receptor Fragments by Cell Surface Proteolysis

Materials

TBASW: ASW buffered with 50 mM Tris-HCl, pH 8
Trypsin
Soybean trypsin inhibitor
Aprotinin (Trasylol)
Column (1.25 in × 15 in) containing Sepharose CL-4B
Sepharose column buffer: 0.5 M NaCl, 0.02% (w/v) sodium azide, 10 mM Tris-HCl, pH 8
DEAE (DE-52) anion-exchange column (1.5 in × 6 in)
DEAE buffer: 0.02% (w/v) sodium azide, 10 mM Tris-HCl, pH 8

Procedure

1. Remove the jelly coats from the eggs as described above, wash them, and make a 20% suspension in TBASW containing trypsin at a final concentration of 100 μg/ml suspension, at room temperature. Stir gently for 5 min.
2. Stop the reaction by adding SBTI (3 mg/mg trypsin) and aprotinin (12 trypsin inhibitory units/100 ml of suspension).
3. Allow the eggs to settle on ice, and collect the supernatant containing the proteolytic digest.
4. Centrifuge the proteolytic digest at 10,000 g for 20 min at 4° to remove the remaining particulate material. Centrifuge the supernatant for 1 hr at 100,000 g at 4°.
5. Concentrate the 100,000 g supernatant by ultrafiltration on a YM 30 membrane. (Amicon, Danvers, MA).
6. Chromatograph the concentrated supernatant on a Sepharose

CL-4B column. The proteolytic fragments of the receptor will be in the excluded volume of the column.

7. Pool the fractions containing the receptor and dialyze them against 10 mM NaCl in DEAE buffer.

8. Chromatograph the receptor fragments on a DEAE-cellulose anion-exchange column. Wash the column with at least 10 column volumes of 10 mM NaCl in DEAE buffer, and elute the receptor fragments with a gradient consisting of equal volumes of 10 mM NaCl and 1 M NaCl in DEAE buffer.

9. Receptor fragments elute as two broad peaks between 130–300 and 400–610 mM NaCl.

Bioassays

Fertilization Assay

The fertilization assay is used to determine the ability of receptor preparations to compete for bindin exposed on the acrosomal filament of acrosome-reacted sperm, thus reducing the total number of eggs fertilized by a limiting concentration of sperm.[8] To ensure that any effect on fertilization is due to a competition for bindin, and not to an inhibition of the ability of the sperm to undergo the acrosome reaction, the sperm are exposed to the competitors at the same time that they are induced to undergo the acrosome reaction. The sperm are subsequently examined for the presence of an acrosomal filament by electron microscopy as described by Decker *et al.*[35] Variations of this assay can be used to determine the ability of antibodies or other inhibitors to affect fertilization. A detailed description of the fertilization assay is shown below.

Materials

ASW (see above), at 10–14° for *S. purpuratus,* and at 20° for *Arbacia punctulata*
CaCl$_2$ (1 M)
Jelly coat: Concentrated, crude in ASW
Test tubes: 12 × 75 mm glass or polypropylene
Hoechst dye 33342 (Calbiochem-Behring Corp., La Jolla, CA)

Gametes

Eggs

1. Collect eggs into ASW. Allow them to settle, and make a 10% suspension in fresh ASW.

[35] G. L. Decker, D. B. Joseph, and W. J. Lennarz, *Dev. Biol.* **53,** 115 (1976).

2. Remove the jelly coats as described above and wash the eggs three times. After the jelly coats have been removed from the eggs, it is essential to treat them gently and to work quickly, because they are more fragile than intact eggs.
3. Make a 1% suspension in ASW and keep at the appropriate temperature (10–14 or 20°) until use.

Sperm. Collect sperm undiluted (dry) and store on ice. These sperm will retain their ability to fertilize for at least 2 days.

Procedure for Inducing Acrosome Reaction

1. Make a stock suspension by diluting the sperm 1 : 1 in ice-cold ASW, and keep the suspension on ice. These sperm will retain their ability to fertilize for a minimum of 2 hr.
2. Dilute a small amount of the sperm stock suspension (5–10 μl) to approximately 1 : 1000. To achieve this dilution by the method that will be used when performing a fertilization assay (see below), use a series of two dilutions: the amount of ASW in the first tube is varied as necessary, and the second tube always contains 100 μl of more than enough jelly coat (6 mol fucose equivalents) to induce the acrosome reaction in all those sperm capable of responding (90–100% for *S. purpuratus* and 60–80% for *A. punctulata*). In the case of *A. punctulata,* it is necessary to supplement the jelly coat with Ca^{2+} to a final concentration of 36 mM. Allow the sperm to remain in the tube containing jelly coat for 15–20 sec.
3. Stop the reaction by adding an equal volume of 6% (v/v) glutaraldehyde in ASW, Tris-HCl, pH 8. The number of sperm that have undergone the acrosome reaction can be monitored by electron microscopy as described by Decker *et al.*[35]

Procedure for Fertilization Assay

1. Place 0.5 ml of the 1% egg suspension into a 12 × 75 mm tube (tube 1), 0.5 ml ASW in a second tube (tube 2), and 100 μl ASW with jelly coat and extra calcium, if needed, in a third tube (tube 3).
2. Make a 1 : 1 sperm suspension in ASW and dilute it so that only 70–90% of the eggs will be fertilized. Place 10 μl of the 1 : 1 sperm suspension into tube 2 and mix quickly and thoroughly, but gently.
3. Take 10°μl of the sperm from tube 2 and place it in tube 3 to induce the sperm to undergo the acrosome reaction. Mix quickly and thoroughly, but gently. After 15 sec, remove 10 μl of acrosome-reacted sperm and place it in tube 1 to fertilize the eggs. Mix gently. Incubate for 5 min at the appropriate temperature, mixing gently as needed to keep the eggs from settling. Because sperm start to die quickly after being diluted and undergoing the acrosome reaction

quickly after being diluted and undergoing the acrosome reaction (about 30 sec), individual tubes 2 and 3 will be necessary for each tube of eggs to be fertilized. In addition, each tube containing eggs should receive the acrosome-reacted sperm at exactly the same number of seconds (usually 20) after the sperm are introduced into tube 3.

4. Stop the reaction by adding an equal volume of ice-cold 2% (v/v) glutaraldehyde in ASW, pH 8. After fixing for 30 min on ice, wash out the glutaraldehyde by resuspending the gametes in fresh ASW.

5. Successful fertilization is scored by the presence of fertilization envelopes. Each point should be done in duplicate tubes and a minimum of 100–200 eggs should be counted for each tube.

6. If fertilization does not fall within the 70–90% rate, adjust the volume of ASW in tube 2 and repeat the experiment. This volume can vary from 0.1 to 5 ml, depending on the sperm sample. Once the correct dilution of sperm has been determined, potential competitors for bindin can be added to tubes containing jelly coat, taking care to keep the total volume constant ($100 \mu l$) so the final sperm dilution is unchanged.

Testing Anti-Receptor Antibodies in Fertilization Assay. This fertilization procedure is similar to the one outlined above, with the following modifications.[32]

1. After performing step 1 as outlined above, add antibody to the tube containing the 1% suspension of dejellied eggs (tube 1), and incubate them at the appropriate temperature for 30 min. Shake the eggs gently every few minutes to keep them well distributed in the tube.

2. To remove excess antibody, dilute the eggs in a 10-fold volume of ASW, allow them to settle, and remove the supernatant. Repeat this procedure twice.

3. Add ASW to the washed eggs to bring the volume to 0.5 ml.

4. Proceed with the fertilization assay, starting with step 2 above.

5. As a control, use preimmune antibodies at the same concentration as the anti-receptor antibodies. As an additional control, if using intact IgG, it is necessary to use calcium ionophore A23187 ($30–50 \mu M$) to activate antibody-treated eggs. This will help determine whether any observed inhibition of fertilization is simply due to the inability of the egg to elevate a fertilization envelope because of antibody cross-linking of the vitelline layer, or to the masking of receptor sites by the anti-receptor antibody.

Sperm–Egg Binding Assay. The sperm–egg binding assay is designed to test the ability of receptor preparations, antibodies, and so on to inhibit the ability of sperm to bind to the vitelline layer, a prerequisite for sperm – egg fusion.[36] The basic structure of this assay is as outlined above for the fertilization assay, with the following modifications.

1. Prepare tubes as outlined in step 1 for the fertilization assay.
2. Dilute the sperm so that 100% of the eggs fertilize, which requires that approximately 20 – 30 sperm be bound per egg at the end of step 4 below. For diluting the sperm and inducing the acrosome reaction, follow the same procedure as for the fertilization assay.
3. Add the acrosome-reacted sperm to the eggs. Stop the reaction 20 sec after the addition of sperm by adding 0.5 ml of 6% (v/v) glutaraldehyde in ASW, Tris-HCl (pH 8). Incubate on ice for 30 min.
4. Allow the eggs to settle and resuspend them in 100 – 200 vol of MFASW. Gently hand centrifuge the eggs and remove the supernatant to remove unbound sperm. Repeat the washing procedure.
5. Count the number of eggs that have at least one sperm bound to their point of largest diameter when examined microscopically at magnification of ×100 or ×200, and express the number as the percentage of the total number of eggs counted. Alternatively, count the number of sperm bound per egg. As before, a minimum of 100 – 200 eggs should be counted per tube and all determinations should be performed in duplicate.

Sperm – Egg Fusion Assay. The sperm – egg fusion assay is designed to determine the ability of various reagents to inhibit the actual fusion of the sperm and egg rather than their binding, or the ability of the egg to undergo cortical granule exocytosis and fertilization envelope elevation in response to fusion with sperm.[37]

1. Prepare tubes 2 and 3 as described in step 1 of the fertilization assay.
2. Incubate a 10% suspension of dejellied eggs for 30 – 60 min in a solution of Hoechst dye 33342 (10 μg/ml) in ASW, at the appropriate temperature.
3. Wash the eggs twice by repeated settling and resuspension in fresh ASW. Make a 1% suspension of eggs in ASW and aliquot 0.5 ml/tube.
4. Dilute the sperm so that 100% of the eggs fertilize. The procedure for diluting the sperm was outlined in steps 2 and 3 of the fertilization assay.
5. Add acrosome-reacted sperm to the eggs. After a 5-min incubation, fix the sample by adding 0.5 ml of 2% (v/v) glutaraldehyde in

[36] W. H. Kinsey, J. A. Rubin, and W. J. Lennarz, *Dev. Biol.* **74**, 245 (1980).
[37] R. E. Hinckley, B. D. Wright, and W. J. Lyn, *Dev. Biol.* **118**, 148 (1986).

ASW, 10 mM Tris-HCl, (pH 8.35). Wash the samples twice in 4 ml ASW, 10 mM Tris-HCl (pH 8.35) and view the eggs as soon as they have settled. Fluorescence is visualized using a UV-1A filter (Nikon).

Bindin Assay

The bindin assay is not a bioassay because it does not use live gametes, thus allowing the testing of the receptor activity of various preparations even when sea urchins are not in season. The assay takes advantage of the ability of bindin to quantitatively bind receptorpreparations *in vitro*.[8]

Materials

ASW
Polypropylene tubes (12 × 75 mm and 5.7 × 46 mm)
Sucrose (20%, w/v) in ASW
Triton X-100 (2.25%, w/v) in ASW
[125]I-Labeled sperm receptor, 2 × 10^7 cpm/ml in ASW[8]
Bindin, 1 mg/ml in ASW[6]

Procedure

1. In a 12 × 75 mm tube, incubate the ^{125}I-labeled sperm receptor (10 μl, 200,000 cpm) with 10 μg of bindin for 20 min at room temperature in a total volume of 50 μl. To test the ability of unlabeled receptor preparations to compete for bindin, preincubate (also for 20 min) the competitor and bindin in a total volume of 40 μl before adding the ^{125}I-labeled sperm receptor. Shake gently during the incubations (60–80 rpm).
2. Add 100 μl Triton X-100 and incubate, shaking an additional 5 min.
3. Layer 50 μl over each of two 100-μl sucrose cushions in 5.7 × 46 mm tubes.
4. Centrifuge the samples at 10,000 g for 1 min at room temperature and quickly freeze them on dry ice.
5. Cut away the bottom 3–5 mm of each tube containing the bindin-receptor pellet with a razor blade and count it in a γ counter.

Acknowledgments

We are indebted to Mrs. Lorraine Conroy for editorial assistance, and to Dr. Kathy Foltz for useful discussions and advice. Work from the authors' laboratory was supported by National Institutes of Health Grant HD 18590 to W.J.L.

[22] Membrane Area and Electrical Capacitance

By RAYMOND T. KADO

Introduction

Cell membrane elements such as channels, pumps, and exchangers may be directly studied through their electrophysiological effects on the membrane. Conversely, the equally important processes of exocytosis and endocytosis occur without producing clear electrical signals. Because the cell membrane is able to maintain an electrical potential difference, however, it exhibits the property of electrical capacitance, which is dependent on surface area. As early as 1776, long before the nature of electricity was understood, Cavendish wondered whether the thin membranes of the cells in the electric organ of *Torpedo,* the electric ray, might not function like the glass walls of Leiden jars, which were used in that time to store electricity.[1]

In the early twentieth century electrophysiology concerned itself mainly with biological capacitance, as this electrical property began to be understood. The early developments are well documented by K. S. Cole, whose book begins with a chapter on membrane capacitance.[2] In 1937, Cole reported measuring a two to three times greater capacitance in fertilized sea urchin egg suspensions than was obtained in nonfertilized suspensions. The same kind of results were also reported some years later,[3] but neither author attributed the capacitance change on fertilization to cortical granule exocytosis. These measurements were done on large volumes of eggs, so that the individual egg could not be observed.

Based on morphology, it is now well established that at fertilization, cortical granule exocytosis physically adds the vesicular membrane of the granule to the egg plasma membrane in many species. Capacitance measurements of single sea urchin eggs at fertilization show a doubling of the capacitance, indicating a near doubling of membrane area with the exocytosis of cortical granules. These results explain earlier observations and provide some insight into the rate of the process.[4] Capacitance measurements also show that the disappearance of microvilli from the *Xenopus laevis* oocytes during meiotic maturation results in up to a fivefold de-

---

[1] H. Cavendish, *Philos. Trans. R. Soc. London* **66,** 196 (1776).
[2] K. S. Cole, "Membranes Ions and Impulses." Univ. of California Press, Berkeley, 1968.
[3] T. T. Iida, *J. Fac. Sci., Tokyo Univ., Zool.* **5,** 141 (1942).
[4] L. A. Jaffe, S. Hagiwara, and R. T. Kado, *Dev. Biol.* **67,** 243 (1978).

crease in membrane capacitance.[5] With newer capacitance measuring techniques, it may be possible to observe the fusion of the sperm to the egg[6] as well as to observe the capacitance changes as secretory vesicles attach to the plasma membrane and open.[7] It would appear that the tools are now available to address explicitly the question of how exocytosis is controlled at the level of the single vesicle.[8]

To study exo- and endocytotic processes, it is necessary to have a parameter quantitatively related to its progress under different experimental conditions. Such information can be provided by a continuous measure of the membrane surface area as membrane is added or subtracted by the exo- or endocytotic process. This chapter describes electrical capacitance and how it can be used to measure membrane surface area.

Basic Electrical Properties

The electrical force is one of the fundamental forces we know in nature. We describe this force in units of volts (V) and the rate of charge movement produced by this force in a conductor, the current, in amperes (A). The movements of charges or their accumulation give rise to three properties: resistance, capacitance, and inductance. The electrical resistance, often described as equivalent to the property of mechanical friction, is defined by Ohm's law as

$$R = V/I \qquad (1)$$

where V (in volts) is across R and I (in amperes) is the current in R. The unit for resistance is the ohm (Ω); conductance (G) is the reciprocal of the resistance, $1/R$, with siemens (S*) as units.

All the electrical energy, the voltage and current, in a resistance is liberated instantly as heat; the resistance is a purely dissipative property. This is the property responsible for producing light in a light bulb by heating the filament. The resistance property is ubiquitous; it is present everywhere charges can be moved.

The two other properties, capacitance and inductance, are not dissipative and arise from the storage of electrical energy in the electrostatic and electromagnetic fields produced by the electrical current. The inductance property, arising from the magnetic field produced by current, is not likely to have an influence in biological systems, where currents tend to be too

[5] R. T. Kado, K. S. Marcher, and R. Ozon, *Dev. Biol.* **84**, 471 (1981).
[6] D. McCulloch and E. L. Chambers, *40th Annu. Meet. Soc. Gen. Physiol.*, p. 38a (1986).
[7] E. Neher and A. Marty, *Proc. Natl. Acad. Sci. U.S.A.* **79**, 6712 (1982).
[8] W. Almers, *Annu. Rev. Physiol.* **52**, 607 (1990).

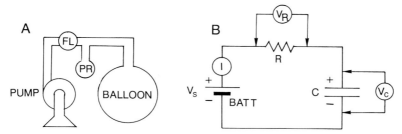

FIG. 1. The balloon analogy for electrical capacitance. (A) The pump takes air in through the funnel and moves it to the balloon. The rate of air flow (FL) is measured, as is PR, the pressure in the pipe leading to the balloon. (B) The battery (BATT) with voltage V_s moves charges in the circuit. The rate of charge movement is measured by the ammeter (I). The current passing through the resistance (R) produces an IR voltage (V_R). The charges move into C, where they will accumulate and produce a voltage V_c.

small to produce appreciable magnetic fields. Such fields can be measured, but only by using extremely sensitive instruments. Capacitance, on the other hand, does influence cellular electrical processes, especially in neurons.

Capacitance

The effect of capacitance in electrical circuits is closely analogous to that of the ordinary rubber balloon in a system of pipes. In Fig. 1, the following substitutions are made: electrical energy source (battery, V_s) for the pump, wires and a resistance (R) for the pipe, and a capacitance (C) for the balloon. Electrically, the upper and lower conductors of C are analogous to the inside and outside of the balloon. The battery "pumps" electrons from the upper conductor of C to the lower to create a difference in quantity of charge. The potential in the upper plate becomes positive and the lower becomes negative. The potential difference is the voltage at C (V_c), analogous to the difference in pressure from inside to outside the balloon. The capacitance property can be shown between any two conductors with different charges, as long as somewhere they share a common reference. This capacitance is known as "stray capacitance," and is even more omnipresent than resistance.

Electrical Definition of Capacitance

The capacity to hold charge in a capacitor is related to the quantity of its stored charge Q and its voltage V_c by the relation:

$$C = Q/V_c \qquad (2)$$

where Q is in coulombs and V_c is the voltage (in volts) between the two conductors. The physical unit for C is coulombs per volt and is named the farad (F_d). One farad is really a large capacitance: it stores 1 coulomb (10^{19} more electrons or monovalent ions on one side than the other) with only 1 V of electrical force. Most capacitors used in electronic circuits are in the micro farad (10^{-6}) range, and most cells are found to have from hundreds of nanofarads (10^{-9}) to picofarads (10^{-12}) of capacitance, depending on their size. The capacitance of vesicles is in the femtofarad (10^{-15}) range.

As shown in Fig. 1B, V_c produced across the capacitor by the accumulating charge has the same polarity as V_s. Because current in the capacitance, $I_{c(t)}$, is moved by the voltage difference between V_{sb} and V_c, as V_c increases the difference becomes smaller and $I_{c(t)}$ decreases. The process will have an exponential time course with a characteristic rate constant, just as for any process in which the quantity of product, V_c, limits the rate of production. The current in the capitance will decrease as

$$I_{c(t)} = I_{(t=0)} e^{-t/T} \qquad (3)$$

where $I_{(t=0)}$ is the current at the instant the source is connected to the uncharged capacitance and T is called the time constant, given by RC (R in ohms and C in farads). The time constant is in seconds when the physical units for ohms and farads are multiplied.

It should be noted that an uncharged capacitance, having equal numbers of free charges on both sides, behaves electrically like a conductor that has equal numbers of free charges everywhere. Because C behaves like a piece of wire, the current $I_{c(t=0)}$ will be just V_s/R; all of V_s will be dropped at R in Fig. 2A; V_c will be zero and will increase with time as

$$(V_c)_t = V_{max}(1 - e^{-t/T}) \qquad (4)$$

where V_{max} is the maximum voltage to which C can be charged by V_s, the voltage of the source. If R is made smaller V_c can be brought more quickly to V_s, because T becomes shorter. The initial current will be larger but will decay faster and, interestingly, the total charge Q will stay the same, because the same change in V_c was produced in the same C. These properties are illustrated in Fig. 2.

Dielectric

Any insulator separating the conductors of a capacitance is known as the *dielectric*. The dielectric is characterized by an empirically determined constant called the dielectric constant (ϵ). Some standard ϵ values are 1.0 for a vacuum, about 80 for water, and about 2 to 3 for lipids. In the case of cell membranes, a satisfactory value has not been found for ϵ, in spite of

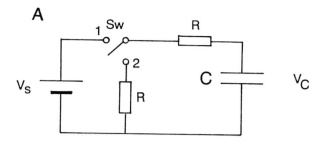

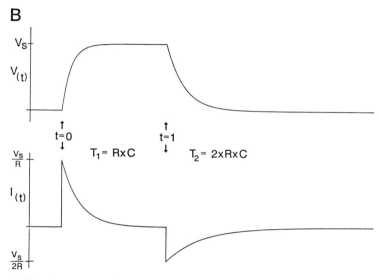

FIG. 2. The time course of V_c is exponential on charging and discharging and the time constant depends on resistance. (A) When switch Sw is at position 1, V_s will charge C through a resistance R. Once fully charged, C can be discharged by moving Sw to position 2, completing the path across C through a second R. C discharges through a resistance of $2R$. (B) The time courses of V_c [$V_{(t)}$] are exponential in both directions but have different T values, as seen in the upper trace. The current in C [$I_{(t=0)}$], Sw at 1, has a peak magnitude of V_s/R and decreases with time constant T_1. On discharge, Sw at 2, peak $I_{(t=0)}$ is $V_s/2R$ or one-half of the charging $I_{(t=0)}$, but decreases with a time constant twice as long, T_2. The total charge moved is the same on charging and discharging as can be found by integrating the two currents.

the large number of capacitance measurements that have been made. This is primarily due to the lack of precise knowledge of the physical dimensions of the membrane. However, a "rule of thumb" constant has been widely adopted: the *specific membrane capacitance* should be about $1 \mu F/cm^2$. The small capacitances measured for small cells and vesicles also give specific membrane capacitances near 1 $\mu F/cm^2$.[7] Larger values are obtained, but they are usually attributed to membrane extensions that are not accounted for in the surface area measurement.[5]

In some cells membrane capacitance is found to depend on the membrane voltage. This unexpected dielectric property of cell membranes is due to the movement of charges within the membrane, and is the subject of two other kinds of studies based on capacitance: one concerns the charges displaced as charged molecules are moved in the membrane by voltage[9] and the other is the possibility that mobile charges in the channel proteins themselves may be displaced by voltage, for example, to change the state of channel molecules.[10] Both kinds of charge movements are dependent on membrane voltage, that is, they are nonlinear, usually increasing with voltage up to some maximum. The charge movements produced by membrane voltage for charging C_m, on the other hand, stay proportional to voltage (linear). This difference allows separation of the two kinds of membrane charge movements as, for example, in muscle,[11] and could become a method for studies on the behavior of molecules in other kinds of membranes.

Membrane Resistance and Capacitance

The many ionic pathways of the membrane are protein molecules that traverse the membrane, and the proteins are thought to be held in a lipid matrix. The membrane can therefore be taken to be a sort of mosaic of insulating and conducting regions. These regions can be modeled as parallel unitary capacitances and resistances, because the ends of the unitary elements are all uniformly connected by the inside and outside media, as illustrated in Fig. 3A. This mosaic model is valid assuming that the same voltage difference is found across all parts of both surfaces at all times. The many capacitances and resistances of the mosaic can therefore be treated as a single capacitance value C_m in parallel with a single resistance value R_m, as shown in Fig. 3B.

Measurement Methods

The basic rules for measuring voltages and currents in a circuit containing electrical elements are (1) voltages must be measured from one end of the elements to the other; and (2) currents must be measured in the path

[9] K. Asami, Y. Takahashi, and S. Takashima, *Biophys. J.* **58,** 143 (1990).
[10] S. Duane and C. L. H. Huang, *Proc. R. Soc. London, Ser. B* **215,** 75 (1982).
[11] C. L.-H. Huang and L. D. Peachey, *J. Gen. Physiol.* **93,** 565 (1988).

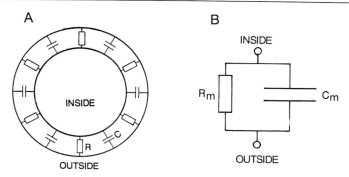

FIG. 3. The cell membrane as a spherically distributed electrical circuit. (A) Cell membrane in cross-section, represented by resistances (R) and capacitances (C) connected in parallel by the inside and outside media. (B) The electrical equivalent is a simple parallel circuit with one resistance for all the Rs and one capacitance for all the Cs of the membrane. The Cs simply add their capacities, but the Rs add in parallel, an operation most conveniently done using the conductances, as

$$G_m = G_1 + G_2 + \cdots + G_n$$

Because no known cell is really a pure sphere, and neither medium is perfectly conducting, this representation is idealized. However, it is a good starting point.

containing the element (in series). Making such measurements in cells poses several problems; one side of the membrane faces the inside of the cell, neither side is facing metallic electrical conductors, and cells are small. Electrical access to the inside face of the membrane can be obtained with fine, membrane-penetrating glass microelectrodes (Fig. 4A) or with a "patch" electrode isolating a ruptured patch of membrane (Fig. 4B). The methods and details for their application are described in Purves[12] and Standen et al.[13] for the microelectrodes and in Sakmann et al.[14,15] for the patch techniques.

With both techniques, the electrodes are filled with a cytoplasm-compatible electrolyte. Potassium chloride (3 M) is used for microelectrodes in which the fine opening at the tip allows only the small leakage of KCl that can be tolerated by most cells. More complex solutions are required for filling the patch pipette especially when using the whole-cell patch configu-

[12] R. D. Purves, "Microelectrode Methods for Intracellular Recording and Ionophoresis." Academic Press, London, 1981.
[13] N. B. Standen, P.T.A. Gray, and M. J. Whitaker, eds., "Microelectrode Techniques." Company of Biologists Ltd., Cambridge, 1987.
[14] O. P. Hamil, A. Marty, E. Neher, B. Sakmann, and F. J. Sigworth, *Pflügers Arch.* **391,** 85 (1981).
[15] B. Sakmann and E. Neher, eds., "Single Channel Recording." Plenum, New York, 1983.

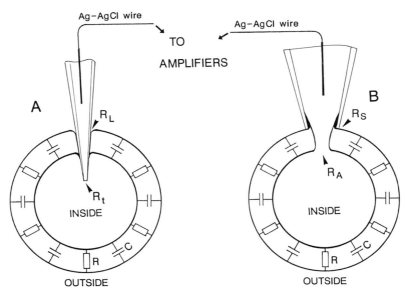

FIG. 4. Accessing the inside of the living, functional cell. (A) A glass micropipette, open at the tip, penetrates the membrane to contact the cytoplasm through the tip resistance R_t. At the penetration site, a perfect seal with the membrane is not obtained and a leakage pathway is formed (R_L). R_t will be in series with the membrane and R_L will be in parallel (see text). (B) An almost perfect seal of the membrane to the pipette is formed if a clean portion of membrane is sucked into the clean tip of the pipette. The leakage path between membrane and glass (R_s) is small. Once the tip is sealed to the membrane, the enclosed patch is ruptured, making an opening (resistance R_A) to establish contact with the cytoplasm. In this case, R_A is in series and R_s is in parallel with the membrane.

ration in which rapid (within minutes) exchange between cytoplasm and pipette contents can take place through the rupture in the membrane. Both kinds of electrodes have resistances in the megaohm (MΩ) range at the fine tips. The tip resistance in microelectrodes (R_t) is about 50 to 80 MΩ.

Both kinds of electrodes also contain an AgCl-coated silver wire that serves as a nonpolarizing transition from ionic conduction in the electrolyte to electron conduction in the wire. The AgCl coating on the silver wire allows transfer of electrical energy in the form of electrons in the wire to movement of Cl$^-$ in the electrolyte and vice versa without significant change of potential at the interface. The wire then leads to the measuring instrument. An electrical connection to the outside of the membrane is made by another Ag–AgCl wire, either placed directly in the bathing medium, or indirectly through an agar-filled tube to prevent contact of the wire with the medium, which might be changed during the experiment.

This external connection is also referred to as the "ground," because it is connected to the ground side of the apparatus. Currents are measured in series with the injecting electrode by using the electronics, which provide an output voltage proportional to the current; for example, 1 mV at the monitor output equals 1 nA injected.

The membrane resting potential (V_{RP}) is measured as soon as the microelectrode penetrates the membrane, or the patch is ruptured and contact is made with the inside of the membrane. Although V_{RP} itself is of interest, more interesting observations can be made once everything is in place to make such measurements. Injecting a current, which must leave the cell through the membrane, as the membrane current (I_m), is a basic working philosophy for electrophysiology. The practice can become complex because the membrane is not entirely passive.[16]

Input Resistance

The measured resistance of a cell is not that of the membrane (R_m) alone. In the case of cells impaled with microelectrodes, the membrane does not seal perfectly onto the exterior of the glass and the resulting spaces act as a leak path for I_m, giving rise to a leak resistance (R_L) or conductance (G_L) (see Fig. 4A). The measured resistance is the parallel combination of R_L and R_m and is called the input resistance (R_i) or conductance (G_i). Poor impalements, pipettes that are too large, or relative movements between microelectrode and cell will result in low R_L and low R_i. With the patch technique, leaking is greatly curtailed because the membrane can be almost perfectly sealed to the interior of the pipette (Fig. 4B). The resistance of the seal (R_s) is on the order of gigaohms (10^9 Ω) before rupturing the patch. The access resistance R_A is usually small relative to R_m, and the measured R_i of the cell will be nearly equal to R_m. To obtain gigaohm seals, the membrane must be free of debris and the interior of the pipette must be clean.

Morphological Considerations

Eggs, unlike other cells, are nearly perfect spheres and V_m can be expected to be the same everywhere at any given time. This ideal condition cannot be assumed in cells such as oocytes, in which membrane tortuosities such as long microvilli or follicular cells, electrically coupled through gap junctions, may be present, or in cells such as neurons, which have

[16] T. G. Smith, H. Lecar, S. J. Redman, and P. W. Gage, eds., "Voltage and Patch Clamping with Microelectrodes." Williams & Wilkins, Baltimore, MD, 1985.

extensively ramified dendrities,[17] or muscle with their T-tubule system.[18] Such remote membranes are isolated to different degrees from the I_m injection and V_m measuring sites in the cell by internal resistances. The measured V_m responses to I_m will be a complex average of the various contributing parts.

To illustrate this complication, imagine invaginating the sphere shown in Fig. 3A so that part of the outside surface is essentially pinched off from the rest. The outside surface in the cul-de-sac will no longer be at the same potential as those parts still freely communicating with the bathing medium because the path for I_m in the invagination is longer and thinner.

Identifying Methods

Two categories of capacitance masurement methods can be identified: the direct current (DC) methods, which require interruption of other measurements to make the capacitance measurement, and the alternating current (AC) methods, which allow simultaneous recording of membrane potential and other parameters as well as membrane capacitance. Both methods can be applied in current or voltage clamp measurements by using microelectrodes or patch electrodes.

Direct Current or Time Constant Method

One way to avoid the problem of access to the membrane through high electrode resistances is to use the current clamp technique to reveal the time-dependent effects of C_m. The current clamp shown in Fig. 5A supplies an I_m that will not change magnitude regardless of the voltage or resistance of the membrane, within limits impared by the electronics. I_m changes produce V_m changes (ΔV_m) that can be measured on the recorded V_m. The polarity of ΔV_m will depend on the direction of I_m; positive for an *outward current* (positive charges moving outward across the membrane) and negative for an *inward current.*

Ideally, with both I_m and R_m remaining constant, V_m will change with a single exponential time course as shown in Fig. 5B. For I_m at $t = 0$, there will be no change in the membrane resting potential V_{RP}, because I_m is carried entirely by C_m. As C_m charges, the membrane potential exponentially changes from V_{RP}. After six to eight time constants, V_m will plateau at $V_{m(max)}$ at this time, C_m takes no more charge, and all of I_m now passes through R_i. Dividing $V_{m(max)}$ by I_m thus yields R_i.

[17] D. H. Edwards, Jr. and B. Mulloney, *J. Physiol. (London)* **348**, 89 (1984).
[18] M. F. Schneider and W. K. Chandler, *J. Gen. Physiol.* **67**, 125 (1976).

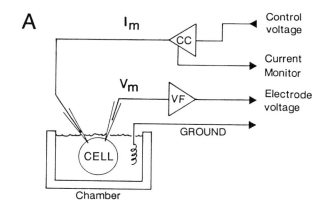

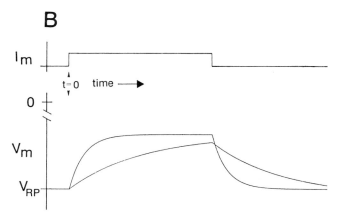

FIG. 5. Current clamping a cell. (A) The cell, in a chamber filled with bathing medium, is impaled with two microelectrodes, one for current (I_m) and another for recording the membrane potential (V_m). The outside of the cell and the bathing medium are electrically connected (GROUND) to the instrumentation. The intensity of I_m is regulated by the current clamp amplifier (CC) in proportion to the applied control voltage, and a monitoring signal proportional to I_m is made available at the current monitor output for recording. Membrane voltage V_m is recorded by a separate amplifier called a "voltage follower" (VF), because it can respond 100% to V_m in spite of the large microelectrode R_t. VF also has adjustments used to annul the intrinsic potential at the tip of the microelectrode, and some stray capacitance effects. (B) The upper trace is a square pulse of I_m injected by CC at time $t = 0$. The lower traces show the exponential V_m responses when the time constant of the cell is short, relative to the length of the I_m pulse, and when T is much longer than the pulse. The former V_m trace can be used for R_i and C_m determinations, but not the latter.

At time t, at which V_m is just $0.63\ V_{m(max)}$, t is just equal to the time constant T for the membrane and because $T = R_iC_m$,

$$C_m = t\ (\text{sec})/R_i\ (\text{ohms}) \tag{5}$$

where C_m will be in farads (F_d). It may appear strange that seconds divided by ohms will give farads, but it is because the farad is the name given to coulombs per volt and the ohm is volts per coulomb per second.

This method can provide rapid estimations of C_m for any cell already impaled by microelectrodes for potential measurements and current injection. With two independent microelectrodes, fairly large currents can be injected into eggs, resulting in a ΔV_m that is easier to read. However, I_m should not be so large that V_m activates voltage-dependent channels, resulting in voltage-dependent changes in R_m. The V_m-dependent R_m changes distort the rising as well as the falling phases of ΔV_m, the latter as the channels return to their normal conductance states at V_{RP}. For this reason, the method is difficult to apply in excitable cells, in which channel thresholds are near the resting potential.

One test for a ΔV_m that is too large is to apply the same I_m in the inward and outward directions; the resulting $\pm \Delta V_m$ should be identical in magnitude and time course for a purely passive membrane. Because most channels are voltage sensitive, they behave differently at different voltages. A further test is to plot the logarithm of ΔV_m against time. If the resulting plot is a straight line, the membrane has only one time constant and is behaving like the cell shown in Fig. 2A. If the plot deviates from a straight line, it may be that R_m did not remain constant or that some parts of the membrane were not equally accessible for I_m. If a smaller I_m gives a straight line, then R_m has been changed by ΔV_m. If the smaller $\pm I_m$ still produces the same deviation from a straight line, the membrane is not uniformly participating everywhere. This test is a good indicator of the presence of morphologically complex parts of the membrane.[17]

The current clamp method can also be applied by using a single impaling microelectrode (with a suitable VF) or a whole-cell patch electrode. The voltage drop at the tip of the electrode must be carefully eliminated with the bridge to avoid calculating $R_i + R_t$ as the R_i for the cell. This drop, equal to R_tI_m, must be subtracted from the measured voltage to obtain the true membrane potential (V_m). Because R_t has only the stray capacitance of the pipette associated with it, its T will be short. The voltage drop at R_t will be fast, almost the same as I_m, whereas V_m will have an exponential response. In the recorded V_m, the effect of R_t is seen as an abrupt change preceding the slower changes due to R_m and C_m. These jumps in V_m will be produced by any abrupt changes in I_m. Commercially available voltage followers for use with microelectrodes are equipped with a bridge circuit for this purpose. With proper care these methods can also provide good measurements of C_m.

Because of dilution in the tip of the microelectrodes, R_t arises in an osmotic and ionic gradient. I_m must be kept small enough that R_t does not change during current injection. Such changes cannot be compensated for with the bridge, because they will be a part of the ΔV_m response. This effect of I_m on R_t is accentuated as the tip is made smaller, or the resistance is made higher for any other reason. R_t changes with current can also be tested by $\pm I_m$ injection, because they are due to rectification [the same current usually produces a smaller V_{R_t} in the positive (outward) than in the negative (inward) direction] in the tip. The same precaution must also be observed with the whole-cell patch, in which the hole in the patch tends to change its apparent size, thus changing R_A.

Because reducing I_m reduces ΔV_m, how small a ΔV_m is acceptable? This will be determined by the amount of noise present in the V_m trace. Because of the high resistance of microelectrodes, their intrinsic peak-to-peak noise can be in the millivolt range. This noise is usually of high frequency, that is, much faster than the ongoing process, and can be reduced by low pass filtering (see Alternating Currents, below). Filtering must be done with care because too much filtering, while giving a clean trace, will also change the recorded time course of ΔV_m.

Voltage Clamp Transient Methods

Voltage clamp means that V_m is forced, by a system of amplifiers, to be equal to the experimentally desired value (V_d). V_d is composed of V_{RP} plus the desired voltage change. The clamping system (Fig. 6A) does this by automatically providing an I_m proportional to the difference ($V_d - V_m$). When $V_m = V_{RP}$, I_m will be zero, I_m, being the result of membrane channels, gives the status of the channels at that V_m. It can be seen in Fig. 6B that V_m changes only as fast as C_m can be charged, and that I_m corresponds to V_m/R_m only after the current has become steady. To study the channel kinetics at different V_m, it is necessary that the clamp bring V_m to V_d as rapidly as possible.

As was shown in Fig. 2B, the peak charging current transient is limited by series resistance and the voltage applied. In the case of the voltage-clamped cell, the series resistance is in the current electrode. Conveniently, the voltage clamping system can transiently and greatly increase the initial current by increasing its output voltage, because at the instant of V_d change, $V_d - V_m$ is large. The large I_m produced at the step change in V_d is called the *capacitive transient* current and is necessary to bring V_m to V_d within micro- or milliseconds. The advantage of this transient is that it is

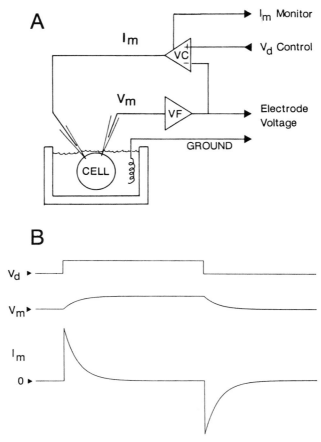

FIG. 6. Voltage clamping with two electrodes. (A) The clamping current I_m, supplied by the voltage clamp amplifier (VC), is determined by the amplified difference $V_d - V_m$; its intensity can be measured at the I_m monitor output. The control voltage (V_d) consists of a DC voltage at which V_m will be held (usually V_{RP}) and the desired ± step change. Membrane voltage is measured by the same kind of VF as in Fig. 5, but its output is also connected to the negative input of VC. (B) The upper trace is a sample V_d, and the second trace shows the real V_m response. The slow response of V_m to V_d is explained by the behavior of I_m seen in the third trace (recorded at low sensitivity to show the transients; I_m due to R_m is not resolved). When V_m is well clamped, the ± I_m transients will contain the same quantity of charge, regardless of how R_m and R_t might have changed during the pulse.

dependent only on C_m, not on R_i *and* C_m, as in the time constant method. The area under the capacitive current transient is the total quantity of charge Q needed to change V_m to its new value, and C_m can be obtained from Q/V_m.

Cortical granule exocytosis has been monitored in *Xenopus* eggs, using the capacitive transient method.[19] C_m is computed by digitizing V_m and the capacitative transients into a computer. However, voltage clamp of such large cells severely taxes the clamping system. The major limitation is imposed by G_L, due to the large current microelectrode needed to accommodate the large currents. The activation of membrane conductances as well makes extensive correction of the data necessary.[19] In view of the complications, this is not likely to be the method of choice for the two-electrode voltage clamp of large cells.

When used with the whole-cell patch in voltage clamp, it can yield good results if the cell is small and R_A is not so large as to introduce serious errors in the measurement of V_m. The capacitance of the electrode must be carefully compensated to zero before rupturing the patch.[20] The whole-cell patch will have the same difficulties as the two-microelectrode method with large cells, such as the *Xenopus* oocyte. Using a computer to collect and analyze the data is essential for capturing and integrating the current transients.

Ramp Voltage Clamp Method

This method is also independent of R_m and has been used by Moody and Bosma[21] to measure C_m in starfish oocytes and eggs. The current in a capacitance is dependent on both its value and the rate of change of voltage. From Eq. (1), the rate of change of voltage can be expressed as

$$dV_c/dt = C \, dQ/dt \qquad (6)$$

and because $dQ/dt = I_{(t)}$,

$$I_{(t)} = 1/C \, dV_{C_m}/dt \qquad (7)$$

Voltage clamping with V_d changing at a constant rate (dV_{C_m}/dt = constant) produces an $I_{(t)}$ that is constant throughout the duration of each ramp (see Fig. 7). For a triangular V_d, the current supplied by the clamp system need only be constant to keep V_m changing at a constant rate in either direction. A positive ramp V_d produces a positive I_{C_m}, and a negative one produces a

[19] A. Peres and G. Bernardini, *Pflügers Arch.* **404**, 266 (1985).
[20] M. Lindau and E. Neher, *Pflügers Arch.* **411**, 137 (1988).
[21] W. J. Moody and M. M. Bosma, *Dev. Biol.* **112**, 396 (1985).

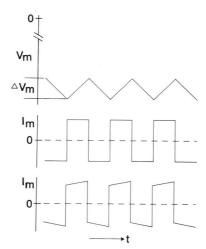

FIG. 7. Using a ramp control voltage in voltage clamp. The first trace represents a V_d held at a negative V_{RP} with a triangular wave of peak-to-peak amplitude ΔV_m. The resulting square pulses of I_m, positive and negative around zero, are seen in the second trace. These I_m pulses correspond to the positive- and negative-going parts of the triangular ΔV_m and C_m as

$$C_m = I_m / 2\Delta V_m / \Delta t$$

where I_m is measured from positive to negative at the time ΔV_m changes direction.

negative I_{C_m}, even if V_m does not reverse in polarity. Current measured from top to bottom of the square pulses must be divided by two for C_m determination. Because there is no stepwise change in V_d, there is no capacitive transient current.

This method is more flexible and direct than the capacitive transient method, as practically no computation is involved. A 1-V/sec ramp will produce 1 nA of current in each direction for 1 nF of C_m. Although a 1-V/sec ramp can be obtained by many combinations of maximum V with variable periods, certain limitations must be observed. If V_m becomes too large, because dV_m/dt continues too long, membrane conductances will be activated and their currents will add to I_{C_m}. If R_i is small, because of a large leak, a current proportional to dV_m/dt will be added to I_{C_m} as shown in Fig. 7. If dV_m/dt is too fast, I_{C_m} will not be constant; it will simply follow dV_m/dt and eventually become just the capacitive transient obtained with the voltage step. The ramp parameters are best chosen to obtain a square I_m with the cell being used. The voltage clamp system must be able to clamp the cell adequately, and the ramp must be generated without deviation from linearity or fast transients, especially at the points where V_d changes directions. Such fast transitions will produce capacitive transients.

Direct Current Methods: Implications

The DC methods described above suffer from the fact that membrane potential recording must be interrupted during the V_m or I_m pulse or ramp to measure C_m. In many applications, the lengths of time over which the membrane potential or ionic changes take place are long and the interruptions will not contaminate the other data. However, being discrete, they cannot provide measurements for rapidly changing C_m, such as might occur during fusion.

Time and Frequency Domains

Thus far the effects of capacitance on voltage or current have been described, using time as the other, independent variable. This method of interpreting voltage and current to deduce capacitance is also known as working in the *time domain*. Another method is working in the *frequency domain*. In this domain, the independent variable is the sinusoid. A sinusoidal wave has the property of purity; a 100-cycle/sec (hertz, Hz) wave contains energy only at that frequency (f) and no other. There can be adjacent waves at 100.1 Hz or 99.8 Hz, but these are also pure waves in their own right and are not part of the 100-Hz wave. Sinusoids can be added together linearly to make one wave containing both or even many; they can each be recovered without losing their identity. They do not change their form on passing through any linear element, but they may be shifted in phase, that is, shifted in time. Linear elements are a good amplifier, as is resistance or capacitance which do not change the form of the sine wave. Sine waves interacting in nonlinear elements (a diode or even a rectifying R_t) result in their energy being divided among many different harmonics, thus making each one lose at least part of its identity.

A powerful method available in the frequency domain is known as Fourier analysis, which transforms complex signals into a broad spectrum of complexly related pure sine waves. Fourier analysis has been adapted to the digital computer in the form of the fast Fourier transform, or FFT.[22] The FFT, although widely available, is not a word processing software. Its ability to deal rapidly with a large number of sinusoids at the same time has been applied mostly to analyzing complexly organized cells.[23]

There is an absolute equivalence between the two domains, whereby one does not provide a more "true" deduction than the other. The choice of working in either domain depends on the problem at hand.

---

[22] E. O. Brigham and R. E. Morrow, *IEEE Spectrum, December,* p. 63 (1967).
[23] C. Clausen and J. M. Fernandez, *Pflügers Arch.* **390,** 290 (1981).

Alternating Currents

Sine wave voltages or currents (AC) change amplitude constantly with time and periodically reverse their direction as shown in Fig. 8 (solid line). This means that if we wait through the period of one cycle, everything will repeat itself (see also Vectors, below). A capacitance with an applied AC voltage (v_c) can never become fully charged and so the current (i_c) is also sinusoidal, but is not *in phase* with the voltage. (Note the use of lower-case v and i for alternating currents.) The dashed line in Fig. 8 shows that i_c is maximal when v_c is zero, and zero when v_c is maximal, as it was in the time domain in Fig. 2B. In vectorial terms, current is orthogonal to the voltage and leading by 90° or $\pi/2$ radians. For an applied AC voltage, the magnitude at any time t, $v_{(t)}$, is given in the simplest case by

$$v_{c(t)} = V_M \sin \omega t \tag{8}$$

and for the current as

$$i_{c(t)} = I_M \cos \omega t \tag{9}$$

where $\omega = 2\pi f$ and V_M and I_M are the maximal or peak values. ω is also called the angular frequency, with units of radians per second, and is an important parameter in the frequency domain.

Because the phase relations for v_c and i_c are fixed for any f, it can be shown that a capacitive parameter is obtained that is dependent both on frequency and capacitance. This is the capacitive reactance X_c, determined as

$$X_c = 1/\omega C \tag{10}$$

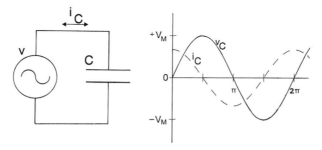

FIG. 8. A capacitance with applied sine wave voltage. A sine wave generator v is directly connected to a capacitance C, producing a current i_C. In a resistance, at every instant in time, current and voltage magnitudes are strictly proportional to R, regardless of the wave form. In the capacitance, current is dependent on the rate of change of voltage (Fig. 2). This property results in a 90° phase difference for sinusoidal i_C and v_C, as seen in the traces.

where C is in farads. X_c is given in ohms and can be used as a nondissipative AC resistance, that is $v_c = i_c X_c$. Note that X_c varies inversely with f and C, therefore for a constant AC current applied at a fixed frequency, v_c will vary inversely and proportionately with capacitance as $v_c = i_c / \omega C$.

The current in a parallel RC circuit, such as in Fig. 3B, divides between the resistance, where it will be in phase with the voltage, and in the capacitance, where it will lead voltage by 90°. The vector sum of these currents will be neither in phase nor orthogonal to the applied voltage. The circuit is no longer purely resistive or capacitive, but is an *impedance (Z)*. Z_m can be used like an AC resistance in making Ohm's law manipulations of AC currents and voltages:

$$i = v/Z \tag{11}$$

Z_m for a membrane can be found by vectorially adding the currents and dividing the voltage across R and C by the vectorial sum of the currents:

$$Z_m = v_m/i_m$$

where

$$i_m = i_{R_L} + i_{R_m} + i_{C_m}$$

and because the currents in R and C are orthogonal their AC properties add trigonometrically as

$$1/Z_m^2 = 1/R_i^2 + 1/X_{c_m}^2 \tag{12}$$

or

$$Y_m^2 = G_i^2 + (\omega C_m)^2$$

The use of admittance (Y_m) for $1/Z_m$, conductance G_i for $1/R_i$, and ωC_m for $1/X_{c_m}$ simplifies Eq. (12). The phase angle for Z relative to R can be found by Eq. (13).

$$\text{Phase angle } \theta = \tan^{-1}(X_c/R) \tag{13}$$

A sine wave current in this circuit will produce a voltage with the same phase angle θ with respect to the current.

Vectors

Sinusoidal voltages or currents can be described in three ways: as curves with respect to time, as in Fig. 8; as equations, for example, Eqs. (6) and (7); or as *vectors*. A vector is a point in a four-quadrant, two-dimensional space. The point is rotating around the origin at ω radians/sec and its distance from the origin (center) represents the maximum magnitude. If more than one voltage or current are involved, they can all be represented on the same vector diagram by different points. All the points will be rotating together at the same speed, if they all have the same f. Lines drawn from the origin to each point graphically describe the position of each at the same instant in time. Their angular relations in the diagram are their phase relationships as they rotate. Vectors are manipulated using trigonometric rules to produce "resultant" vectors.

The principal use of vectors is to illustrate the phases of complexly related sine waves that, if drawn as sinusoids, will become difficult to interpret. The phase and magnitude relations for voltage and current in a parallel RC circuit are illustrated in Fig. 9A. The vector diagram in Fig. 9B illustrates the changes in magnitudes and phase introduced when a resistance is added between the RC circuit and the voltage source. Vector diagrams help to rapidly visualize changes in phase relations introduced by changing the magnitude of any component(s).

Alternating Current Methods

A way to simultaneously record membrane potentials and C_m is to use one of the alternating current (AC) methods in current or voltage clamp. An AC (sinusoidal) signal is added to the control voltage or current before applying it to the clamping system. The resultant signal from the membrane will contain both DC (zero frequency) membrane voltage and AC components produced by i_m. The two components can be separated by appropriate filtering and analyzed for V_m, C_m, and R_m. Because AC signals can be greatly amplified before being measured, small voltages and currents can be used at the level of the membrane. Depending on the type of amplifier, the magnitudes can be on the order of microvolts and picoamperes (10^{-15}A).

In addition to measurements of magnitude, AC signals also allow measurement of the phase relation of the returning signal and the applied current or voltage. The most sensitive measurements up to now are made by phase detection. Phase detection is analogous to putting a shutter in front of the rotating vector, then opening it just at the instant when the desired component will be in front of the shutter.

Alternating Current Clamp Method

The frequency-dependent property of X_{C_m} is used to make the simplest kind of AC measurements of C_m. The fact that an AC current injected into a cell divides between R_i and C_m was recognized and used by Hagiwara and

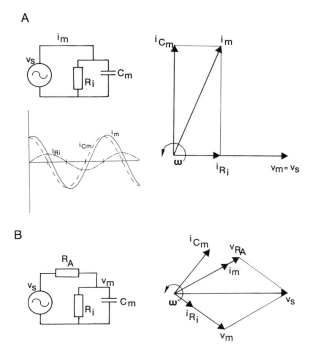

FIG. 9. Behavior of AC currents and voltage in parallel RC circuits. (A) The imaginary AC generator supplies a current i_m to R_i and C_m of a membrane. As i_m divides into two components, each must obey their respective phase relationship with the voltage across each element. The currents are shown in the sine wave diagram, where i_{C_m} is about 2.5 times greater than i_{R_i} and v_S is not shown, but is in phase with i_{R_i}. These phase and magnitude relations are shown in the form of a vector diagram (see text) in which $v_m = v_s$ is used as the reference vector. (B) A resistance R_A is added to the circuit, which reduces I_m to half its former value. The effects of adding R_A on the magnitudes and phases are seen in the vector diagram. Again, using v_S as the reference, the introduction of the voltage drop v_{R_A}, in phase with i_m, causes a clockwise (retarding) rotation of all the phases relating to the membrane. The ensemble rotation does not change the proportions or phases for i_{C_m} and i_{R_i} with v_m, but their relations to v_S no longer resemble that in (A). The capacitance current is still 90° leading v_m, but not v_S.

Saito[24] to determine C_m in neurons. A constant current i injected into a cell produces an alternating voltage (v_m), proportional to Z_m across the membrane. At high f, X_{C_m} can be made to be much smaller than R_i, Z_m equals X_{C_m} [see Eq. (12)], and almost all of i_m will be carried by C_m. The AC membrane voltage, v_m, will be equal to $i_m X_c$. X_{C_m} can then be found from

[24] S. Hagiwara and N. Saito, *J. Neurophysiol.* **22**, 204 (1959).

v_m/i_m. Because i_m is held constant by the clamp, v_m will change proportionately with changes in X_{C_m}.

In a cell of 100-μm diameter C_m should be about 0.3 nF and, at $f = 200$ Hz, X_{C_m} will be 2.7 MΩ and R_m may be 1 GΩ. If R_L is 3 MΩ, due to poor impalement, Z_m will be near 2 MΩ and, for an i_m of 1 nA, v_m will be 2 mV. The current will be about equally divided between R_i and X_{C_m}, and any changes in either will result in changes in v_m. If the impalement of the cell is good and R_i is nearer to 200 MΩ, and most of i_m will go through X_{C_m}, Z_m will be 2.69 MΩ. For an i_m of 1 nA, v_m will be nearly equal to the 2.7 mV expected for an X_{C_m} of 2.7 MΩ. Thus, with R_i much larger than X_{C_m}, v_m will be proportional to $1/C_m$ and faithfully reflect its changes. If R_m decreases significantly, as it does during egg activation, R_i may again be near X_{C_m} and v_m will no longer be dependent only on C_m. The way to avoid this is to choose f so that at the lowest expected R_m, X_{C_m} will still be at least 10-fold smaller than R_i.

In general, v_m will be less than a millivolt for i_m, which can be easily injected because X_{C_m} is deliberately made small relative to R_i. This does not present a problem, as v_m can be amplified easily before being measured. The amplifier can also be one of those that can be narrowly tuned to the frequency being used, which will help to reduce the noise. Measuring the magnitude of v_m can be done with one of the commercially available "true root-mean-square (rms) voltmeters." Such voltmeters measure the true rms value of an applied AC signal and output an exactly equal DC voltage that can be continuously recorded. The same voltmeter connected to the current monitor output will also provide a direct reading of i_m. The connection for the two signal pathways is shown schematically in Fig. 10.

The output from the voltmeter will be proportional to X_{C_m}, so that an increased C_m will result in decreased voltmeter DC output and vice versa. This signal should be inverted before recording or connected in an inverted fashion to the recorder to obtain a trace that goes upward for an increased C_m. The recorded DC trace will be proportional to C_m as long as X_{C_m} remains smaller than R_i and can be calibrated directly in units of F_d.

The DC V_m recordings can be made entirely free of AC components without losing any of their slower changes by suppressing the AC with a filter with adjustable cutoff frequency (f_0). Such filters rapidly attenuate signals with frequencies above f_0 and are also available commercially. There is a lower limit to setting f_0 because the DC signals also change in time (i.e., have frequency domain components). However, to satisfy the small X_{C_m} requirement, f is usually much higher than the highest frequency component in the membrane potential data. Therefore setting f_0 at one-half f or less should eliminate the AC component without affecting membrane potential changes.

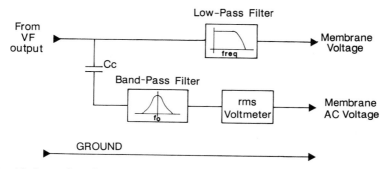

FIG. 10. Separation of the DC components from the AC. The signal from VF contains both DC membrane potential information and the AC signals produced by the injected AC. The DC components are obtained by a low-pass filter, which allows only low frequencies to pass on to its output. The same VF signal is also coupled through a capacitor (Cc), and its X_C allows only the higher frequency components to enter another filter. This filter is tuned to give maximal output for a narrow band of frequencies centered on the frequency (f) being used for measurements. The output of the tuned filter is led to a true rms voltmeter, which usually has a digital display and provides a DC voltage equal to the rms value of the AC signal. Both outputs are available for recording.

This method works well for measurements made with two impaling electrodes in sea urchin eggs,[4] as well as in frog eggs.[25] Shielding is needed between the two electrodes to prevent coupling through the stray capacitance between the pipettes, especially with the larger cells in which i_m will be greater. This method can also be used with a single impaling microelectrode, or in the whole-cell patch, if the AC voltage drop at R_t, or R_A, caused by i_m can be balanced to zero and does not change during the recording period.

Alternating Current Voltage Clamp Methods

Capacitance can also be determined for a membrane in AC voltage clamp by measuring the i_m needed to clamp v_m to an AC v_d. If f is selected as for the current clamp method, i_m will be exclusively in X_{C_m}, so that measurement of i_m, and knowing v_m, allows the calculation of X_{C_m}. Also, as in the current clamp method, i_m can be amplified, measured, and recorded. In this case, however, changes in i_m will be directly proportional to changes in C_m because $i_m = v_m \omega C_m$.

The cell described above, clamped to a v_m of 10 mV, at $f = 200$ Hz, will give a current of 3.7 nA. This current will appear at the I_m monitor output

[25] L. A. Jaffe and L. C. Schlicter, *J. Physiol. (London)* **358**, 299 (1985).

as 3.7 mV (at 1 mV = 1 nA). When amplified 1000 times to 3.7 V, it is easily measured with an AC voltmeter. Because X_{C_m} equals v_m/i_m, X_{C_m} = 2.7 MΩ; then from $C_m = 1/2\pi f X_{C_m}$, a value of 0.294 nF is found. Notice that this C_m estimate is nearly correct only when R_i is 200 MΩ and i_{R_i} is small compared to i_{C_m}.

Phase Detection

The use of the voltage clamp provides another way to measure the magnitudes of R_i and X_{C_m} independently, using the phase difference between the currents in R_i and C_m. The current in R_i must be in phase with v_m, and i_{C_m} must lead v_m by 90°. The clamping current i_m is the vector resultant of the two current components, as shown in Fig. 9. An AC signal that is the resultant of two components with a fixed phase relation can be decomposed into the two original components by sampling the resultant at the right time in each cycle. This is illustrated in Fig. 11, in which two sampling periods, A and B, are positioned in time on i_m to sample the peak magnitudes of i_{R_i} and i_{C_m}. Using two independent systems for the sampling provides independent outputs for the A and B samples. Because A samples only i_{R_i} and B samples only i_{C_m}, their respective outputs will be proportional to G_i and C_m, and if v_m is known, their absolute values may be calculated. This way of detecting components in a resultant signal is known as *phase detection*. Instruments that take an AC signal at its input

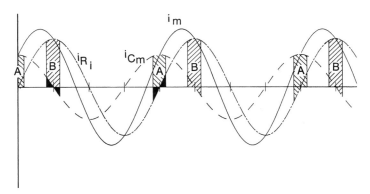

FIG. 11. Phase detection of two components in one sine wave. The trace i_m is the vector sum of two sine waves, i_{C_m} and i_{R_i}. Two sampling pulses, A and B, are positioned on i_m to coincide with the peaks of i_{C_m} and i_{R_i}. Note that for both pulses the sampling includes part of the other current as it passes through zero (blackened regions for B and A pulses). Thus each sample contains the signal only from its own component, and the contribution from the other is "autocancelled."

and outputs one or two phase-detected signals are known as "phase lock amplifiers" and are commercially available.

The problem here is to find the correct instant at which to do the detection. If a phase-accurate v_m measurement can be made, i_{R_i} is detected by putting A in phase with v_m and B exactly $90°$ earlier for i_{C_m}. The exact phase of v_m usually cannot be known, but if one of the current components is spontaneously changing magnitude the detection phase can be adjusted on that component until the pulsations just disappear. Detection is then taking place at the time the component is going through zero. (Note that adjusting the phase beyond this point reverses the direction of the pulsations.) A precise $90°$ shift from this null phase will put the detection at the peak of that component (compare the A and B sampling periods in Fig. 11). This technique is used to set the phase for detecting an electromagnetically induced pulsatile blood flow signal.[26]

Usually neither i_{C_m} nor i_{R_m} are pulsatile, but pulsations can be introduced as small changes in capacitance or resistance (manually, in the compensating electronics) to set the detection phase, before switching, by $90°$.[7,25,26] The AC voltage-clamped membrane current i_m, consisting of i_{R_i} and i_{C_m}, will maintain their phase relations to v_m regardless of changes in their individual magnitudes. Thus, in principle, R_m and C_m and their changes may be determined from the measured currents, at their respective phases, and v_m.

In the whole-cell patch configuration, v_m is taken as the voltage measured in the pipette (v_p). However, as connection to the cell interior is through R_A, its effects must be kept in mind. If the clamping i_m is small, so that the voltage drop at R_A is negligible, v_m will be nearly the same in magnitude and phase as v_p and small changes in R_A will have little effect. In the case of large cells, the larger i_m will drop a significant voltage at R_A. Changes in R_A will change the magnitude and phase of v_m with respect to v_p, as in Fig. 9B. A phase setting for i_{R_m} and i_{C_m}, established before the change in R_A, will no longer be correct. This problem has been extensively analyzed[27,28] and has led to the use of the DC constant current method instead in a complex cell.[29]

The other patch configurations for which the membrane is not ruptured, but instead is sealed to the end of the pipette,[30] offers a different situation.[30] In these configurations the gigaohm seals cause practically all of the current in the pipette to pass through only the piece of membrane

[26] A. Kolin and R. T. Kado, *Proc. Natl. Acad Sci. U.S.A.* **45**, 1312 (1959).
[27] E. Neher and A. Marty, *Proc. Natl. Acad. Sci. U.S.A.* **79**, 6712 (1982).
[28] C. Joshi and J. M. Fernandez, *Biophys. J.* **53**, 885 (1988).
[29] K. Narita, T. Tsuruhara, J. H. Koenig, and K. Ikeda, *J. Cell. Physiol.* **141**, 383 (1989).
[30] Y. Maruyama, *J. Physiol. (London)* **406**, 299 (1988).

sealed to the tip. The currents will be small, because the area of sealed-off membrane is small. The voltages v_p and v_m will be nearly equal. Because the area of the isolated patch is only a few square micrometers, the capacitances will be near 10^{-14} F_d or about 10 femtofarads (fF_d). The attachment of a vesicle to this patch of membrane will constitute an appreciable change in area.

Contrast this situation to that of measurements in the entire cell, in which the contribution of one vesicle or sperm to the total area will be miniscule. The patch approach allows one to resolve the capacitance added by single secretory granules from chromaffin cells.[7] The fusion of one sperm to a patch of egg membrane, even with a somewhat leaky seal, may be followed.[6] The fact that i_C in a capacitance depends on the series resistance makes it possible to follow the widening of the pore as an erythrocyte begins to fuse with another cell.[31] The finding of different capacitance sizes during exocytosis has led to the suggestion that some form of compound exocytosis may also occur.[32] Endocytosis following exocytosis has also been followed and the results indicate different regulatory mechanisms.[33]

Method of Choice

Several methods have been described for making capacitance measurements of cell surface area. How to decide which method to use depends, as usual, on many factors. Perhaps a good starting point is the size of the cell and whether capacitance is to be followed in the entire cell. Measuring the capacitance of the entire membrane in cells larger than about 100 μm is probably best done using the two microelectrode techniques. Because I_m injection and V_m measurement are made by independent electrodes, the data will be less complicated to interpret than with the whole-cell patch. For cells smaller than about 100 μm, but larger than about 30 μm, the single microelectrode may be better than the whole-cell patch, if the recording must be made over many hours. In cells under 30 μm, R_L, due to an impaling microelectrode, will cause the time constant to be short, making measurement difficult. In these cases, the whole-cell patch will be obligatory. To resolve the C_m changes produced by vesicles fusing with the plasma membrane, whole-cell patches for small cells with large vesicles or cell-attached patches to the plasma membrane are the only methods presently available.[34]

[31] A. Ruknudin, M. J. Song, and F. Sachs, *J. Cell Biol.* **112**, 125 (1991).
[32] A. E. Spruce, L. J. Breckenridge, A. K. Lee, and W. Almers, *Neuron* **4**, 643 (1990).
[33] G. Alvarez De Toledo and J. M. Fernandez, *J. Gen. Physiol.* **95**, 397 (1990).
[34] W. Almers and E. Neher, *J. Physiol. (London)* **386**, 205 (1987).

The descriptions of the membrane properties given above are in their simplest forms and many details for the application of the methods have been omitted. However, it appears that the tools are nearly at hand with which to probe directly the ongoing fusion processes that underlie endo- and exocytotic mechanisms in all cells. The methods are not yet finalized, nor are the data always easily interpreted, but the work already done strongly suggests that determination of membrane surface area by measuring one of its fundamental electrical properties is a valid approach. In the not too distant future these methods should become more easily and widely used.

Section VI

Introduction of Macromolecules into Cells by Membrane Fusion

[23] Intracellular Delivery of Nucleic Acids and Transcription Factors by Cationic Liposomes

By Nejat Düzgüneş and Philip L. Felgner

Liposomes of various types have been used for the delivery of nucleic acids into cultured cells.[1] Earlier studies involved the encapsulation of the nucleic acids inside liposomes and the delivery of liposomes most likely via an endocytotic pathway.[2] Liposomes containing the cationic lipid N-[1-(2,3-dioleyloxy)propyl]-N,N,N-trimethylammonium (DOTMA) have been found to mediate the efficient delivery of DNA[3] and RNA[4] into cells by forming a positively charged complex with the nucleic acids. Although the mechanism of delivery is not well understood, DOTMA-containing liposomes have been shown to undergo fusion with negatively charged liposomes[5] and cells.[3,6]

DOTMA liposomes have also been used to deliver purified transcription factors into cells, mediating the expression of specific genes.[7] Procedures for transfection and delivery of regulatory proteins are described below.

Delivery of Nucleic Acids

Cells to be transfected are plated on 60-mm diameter plastic tissue culture plates at a cell density of 0.5×10^6/plate, and incubated overnight to become approximately 80% confluent. The transfection procedure may cause toxicity in some cell lines if performed at lower cell densities. The culture medium is composed of Eagle's or Dulbecco's minimal essential medium (DMEM) with 10% (v/v) fetal bovine serum and Fungi-Bact solution (penicillin, streptomycin, and Fungizone; Irvine Scientific, Irvine,

---

[1] R. M. Straubinger and D. Papahadjopoulos, this series, Vol. 101, p. 512.

[2] R. M. Straubinger, K. Hong, D. S. Friend, and D. Papahadjopoulos, *Cell (Cambridge, Mass.)* **32,** 1069 (1983).

[3] P. L. Felgner, T. R. Gadek, M. Holm, R. Roman, H. W. Chan, M. Wenz, J. R. Northrop, G. M. Ringold, and M. Danielson, *Proc. Natl. Acad. Sci. U.S.A.* **84,** 7413 (1987).

[4] R. Malone, P. L. Felgner, and I. Verma, *Proc. Natl. Acad. Sci. U.S.A.* **86,** 6077 (1989).

[5] N. Düzgüneş, J. A. Goldstein, D. S. Friend, and P. L. Felgner, *Biochemistry* **28,** 9179 (1989).

[6] K. Konopka, L. Stamatatos, C. E. Larsen, B. R. Davis, and N. Düzgüneş, *J. Gen. Virol.* **72,** 2685 (1991).

[7] R. J. Debs, L. P. Freedman, S. Edmunds, K. L. Gaensler, N. Düzgüneş, and K. R. Yamamoto, *J. Biol. Chem.* **265,** 10189 (1990).

METHODS IN ENZYMOLOGY, VOL. 221

CA). To ascertain the complete absence of bacteria or fungi, an aliquot of cells may be incubated periodically in medium without the antibiotics, because undetectable numbers of microorganisms may still grow in the presence of antibiotics. The cells are washed three times with phosphate-buffered saline (PBS). Thirty micrograms of Lipofectin reagent (Life Technologies, Inc., Gaithersburg, MD) in 1.5 ml Opti-MEM I medium (Life Technologies) is mixed with 5 μg DNA in 1.5 ml Opti-MEM I. Different amounts of lipofectin and DNA may be optimal for different cell lines. The lipofectin reagent and DNA may also be added sequentially (but without washing away) to culture medium consisting of Opti-MEM I. The cells are incubated for 6 hr in a humidified CO_2 (5 – 10%) incubator. Three milliliters of medium containing 20% (v/v) FBS is added to the plates, and the calls are incubated for another 24 – 48 hr.

When preparing complexes of DNA and Lipofectin, it is essential to maintain a net positive charge on the complex. Optimal transfection occurs when the ratio of the molar equivalent positive charge contributed by DOTMA to the molar equivalent negative charge contributed by the nucleic acid is in the range 1.1 – 2.5. The corresponding ratio of the weight of the lipid (i.e., Lipofectin) to the weight of the nucleic acid is about 4 – 10.

The above procedures should be optimized for each cell line used. It is first necessary to determine the toxic levels of Lipofectin. The Lipofection experiments can then be performed at about half the toxic concentration, with varying concentrations of nucleic acid in different culture plates to produce the maximal transfection. When quantities of nucleic acid (e.g., plasmid) are limited, a reasonable amount of plasmid is added to several culture plates, and the amount of lipofectin may be varied below the toxic range.

Lipofectin-mediated transfection has been used to deliver plasmids such as pRSV-CAT, using chloramphenicol acetyltransferase activity as a marker for the delivery of the plasmid,[4] and pSV2-LacZ, using β-galactosidase activity as a marker.[8]

Delivery of Transcription Factors

Purified transcription factors can be delivered to cultured cells by forming a complex of the protein, the reporter plasmid containing the response element, and DOTMA liposomes. Although the commercially available lipofectin reagent can be used directly for this application, the use of pure DOTMA (obtained from Syntex Research, Palo Alto, CA) liposomes results in substantially higher levels of delivery.[7] Liposomes are

---

[8] J. Felgner, R. Kumar, R. Border, and P. L. Felgner, *J. Cell Biol.* **111**, 381a (1990).

prepared by sonication under an argon atmosphere (produced by purging the glass tube with argon for 15 sec), in a bath-type sonicator (Laboratory Supply Co., Hicksville, NY), following the formation of a dry film of lipid by rotary evaporation.[9]

Cells (e.g., CV-1 or HTC) are plated at a density of 10^6/100-mm diameter plastic Petri dishes in DME H-21 medium containing 10% (v/v) FBS, and incubated overnight in a CO_2 incubator at 37°. The cells are treated for 1 hr with 100 μM chloroquine, washed twice with PBS, and a mixture of liposomes, transcription factor, and reporter plasmid is added to the cells in DMEM without serum, as described below. The optimal concentrations of the transcription factor and plasmid to be used are determined empirically. The transcription factor (20–125 μg) is mixed gently with reporter plasmid (0.5–2.5 μg) and DOTMA liposomes (25 nmol) in 1 ml of DMEM. This mixture (1 ml) is placed in each petri dish containing the cells, and 4 ml of DMEM with 100 μM chloroquine and 0.05% (v/v) gentamicin is added immediately afterward. After the cells are incubated for 5 hr at 37°, 10 ml of DMEM/10% (v/v) FBS is added and the cells cultured for another 12 hr. The cells are then washed twice with PBS, and incubated in DMEM/10% (v/v) FBS for 24–36 hr. The transcription factor used in our studies is the glucocorticoid receptor derivative T7X556, the segment of the glucocorticoid receptor from amino acid 407 to 556 (with additional nonreceptor amino acids at the C and N termini added during expression in *Escherichia coli*), which includes the DNA-binding region but lacks the hormone-binding region.[7] The reporter plasmids contain (1) the glucocorticoid response element, (2) a promoter, and (3) the gene encoding chloramphenicol acetyltransferase (CAT). For example, the plasmid GTCO[7] contained a 46-base pair synthetic glucocorticoid response element (GRE)[10] fused to the herpes simplex virus thymidine kinase promoter. To determine the level of transcription, the cells are washed twice with PBS, scraped from the plate with a rubber policeman or Teflon cell scraper, centrifuged (1000 g, 5 min, 4°), resuspended in a small volume (100–200 μl) of 250 mM Tris (pH 7.5), freeze-thawed three times, incubated at 65° for 10 min, and centrifuged again (12,500 g, 10 min, 4°). The protein concentration in the supernatant is determined and 20 μg of supernatant protein is placed into a standard CAT assay.[11]

This method has been used to show that the transcriptional regulator T7X556 localizes in the nucleus following intracellular delivery, and that it

[9] N. Düzgüneş and J. Wilschut, this series, Vol. 220 [1].

[10] D. D. Sakai, S. Helms, J. Carlstedt-Duke, J. A. Gustafsson, F. M. Rottman, and K. R. Yamamoto, *Genes Dev.* **2**, 1144 (1988).

[11] C. M. Gorman, L. F. Moffat, and B. H. Howard, *Mol. Cell. Biol.* **2**, 1044 (1982).

selectively enhances expression from promoters linked to the glucocorticoid response element.[7] These experiments have also demonstrated that the expression of the transcriptional regulator in bacteria and subsequent biochemical purification does not affect the *in vivo* activities of the molecule, as shown by endogenous expression in mammalian cells.

[24] Microinjection of Macromolecules into Cultured Cells by Erythrocyte Ghost – Cell Fusion

By YOSHIHIRO YONEDA

Introduction

Various methods for introducing macromolecules, such as proteins and nucleic acids (DNA and RNA), into cultured cells have been developed and used to analyze the biological activities of these molecules in living cells. One of the most useful and commonly used methods is microinjection with a microcapillary. By this method, foreign substances can be introduced precisely into single cells. However, this method requires a special apparatus and skillful techniques. An alternative method is the widely used erythrocyte ghost – cell fusion method mediated by hemagglutinating virus of Japan (HVJ, or Sendai virus). This method is easier, and has the special advantage that it can be used for many cells at the same time. Furusawa *et al.* first demonstrated the reliability of erythrocyte ghost – cell fusion with fluorescein isothiocyanate (FITC) as a marker.[1] At first the injection frequency was low, but now it has been increased due to various improvements in the technique. This chapter describes the improved erythrocyte ghost – cell fusion method and its applications in biological studies.

Materials

Phosphate-buffered saline [PBS(−)]: 137 mM NaCl, 2.7 mM KCl, 8.1 mM Na$_2$HPO$_4$, 1.5 mM KH$_2$PO$_4$; pH 7.2

Reverse PBS (rPBS): 137 mM KCl, 2.7 mM NaCl, 8.1 mM Na$_2$HPO$_4$, 1.5 mM KH$_2$PO$_4$; pH 7.2

BSS(−): 140 mM NaCl, 54 mM KCl, 0.34 mM Na$_2$HPO$_4$, 0.44 mM KH$_2$PO$_4$, 10 mM Tris-HCl; pH 7.6

BSS(+): BSS(−) + 2 mM CaCl$_2$

---

[1] M. Furusawa, T. Nishimura, M. Yamaizumi, and Y. Okada, *Nature (London)* **249**, 449 (1974).

HVJ: Culture, concentration, and inactivation by ultraviolet (UV) irradiation of HVJ are carried out by the methods of Okada and Tadokoro.[2] When used, dilute with cold BSS(−)

(1/6)rPBS: Sixfold diluted rPBS containing 4 mM MgCl$_2$

Dialysis tubing (20/32 in)

Round-bottomed tube (15 ml)

Centrifuge tube (15 ml)

Sodium citrate solution (3.8%, w/v)

Autoclave the materials when necessary. PBS(−), rPBS, and BSS(−) can be stocked as 10-fold concentrated solutions.

Methods of Injection

Preparation of Human Erythrocytes

1. Collect 3–4 ml of human blood in a centrifuge tube from the median vein of the forearm with a syringe containing 3.8% (w/v) sodium citrate to prevent blood coagulation.

2. Wash the blood with PBS(−) three times by centrifugation at 2000 rpm for 5 min at 4° each time to remove serum and leukocytes. Leukocytes are collected as a white layer on top of the pellet.

3. Wash the erythrocytes once with rPBS and estimate their packed volume.

4. Add 4 vol of rPBS to prepare a 20% (v/v) erythrocyte suspension.

Notes

1. Blood of other animals (dog, cow, and guinea pig) can be used, but human red blood cells are most effective for introducing macromolecules into cultured cells.[3]

2. Blood supplied by a blood bank can also be used. However, care must be taken to avoid use of blood containing viruses or other pathogens.

3. Erythrocyte suspensions can be stored at 4° for 3 or 4 days if they are sterile, but fresh blood gives the best results.

Introduction of Macromolecules into Erythrocyte Ghosts

1. Transfer 1.5 ml of a 20% (v/v) erythrocyte suspension to a centrifuge tube.

---

[2] Y. Okada and J. Tadokoro, *Exp. Cell Res.* **26,** 108 (1962).

[3] M. Furusawa, M. Yamaizumi, T. Nishimura, T. Uchida, and Y. Okada, *Methods Cell Biol.* **14,** 73 (1976).

FIG. 1. Erythrocyte ghosts during dialysis against hypotonic ($\frac{1}{6}$)rPBS. The solution, containing a mixture of erythrocytes and target molecules, is dialyzed against ($\frac{1}{6}$) rPBS as shown here. The apparatus shown is usually used because it allows aseptic treatment of samples.

2. Centrifuge and remove the supernatant by aspiration. (The packed volume of erythrocytes is 0.3 ml.)

3. Add 1.7 ml of target macromolecules and mix.

4. Put the mixture into a small dialysis tube.

5. Dialyze against 500 ml of hypotonic (1/6) rPBS precooled to 4°, with vigorous stirring of the outer solution at room temperature, or if necessary at 4° for 30 min. This step may be performed in the apparatus shown in Fig. 1.

6. Dialyze the solution against 500 ml of isotonic PBS(−) prewarmed to 37°, stirring vigorously at room temperature for 30 min.

7. Collect the erythrocyte ghosts containing the target macromolecules from the dialysis tubing in a centrifuge tube.

8. Wash the ghosts three times with PBS(−) by centrifugation at 2800 rpm for 5 min at 4° to remove untrapped macromolecules and leaked hemoglobin.

9. Wash the ghosts once with BSS(+) and suspend them in 1.2 ml of BSS(+) to obtain a 20% (v/v) erythrocyte ghost suspension.

Notes

1. At step 5, hemolysis proceeds and the contents of the tubing become transparent. At this time, transient holes in the erythrocyte membrane seem to be formed.[4] During this step, hemoglobin leaks out of the erythro-

[4] P. Seeman, *J. Cell Biol.* **32,** 55 (1967).

cytes and target macromolecules enter them according to their concentration gradients.

2. At step 6, ruptured erythrocyte membranes reseal and the samples (target macromolecules) become trapped in the ghosts.

3. At step 8, after the first centrifugation, the supernatant is red due to leakage of hemoglobin and the ghost pellet is pale red or almost white if the hemolysis takes place efficiently.

4. ^{125}I-Labeled bovine serum albumin (BSA) trapped in the ghosts by this method was found to be quite stable.[5] Thus, the ghost suspension can be stored at 4° for a few days before its use in fusion experiments.

5. The concentrations of macromolecules trapped in the ghosts decrease as their molecular weights increase. For example, the concentrations of immunoglobulin G (IgG) and BSA trapped in ghosts are about one-third and one-half of their original concentrations, respectively.[3,5]

6. The concentration of a foreign substance trapped in ghosts increases linearly as its concentration in the tubing increases.[5] Therefore, higher concentrations of these substances should be used when more molecules are to be introduced into the cells.

Preparation of Target Cells

1. Detach plated cells with trypsin and ethylenediaminetetraacetic acid (EDTA).

2. Collect the cells in a centrifuge tube.

3. Wash the cells three times with PBS(−), by centrifugation at 1000 rpm for 5 min at 4° to remove the culture medium and serum.

4. Wash the cells once with BSS(+) as described in Step 3.

5. Suspend the cells at a concentration of 2.5×10^6 to 1×10^7 cells/ml in BSS(+) and keep them on ice until fusion.

Notes

1. During step 3, count the cell number and estimate the total cell number.

2. Suspension cultured cells can also be used. In this case start at step 2.

Fusion of Erythrocyte Ghosts with Target Cells

Erythrocyte ghost–cell fusion mediated by HVJ is performed by the method of Okada and Murayama.[6]

[5] M. Yamaizumi, M. Furusawa, T. Uchida, and Y. Okada, *Cell Struct. Funct.* **3**, 293 (1978).
[6] Y. Okada and F. Murayama, *Exp. Cell Res.* **44**, 527 (1966).

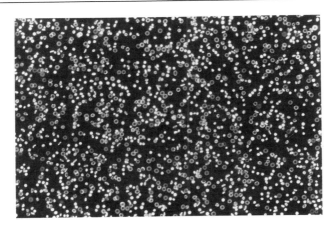

FIG. 2. Erythrocyte ghosts containing FITC–BSA. Erythrocyte ghosts are prepared as described in text, using FITC–BSA. The ghosts with trapped FITC–BSA are observed by fluorescence microscopy.

1. Put into a round-bottomed tube 0.25 ml of cell suspension, 0.25 ml of ghost suspension containing macromolecules, and 0.5 ml of UV-inactivated HVJ [1000 hemagglutinating units (HAU)/ml; Salk[7]] and mix.

2. Keep the mixture at 4° for 10 min with occasional mixing.

3. Incubate the mixture with gentle shaking in a water bath at 37° for 30 min.

4. Add an appropriate buffer (e.g., culture medium) and centrifuge the mixture at 800 rpm for 4 min at 4° to remove free ghosts and HVJ.

5. When necessary, repeat step 4 until scarcely any free ghosts are detectable.

Notes

1. At step 2, cell agglutination can be seen, indicating fusion activity of HVJ. When this is observed, proceed to the next step.

2. At step 3, gentle shaking is necessary for cell fusion because incubation without shaking results in cell degeneration.[8]

3. Centrifugation after fusion to remove free ghosts and HVJ should be performed at 800 rpm for 4 min at 4° because under these centrifugation conditions free ghosts are precipitated only slightly but cells are precipitated effectively. If necessary, transfer the sample from the round-bottomed tube to a centrifuge tube before centrifugation.

[7] J. E. Salk, *J. Immunol.* **49,** 87 (1944).
[8] Y. Okada, *Exp. Cell Res.* **26,** 98 (1962).

4. A hemolysing solution (155 mM NaCl, 10 mM KHCO$_3$, 1 mM Na$_2$-EDTA) can be used to remove free ghosts. After centrifugation of the cell suspension, suspend the pellet in 3 vol of this hemolysing solution, let the suspension stand at 0° for 10 min, add $\frac{1}{10}$ vol of fetal calf serum, and centrifuge. Wash the cells twice with an appropriate buffer.

5. As shown in Figs. 2 and 3, by use of FITC–BSA the injection efficiency can be determined by fluoresence microscopy. By the above protocol, macromolecules can usually be introduced into more than 90% of the cells.

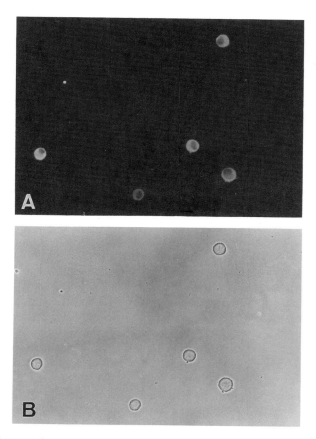

FIG. 3. Mammalian cells fused with erythrocyte ghosts containing FITC–BSA. Human embryonic lung (HEL) cells are fused with erythrocyte ghosts containing FITC–BSA as described in text. Immediately after washing, the cells are observed by (A) fluorescence microscopy and (B) phase-contrast microscopy. The fluorescence of FITC–BSA can be seen in the cytoplasm of HEL cells.

General Comments

The fusion method described here is straightforward, does not require any special apparatus, and is useful for analyses in almost all fields of cell biology. Various types of cultured cells can be used and a large number of cells can be treated at the same time.

When cells without HVJ receptors, such as lymphocytes, are used as targets or HVJ is not available, polyethylene glycol (PEG) can be used. However, the fusion efficiency with PEG is low compared with that with HVJ. Moreover, because PEG has a toxic effect on the cells, washing procedures after cell fusion must be repeated many times.

Schlegel and Rechsteiner reported another kind of erythrocyte-mediated microinjection method.[9] Their method is as simple and widely applicable as our injection method and seems to result in greater trapping of macromolecules in erythrocytes than our method; but our method appears to be better quantitatively. Moreover, by our method, the injection frequency determined with FITC–BSA is almost 100% (at least more than 90%), as shown in Fig. 3.

The concentration of a substance trapped in a single ghost can be controlled by adjusting the concentration of the original solution. Therefore, if cells fused with a defined number of ghosts are collected, cells containing a defined amount of macromolecules can be selected.[10]

When only a small amount of material is available, the procedure may be scaled down. For example, almost the same results as those described above can be obtained by using a mixture of 60 μl (packed volume) of erythrocytes and 200 μl of macromolecules.

This method has the following advantages over the method of microinjection with a microcapillary: (1) As described above, it can be used for quantitative injection of material into cells; (2) no special apparatus and/or technique is needed; (3) materials can be introduced into a large number of cells at the same time; (4) it can be used to introduce material into cells in suspension cultures; and (5) it does not damage the cells appreciably.

On the other hand, this method has the following disadvantages: (1) Erythrocyte membranes, viral envelopes, viral RNA, and residual hemoglobin are cointroduced, and these may have unexpected effects on the cells; (2) a larger amount of test material is required than for the microinjection method with a microcapillary; (3) direct injection into the cell nucleus is not possible; and (4) materials that are insoluble under the hypotonic conditions (low ionic strength) cannot be used.

[9] R. A. Schlegel and M. C. Rechsteiner, *Methods Cell Biol.* **20,** 341 (1978).
[10] E. Mekada, M. Yamaizumi, T. Uchida, and Y. Okada, *J. Histochem. Cytochem.* **26,** 1067 (1978).

Besides proteins, other macromolecules such as tRNA[11] and virus particles[12] have been introduced into cells by this method.

Applications

This method can be used in various kinds of experiments in cell biology. Some examples of studies using this method are described below.

Using fragment A of diphtheria toxin, Yamaizumi et al.[13] clearly demonstrated quantitative introduction of macromolecules into cells by the erythrocyte ghost–cell fusion method. Diphtheria toxin is a single polypeptide consisting of fragments A and B. Fragment B is required for attachment of the molecule to cell surface receptors. Fragment A, once inside the cytoplasm, acts on elongation factor 2, inhibiting peptide chain elongation in translation and thus causing cell death. Erythrocyte ghosts containing a known number of molecules of fragment A of diphtheria toxin with a marker were prepared. A constant amount of FITC–BSA was used as a fluorescence marker. These ghosts were then fused with diphtheria toxin-resistant mouse L cells by the HVJ-mediated cell fusion method. Mononuclear target cells that had fused with only one ghost were selected with a fluorescence-activated cell sorter (FACS) on the basis of their cell size and the fluorescence intensity of FITC–BSA, and their viability was then examined by measuring colony-forming ability. Estimation of the numbers of fragment A molecules introduced into a single cell demonstrated that introduction of one molecule of diphtheria toxin fragment A into a single cell can kill the cell.

Yamaizumi et al.[14] examined whether an antibody can function in a cell by studies using antibody against diphtheria toxin fragment A. Anti-fragment A, which was not effective when added to the culture medium directly, was introduced into diphtheria toxin-sensitive Vero cells or FL (human amnionic) cells. When about 1500 molecules of anti-fragment A antibody were introduced into these cells by the erythrocyte ghost–cell fusion method, the cells became resistant to a mutant protein of diphtheria toxin, CRM176. CRM176 has a completely normal fragment B and its fragment A (fragment A-176) is antigenically identical to fragment A of the wild type but its enzymatic activity is one-tenth that of the wild type. It was

[11] K. Kaltott, J. Zenther, F. Engback, P. W. Piper, and J. E. Celis, Proc. Natl. Acad. Sci. U.S.A. 73, 2793 (1976).
[12] A. Loyter, N. Zakai, and R. G. Kulka, J. Cell Biol. 66, 292 (1975).
[13] M. Yamaizumi, E. Mekada, T. Uchida, and Y. Okada, Cell (Cambridge, Mass.) 15, 245 (1978).
[14] M. Yamaizumi, T. Uchida, E. Mekada, and Y. Okada, Cell (Cambridge, Mass.) 18, 1009 (1979).

calculated from the survival curves of cells containing a definite number of molecules of fragment A-176 that, under the experimental conditions used, about 300 molecules of fragment A-176 entered the cells. These results showed that the antigen–antibody reaction occurred in living cells *(in vivo)* as effectively as in a cell-free system *(in vitro)*. Furthermore, more than 50% of the initial activity of the anti-fragment A antibody was shown to remain even after incubation of the cells containing the antibody at 37° for 20 hr, indicating that the antibody against fragment A retains its function equally well *in vivo* and *in vitro*. The molar ratio of antibody to antigen for neutralization and the functional stability of the antibody in living cells were determined.

The mechanism of protein import into the nucleus was examined by the erythrocyte ghost–cell fusion method. Yamaizumi *et al.*[15] first showed the distribution of ^{125}I-labeled nonhistone chromosomal proteins injected into the cytoplasm of Ehrlich ascites tumor cells. After various times of incubation at 37°, the cells were lysed and separated into two fractions, cytoplasmic and nuclear, and the amount of radioactivity in each fraction was counted. Results showed that nonhistone chromosomal proteins introduced into the cytoplasm rapidly accumulated in the nuclei.

Next, nucleoplasmin, a nuclear protein of *Xenopus* oocytes, was used. Sugawa *et al.*[16] introduced nucleoplasmin or nucleoplasmin chemically conjugated with nonimmune IgG (nucleoplasmin–IgG) into the cytoplasm of mammalian cells by this method, incubated the cells at 37° for 1–2 hr, and then examined the distribution of the proteins. Results showed that the nucleoplasmin–IgG conjugate as well as nucleoplasmin entered the nucleus from the cytoplasm. These results suggested that nuclear proteins have a specific signal to enter the nucleus actively and indicated that this signal was conserved in different species.

Tsuneoka *et al.*[17] obtained further information using nonhistone chromosomal protein high mobility group-1 (HMG-1) and monoclonal antibody against HMG-1. First, they showed that HMG-1 rapidly migrated into the nucleus when injected into the cytoplasm of mammalian cells by erythrocyte ghost-mediated microinjection. A monoclonal antibody against HMG-1, FR-1, was found to inhibit the binding of ^{125}I-labeled HMG-1 to isolated chromatin *in vitro*. They then examined whether FR-1 blocked nuclear transport of HMG-1. When ^{125}I-labeled HMG-1 was introduced with FR-1, the nuclear accumulation of HMG-1 was not inhib-

[15] M. Yamaizumi, T. Uchida, Y. Okada, and M. Furusawa, *Nature (London)* **273**, 782 (1978).
[16] H. Sugawa, N. Imamoto, M. Wataya-Kaneda, and T. Uchida, *Exp. Cell Res.* **159**, 419 (1985).
[17] M. Tsuneoka, N. S. Imamoto, and T. Uchida, *J. Biol. Chem.* **261**, 1829 (1986).

ited. Furthermore, when ^{125}I-labeled FR-1 was cointroduced with HMG-1 into the cytoplasm of human cells, it migrated into the nucleus. These results indicated that HMG-1 has a specific domain responsible for its migration into the nucleus and that this domain is different from that responsible for binding to nuclear components. It was suggested, therefore, that entry of HMG-1 into the nucleus is not due to its ability to bind to nuclear components (chromatin) and simple diffusion.

One of the best characterized nuclear proteins is simian virus 40 (SV40) large T antigen. Studies using an SV40 mutant defective in nuclear transport of this large T antigen showed that its nuclear accumulation requires a short sequence of amino acids. This was confirmed by introducing a point mutation into the large T-antigen gene by mixed oligonucleotide mutagenesis and expressing its product in cells. From these experiments, it was proposed that the sequence Pro-Lys-Lys-Lys-Arg-Lys-Val can act as a signal for nuclear localization. However, because the above experiments were performed by transfection of DNA and study of its transient expres-

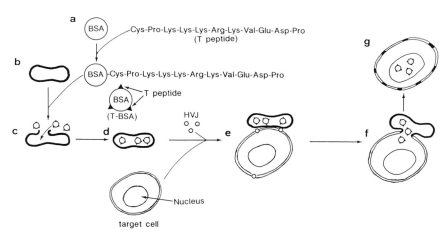

FIG. 4. Diagram of the steps in an experiment on the nuclear transport mechanism, using a synthetic peptide containing the nuclear location signal of SV40 large T antigen (T peptide). (a) Bovine serum albumin (BSA) is treated with the synthetic peptide (T peptide) to make the conjugate as described previously. (b) The conjugate (T-BSA), in which BSA is chemically cross-linked with T peptides, is completed. (c) During dialysis against hypotonic ($\frac{1}{6}$)rPBS, T-BSA enters erythrocyte ghosts through transient holes in their membrane. (d) During dialysis against isotonic PBS(−), the ruptured erythrocyte membrane is resealed and T-BSA is trapped in the ghost. (e) Ghosts containing T-BSA are mixed with target cells and HVJ at 4°, resulting in agglutination. (f) During incubation at 37° for 30 min, the ghosts and target cells fuse and T-BSA trapped in the ghosts enters the cytoplasm of the cells. (g) After incubation at 37° for 1–2 hr, T-BSA migrates into the nucleus. Molecules of the erythrocyte membrane and target cell are intermixed.

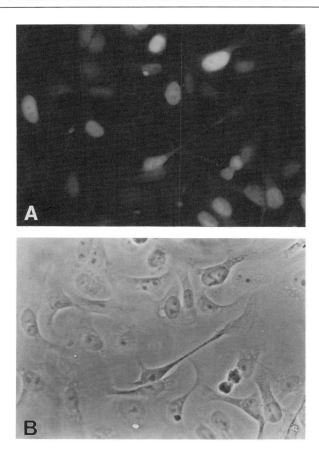

FIG. 5. Nuclear accumulation of T-BSA in mammalian cells. T-BSA is introduced into the cytoplasm of mammalian cells by the erythrocyte ghost–cell fusion method, as illustrated in Fig. 4. The cells are fixed and treated with rabbit anti-BSA as the first antibody and FITC–goat anti-rabbit IgG as the second antibody, and then observed by (A) fluorescence microscopy and (B) phase-contrast microscopy.

sion, the possibility remained that the protein containing the nuclear location signal sequence bound to nuclear components during mitosis when the nuclear envelope broke down. To exclude this possibility completely, Yoneda et al.[18] conjugated foreign proteins chemically with synthetic peptides containing the nuclear location signal sequence and introduced these into the cytoplasm by this method. As shown in Fig. 4, the

[18] Y. Yoneda, T. Arioka, N. Imamoto-Sonobe, H. Sugawa, Y. Shimonishi, and T. Uchida, *Exp. Cell Res.* **170**, 439 (1987).

synthetic peptide containing the nuclear location signal for SV40 large T antigen (T peptide) was conjugated chemically with bovine serum albumin (BSA). Then erythrocyte ghosts containing the conjugates (T-BSA) were fused with human embryonic lung cells. Two hours later the cells were fixed and the subcellular localization of T-BSA was examined by the indirect immunofluorescence method with anti-BSA antibody. The conjugates were found in the nucleus within 2 hr after fusion (Fig. 5). Conjugates of the synthetic peptide with phycoerythrin ($M_r \sim 150,000$) and with secretory IgA ($M_r \sim 380,000$) were also both found in the nucleus shortly after their introduction. These results suggested that the synthetic peptide containing the nuclear location signal sequence directs the transport of proteins into the nucleus.

Thus, by using synthetic peptides containing the nuclear location signal sequence, desired nonnuclear proteins, such as antibody, can be introduced into the nucleus by injecting them into the cytoplasm by the erythrocyte ghost–cell fusion method. One of the disadvantages of the method described above — the impossibility of injecting materials directly into the nucleus — can be overcome.

Acknowledgment

I thank Professor Yoshio Okada (Institute for Molecular and Cellular Biology, Osaka University) for helpful discussions and critical reading of the manuscript.

[25] Introduction of Plasmid DNA and Nuclear Protein into Cells by Using Erythrocyte Ghosts, Liposomes, and Sendai Virus

By Yasufumi Kaneda, Keiko Kato, Mahito Nakanishi, and Tsuyoshi Uchida

Introduction

The best way to study the biological functions of macromolecules (DNA, RNA, and proteins) is to introduce those compounds into living cells. Many methods have been developed to introduce macromolecules into cells, including calcium phosphate[1-3] or DEAE-dextran-mediated

[1] F. L. Graham and A. J. Van der Eb, *Virology* **52**, 456 (1973).

[2] M. Wigler, S. Silverstein, L. S. Lee, A. Pellicer, Y. C. Cheng, and R. Axel, *Cell (Cambridge, Mass.)* **11**, 223 (1977).

[3] C. Chen and H. Okayama, *Mol. Cell. Biol.* **7**, 2745 (1987).

transfection,[4,5] direct microinjection with microcapillaries,[6-8] protoplast fusion,[9,10] polybrene,[11,12] electroporation,[13,14] virus vectors,[15-17] liposomes,[18-22] and erythrocyte (red blood cell, RBC) ghosts.[23-25]

Our efforts have been focused on developing the methods first to introduce exogenous DNA into the cytoplasm efficiently by Sendai virus (hemagglutinating virus of Japan, HVJ)-mediated fusion, and then to transfer the DNA efficiently from the cytoplasm to the nucleus for high-level expression even in cells in the G_0 phase.[26] For this purpose, we developed new vehicles by using HVJ to combine DNA-loaded liposomes with RBC ghosts containing nuclear proteins. By use of these vehicles, referred to here as vesicle complexes, DNA and nuclear proteins could be simultaneously introduced into cultured cells at more than 95% efficiency and the introduced DNA rapidly migrated into the nucleus even of interphase cells.[25] Furthermore, by this delivery system, exogenous DNA was

[4] A. Vaheri and J. S. Pagano, *Virology* **27**, 434 (1965).

[5] D. Warden and H. V. Thomas, *J. Gen. Virol.* **3**, 371 (1968).

[6] M. R. Capecchi, *Cell (Cambridge, Mass.)* **22**, 479 (1980).

[7] E. G. Diakumakos, S. Holland, and P. Pecora, *Proc. Natl. Acad. Sci. U.S.A.* **65**, 911 (1970).

[8] M. Graessman and A. Graessman, *Proc. Natl. Acad. Sci. U.S.A.* **73**, 366 (1976).

[9] W. Schaffner, *Proc. Natl. Acad. Sci. U.S.A.* **77**, 2163 (1980).

[10] M. Rassoulzadegan, B. Bincruy, and F. Cuzin, *Nature (London)* **295**, 257 (1982).

[11] S. Kawai and M. Nishizawa, *Mol. Cell. Biol.* **4**, 1172 (1984).

[12] W. G. Charney, D. R. Howard, J. W. Pollard, S. Sallustio, and P. Stanley, *Somatic Cell Mol. Genet.* **12**, 237 (1986).

[13] E. Neumann, M. Schaefer-Ridder, Y. Wang, and P. Hofschneider, *EMBO J.* **1**, 841 (1982).

[14] U. Zimmermann, *Biochim. Biophys. Acta* **694**, 227 (1982).

[15] S. Karlsson, R. K. Humphries, Y. Gluzman, and A. W. Nienkius, *Proc. Natl. Acad. Sci. U.S.A.* **82**, 158 (1985).

[16] I. M. Shapiro, M. Stevenson, F. Sinangil, and D. J. Volsky, *Somatic Cell Mol. Genet.* **12**, 351 (1986).

[17] A. Oppenheim, A. Peleg, E. Fibach, and E. A. Rachmilewitz, *Proc. Natl. Acad. Sci. U.S.A.* **83**, 6925 (1986).

[18] R. Fraley, S. Subramani, P. Berg, and D. Papahadjopoulos, *J. Biol. Chem.* **255**, 10431 (1980).

[19] T. K. Wong, C. Nicolau, and P. H. Hofschneider, *Gene* **10**, 87 (1980).

[20] M. Schaefer-Ridder, Y. Wong, and P. H. Hofschneider, *Science* **215**, 166 (1982).

[21] M. Nakanishi, T. Uchida, H. Sugawa, M. Ishiura, and Y. Okada, *Exp. Cell Res.* **159**, 399 (1985).

[22] P. L. Felgner, T. R. Gadek, M. Holm, R. Roman, H. W. Chan, M. Wenz, J. P. Northrop, G. M. Ringold, and M. Danielson, *Proc. Natl. Acad. Sci. U.S.A.* **84**, 7413 (1987).

[23] M. Furusawa, T. Nishimura, M. Yamaizumi, and Y. Okada, *Nature (London)* **249**, 449 (1974).

[24] A. Loyter, N. Zakai, and R. G. Kulka, *J. Cell Biol.* **66**, 292 (1975).

[25] R. A. Schlegel and M. C. Rechsteiner, *Cell (Cambridge, Mass.)* **5**, 371 (1975).

[26] T. Uchida, *Exp. Cell Res.* **178**, 1 (1988).

introduced into adult rat liver and highly but transiently expressed in hepatocytes *in vivo.*[27,28]

Thus the vesicle complex is a unique and attractive vehicle that should provide new insights into the functions of cells and macromolecules and allow gene therapy of various diseases. In this chapter, we describe the detailed protocol for preparation of vesicle complexes and the results obtained by the use of this vehicle.

Materials

Sendai Virus (HVJ; Hemagglutinating Virus of Japan).[29] Sendai virus is one of the most important materials for vesicle complexes. Aliquots of the best seed of HVJ (Z strain) are stored in 2-ml polypropylene tubes in liquid nitrogen. The seed is injected into 10-day-old embryonated chick eggs. Sendai virus is harvested from the chorioallantoic fluid of the eggs after a 4-day incubation at 36° and purified by differential centrifugation.[30] Its hemagglutinating activity is determined[31] and it is then stored aseptically at 4°, because the freezing of HVJ must be avoided. Sendai virus is stable in chorioallantoic fluid for at least 3 months, but once purified the fusion activity lasts only 3 weeks. The purified virus is diluted with balanced salt solution (BSS; 137 mM NaCl, 5.4 mM KCl, 0.34 mM Na$_2$PO$_4$, 0.44 mM KH$_2$PO$_4$, 10 mM Tris-Cl, pH 7.6) containing 2 mM CaCl$_2$ to a concentration of 15,000 hemagglutinating units (HAU)/0.5 ml and inactivated with ultraviolet irradiation (100 erg/mm²/sec) for 3 min just before use. It is important to use fresh and intact HVJ for this experiment.

Lipids.[32] For preparation of liposomes, it was important to use purified phosphatidylserine (PS) and egg yolk phosphatidylcholine (PC). Chromatographically pure bovine brain PS (sodium salt) is purchased from Avanti Polar Lipids, Inc. (Birmingham, AL) or Boehringer-Mannheim (Indianapolis, IN). Egg yolk PC, cholesterol (Chol), and bovine brain gangliosides (GS) (type III, containing 20% sialic acids) are purchased from Sigma (St. Louis, MO). Phosphatidylserine, PC, and Chol are each dissolved in chloroform and mixed in a weight ratio of 1 : 4.8 : 2. The mixture is evaporated and the residue stored at −20° under nitrogen gas. Gangliosides are dissolved in tetrahydrofuran, evaporated, and stored at −20° under nitrogen gas. These mixtures can be stored for 1 month at −20°.

[27] Y. Kaneda, K. Iwai, and T. Uchida, *Science,* **243,** 375 (1989).
[28] Y. Kaneda, K. Iwai, and T. Uchida, *J. Biol. Chem.* **264,** 12126 (1989).
[29] Y. Okada, *Exp. Cell Res.* **26,** 98 (1962).
[30] J. Kim, K. Hama, Y. Miyake, and Y. Okada, *Virology* **95,** 523 (1979).
[31] J. E. Salk, *J. Immunol.* **69,** 87 (1944).
[32] Y. Kaneda, T. Uchida, J. Kim, M. Ishiura, and Y. Okada, *Exp. Cell Res.* **173,** 56 (1987).

Erythrocyte Membranes.[32,33] Fresh heparinized human blood (about 20 ml) is diluted with an equal amount of cold phosphate-buffered saline (PBS; 137 mM NaCl, 3 mM KCl, 8 mM Na$_2$PO$_4$, 1 mM KH$_2$PO$_4$) and then spun at 3000 rpm for 15 min in a low-speed centrifuge (Hitachi, Tokyo, Japan) at 4°. Human RBCs are collected and washed twice with PBS. About 5 ml of packed RBCs is suspended in 20–30 ml PBS and dialyzed against 3000 ml of 1:6 diluted PBS for 4 hr at 4°. The dialyzed RBC lysate, to which 5–6 ml PBS is added after dialysis, is centrifuged at 2800 g for 15 min at 4°. The resulting pellet (RBC membranes) is suspended in 20 ml cold PBS. Aggregates, if any, should be removed from the pellet at this step. The membranes are then collected by centrifugation (2800 g, 15 min, 4°). This washing process should be repeated twice. The RBC membranes are kept at 4° until use. They can be stored for 10 days in this state.

Plasmid DNA. Plasmid DNAs are purified by equilibrium centrifugation in cesium chloride containing ethidium bromide. The preparations are suspended in 10 mM Tris-HCl (pH 8.0)–0.1 mM EDTA (final concentration of DNA is 1 mg/ml) and stored at −20°. Before use, the DNA solution is diluted 10-fold with BSS.

Cells. We use mouse L cells deficient in thymidine kinase (Ltk⁻), mouse Ehrlich ascites tumor cells, monkey LLCMK2 cells, and human WI-38 cells. Lymphocytes and lymphoma cells are not available for this delivery system because these cells cannot fuse well with HVJ.[34]

Other Chemicals. Octyl-β-D-glucopyranoside (octylglucoside; Calbiochem Behring Corp. La Jolla, CA) is dissolved in distilled water at a concentration of 30% (w/w) and stored at room temperature. Nonhistone chromosomal protein, high mobility group 1 (HMG-1), is purified from calf thymus as described elsewhere[35] and dissolved in distilled water at a concentration of 10 mg/ml; 250-μl aliquots are stored at −70°.

Apparatus. To prepare liposomes by the reversed-phase evaporation method, a rotary evaporator (Rotavapor R110; Büchi, Flawil, Switzerland) is used.

Preparation of Vesicle Complexes

The procedure for constructing vesicle complexes is shown in Fig. 1.

Preparation of Liposomes Containing Plasmid DNA

Liposomes are prepared by the reversed-phase evaporation method,[36] with a slight modification as follows. Stored GS are dissolved in chloro-

[33] H. Sugawa, T. Uchida, Y. Yoneda, M. Ishiura, and Y. Okada, *Exp. Cell Res.* **159,** 410 (1985).

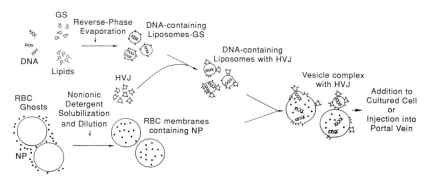

FIG. 1. The procedure for simultaneous introduction of plasmid DNA and nuclear proteins into cultured cells or adult rat liver, using the vesicle complex. GS, gangliosides; NP, nuclear protein. This procedure consists of three steps, as described in text. The first step is the interaction of DNA-loaded liposomes with HVJ. The second step is the construction of RBC ghosts containing nuclear proteins. The final step is the formation of vesicle complexes by mixing DNA-loaded liposome–HVJ complexes with RBC ghosts containing nuclear proteins.

form and added to the mixture of the lipids (PS:PC:Chol, 1:4.8:2) in a weight ratio of 1:1000 and evaporated. Then 7.5 mg of this GS–lipid mixture dissolved in $450 \mu l$ of organic solvents (isopropyl ether:chloroform, 64:36) is mixed with $150 \mu l$ of DNA solution (10–100 μg of plasmid DNA), agitated with a vortex mixer for 30 sec, and sonicated in a bath-type sonicator for 3 sec. The organic solvents are evaporated in a rotary evaporator under high vacuum (700–750 mmHg) at 37°. Finally, 1.5 ml of BSS is added with thorough mixing in a vortex mixer. Those liposomes are now ready for fusion with HVJ. They can be kept at 4° for 1 day before fusion with HVJ. The trapping efficiency of the added DNA in GS-containing liposomes prepared as described above is about 30% with plasmid DNA of less than 10 kbp and about 5% with cosmid DNA of 45 kbp.[32] The DNA trapped in liposomes is intact after incubation with HVJ.[32] In addition, poly(A)$^+$ RNA could also be incorporated into liposomes by this method.[32] Several methods have been reported to remove unencapsulated DNA from liposomes.[20,37] However, we use liposomes without purification in the following experiments because free DNA does not affect the formation of vesicle complexes.

[34] Y. Okada and J. Tadokoro, *Exp. Cell Res.* **32,** 417 (1963).
[35] C. Sanders, *Biochem. Biophys. Res. Commun.* **78,** 1034 (1977).
[36] F. Szoka and D. Papahadjopoulos, *Proc. Natl. Acad. Sci. U.S.A.* **75,** 4194 (1982).
[37] R. Straubinger and D. Papahadjopoulos, "Techniques in Somatic Cell Genetics." Plenum, New York and London, 1982.
[38] Y. Kaneda and T. Uchida, unpublished data (1987).

Preparation of Erythrocyte Ghosts Containing Nuclear Proteins

Proteins cannot be entrapped in liposomes prepared by the reverse-phase evaporation because of the use of organic solvents. Therefore RBC ghosts must be prepared to enclose proteins. Packed RBC membranes (450 μl) prepared as described in the previous section are mixed with 225 μl of proteins (HMG-1; 10 mg/ml) in three microcentrifuge tubes (225-μl total in each tube). Then 6.2 μl of 15% (w/w) octylglucoside is added to each tube twice and the mixtures incubated at 37° for 10 sec. Cold PBS (1.3 ml) is added to the mixture in one tube to dilute the octylglucoside and the diluted mixture is incubated at 4° for 30 min. Then the mixture is centrifuged at 8000 rpm in a Beckman (Fullerton, CA) microfuge B for 15 min at 4°. The pellet is suspended in 1.3 ml BSS with a long Pasteur pipette. At this step aggregates should be removed carefully. Then the suspension is spun as described above. This step is repeated twice. The resulting pellet (about 120 μl, about 9.6×10^8 vesicles) can be kept at 4° for 1–2 hr before fusion with HVJ–liposomes.

The method described above for the preparation of RBC ghosts is a new method involving detergent solubilization–dilution,[32,33] and is different from the traditional RBC ghosts prepared by hypotonic treatment.[23–25] The traditional RBC ghosts may also be used, but their trapping efficiency of proteins is low (about 3% for IgG), and proteins more than 500 kDa cannot be entrapped in the ghosts.[23] On the other hand, the new ghosts are prepared more easily and can entrap almost all proteins efficiently. For example, immunoglobin M (IgM) (900 kDa) can be incorporated into the vesicles at about 10% efficiency.[33] Therefore we use the new RBC ghosts to form vesicle complexes.

Preparation of Vesicle Complexes

DNA-loaded liposomes suspended in 1.5 ml of BSS are mixed with 1.5 ml of UV-inactivated HVJ (about 15,000 HAU/0.5 ml) and incubated at 4° for 20 min and then at 37° for 40 min with shaking to form HVJ–liposomes. About 120 μl of RBC ghosts containing proteins is suspended in the resulting HVJ–liposomes (3 ml) and the mixture is incubated at 4° for 20 min and then at 37° for 3.5 hr with shaking to form vesicle complexes. The disadvantage of the new RBC ghosts is that the fusion efficiency of the vesicle with cultured cells is low (less than 10%) even if HVJ is used,[33] probably because the membrane lipids are lost from the ghost membrane by octylglucoside treatment. Therefore a long incubation (more than 1.5 hr at 37°) of RBC ghosts with HVJ–liposomes is required to supply the membrane lipids and to increase the fusion efficiency.

Introduction of DNA and Nuclear Proteins into Cells by Using Vesicle Complexes

This delivery system is available both for cells in suspension and for cells in monolayers.

Cells in Suspension

Suspension cell lines or cells detached from dishes by trypsinization are collected in a 15-ml conical tube (Falcon, Becton Dickinson, Oxnard, CA) and washed with BSS three times. The number of cells used for fusion with vesicle complexes should be more than 10^6 cells/tube. Cells are then suspended in 0.5 ml of BSS containing 1 mM $CaCl_2$ and mixed with an equal volume of vesicle complex suspension, and incubated at 4° for 5 min and then at 37° for 60 min. After incubation, fresh medium containing serum is added to the cell suspension to stop fusion and centrifuged at 1500 rpm for 5 min in a low-speed centrifuge. Cells are suspended in complete culture medium and inoculated into appropriate dishes.

Cells in Monolayers

Monolayers of cells are washed three times with BSS containing $CaCl_2$ and then 3 ml of vesicle complex suspension is added per 100-mm petri dish. The dishes are incubated first at 4° for 5 min and then at 37° for 120 min. The vesicle complex suspension is then replaced with fresh medium containing serum.

Practical Uses of Vesicle Complex

1. Plasmid DNA containing the thymidine kinase gene *(tk)* of herpes simplex virus and the fluorescent protein phycoerythrin were introduced into mouse Ltk⁻ cells by the vesicle complex. More than 95% of the cells showed red fluorescence in their cytoplasm and incorporated [³H]thymidine into their nuclei.[27] When HVJ was not used, no fluorescence or [³H]thymidine incorporation was observed.

2. Using this delivery system, plasmid DNA was cointroduced with HMG-1 into mouse Ltk⁻ cells.[27]

Localization of the DNA: The DNA was concentrated in the nuclei within 6 hr after cointroduction with HMG-1. In contrast, it took 24 hr for the DNA to enter the nuclei when bovine serum albumin (BSA) was used instead of HMG-1.

Expression of the DNA: The expression of chloramphenicol acetyltransferase gene *(cat)* reached a maximum at 6 hr after cointroduction with

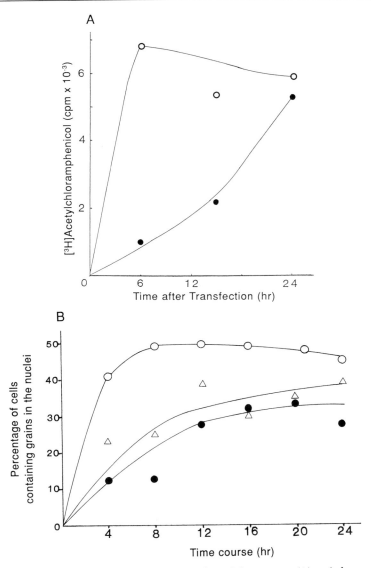

FIG. 2. Influence of HMG-1 on rapid expression of the *cat* gene (A) and *tk* gene (B) in mouse Ltk⁻ cells. (A) pAct-CAT (*cat* gene under the control of the chick β-actin promoter) was introduced into mouse L cells by vesicle complexes containing HMG-1 (O) or nonimmune rabbit IgG (●). *cat* activity was expressed as the radioactivity of [³H]acetylchloramphenicol formed from [³H]acetyl-CoA [M. J. Sleigh, *Anal. Biochem.* **156**, 251 (1986)]. (B) pTK4 (herpes simplex thymidine kinase gene) was introduced with HMG-1 (O), nonimmune rabbit IgG (△), or no protein (●) into mouse Ltk⁻ cells. See Kaneda *et al.*[27] for more detail.

HMG-1 (Fig. 2A). In control experiments in which the *cat* gene was transferred into cells with nonimmune IgG, the *cat* activity was about 14% of that observed with HMG-1 at 6 hr and then gradually increased to almost the same level as that with HMG-1. Similar results were obtained in *tk* gene expression (Fig. 2B).

Stable Transformants: This experiment is now being carried out. As reported previously,[32] stable transformants could be obtained efficiently from various cell lines by the use of HVJ–liposomes with GS. We estimate that the vesicle complex may also be used to obtain stable transformants efficiently.

3. Exogenous DNA [simian virus 40 (SV40) DNA,[27] human insulin DNA[28]] was expressed efficiently but transiently (for about 10 days) in adult rat liver when the vesicle complexes containing the DNA and HMG-1 were injected into the portal vein of adult rat. More recently we succeeded in the expression of the surface antigen of hepatitis B virus in hepatocytes of adult rats.[39] In *in vivo* experiments, HMG-1 is effective for high-level expression of the introduced DNA. Indeed, human insulin gene cointroduced with HMG-1 was transported into the nuclei of rat liver cells about 7.5 times more efficiently than the gene cointroduced with BSA.[28] The amount of transcript of the insulin gene cointroduced with HMG-1 was more than 10 times greater than that of the gene cointroduced with BSA.[28] The level of the human insulin secreted into rat serum corresponded to almost the normal concentration in human serum when cointroduced with HMG-1, whereas when BSA was used instead of HMG-1, the level of human insulin in rat serum was not significant.[28]

Troubleshooting

If all the ingredients of the vesicle complex are carefully prepared as described above, the reproducibility of simultaneous introduction of DNA and proteins is 100%. Nevertheless, several problems may arise. These problems, and our suggestions to solve them, are discussed below.

HVJ and liposomes do not fuse: There are two reasons why this problem may occur:

1. HVJ loses high fusion activity: Check the HVJ fusion activity by cell–cell fusion.[40] In many cases the best way to solve this problem is to change the seed of HVJ and prepare fresh virus solution.

2. Liposomes are defective: Check the trapping efficiency of plasmid

[39] K. Kato, Y. Kaneda, M. Sakurai, M. Nakanishi, and Y. Okada, *J. Biol. Chem.* **266**, 22071 (1991).
[40] Y. Okada and J. Tadokoro, *Exp. Cell Res.* **26**, 108 (1962).

DNA into liposomes.[32] Use fresh lipids. Change the lot of lipids (especially PS and GS). Dissolve all the lipids in tetrahydrofuran before evaporation.

Aggregates appear during incubation of RBC ghosts with HVJ–liposomes: This greatly reduces the efficiency of fusion between RBC ghosts and HVJ–liposomes. Make the HVJ solution fresh. Reduce the HVJ concentration to be used for preparation of HVJ–liposomes. The virus concentration can be reduced to 5000 HAU/0.5 ml to obtain the same results. Remove large aggregates of RBC ghosts before incubation with HVJ–liposomes by nylon mesh (pore size, 49 μm).

Recipient cells are damaged after incubation with vesicle complexes: This trouble has been rare when fibroblast cell lines are used. More than 80% of the cells were viable following treatment with vesicle complexes. But if severe cytotoxicity appears, free HVJ should be removed by sucrose gradient after incubation of liposomes with HVJ. Layer 60, 40, and 30% (w/v) sucrose in an SW 28.1 tube (Beckman) at a volume ratio of 1 : 1 : 8, and overlay 4 ml of HVJ–liposomes solution on top of the gradients. After centrifugation at 22,000 rpm for 4 hr at 4°, HVJ–liposomes remain at the top of the gradients, whereas free HVJ is trapped in the 40% sucrose layer. The clarified HVJ–liposomes are now fused with RBC ghosts to form vesicle complexes.

The introduced DNA does not migrate rapidly: In other words, the effect of nuclear proteins is not detected. Confirm whether the nuclear proteins are intact or not by biochemical analysis. Prepare the nuclear proteins again. We have not encountered such a problem when using HMG-1. We estimate that all nuclear proteins with both DNA-binding domains and nuclear migration domains can be used to transfer the DNA rapidly from the cytoplasm to the nucleus. We have also used DNA-binding proteins of extracts of *Xenopus laevis* oocytes instead of HMG-1. Although these DNA-binding proteins showed the same effects as HMG-1,[27] its reproducibility was about 50%. The mixture of histone H1 and nucleoplasmin (frog nuclear protein) may also be used,[41] but because the ratio of H1 to nucleoplasmin seems to be critical, the use of this mixture is not straightforward. Therefore we recommend HMG-1 as the nuclear protein to be cointroduced with plasmid DNA.

Improvements

The vesicle complex is an attractive vehicle both for *in vitro* and *in vivo* use. But the preparation of the vesicle complex is complicated because the

[41] Y. Kaneda and T. Uchida, unpublished data (1988).

vehicle consists of three kinds of vesicles, that is, HVJ, liposomes, and RBC ghosts. We have succeeded in enclosing DNA and proteins into liposomes without RBC ghosts by vortex mixing. Although the trapping efficiency of the added plasmid DNA is about 5%, the procedure is straightforward. Liposomes containing DNA and proteins are incubated with HVJ, and the resulting HVJ–liposomes can introduce DNA and proteins simultaneously into the same cells as efficiently as the vesicle complex, and are also available for *in vivo* experiments.[39]

Acknowledgments

The authors thank Dr. Y. Okada (Institute for Molecular and Cellular Biology, Osaka University, Japan) for advice and encouragement, and Dr. N. Düzgüneş University of the Pacific, San Francisco, CA) for critical reading of the manuscript.

[26] Delivery of Liposome-Encapsulated RNA to Cells Expressing Influenza Virus Hemagglutinin

By JEFFREY S. GLENN, HARMA ELLENS, and JUDITH M. WHITE

Introduction

The ability to deliver DNA or RNA into living cells provides a powerful tool for studying structural and functional features of a particular nucleic acid molecule *in vivo*. Several well-established techniques are available for promoting the uptake of exogenous DNA.[1-5] In contrast, it has remained more difficult to deliver RNA into large populations of cells, due most likely in part to the greater susceptibility of RNA to degradation. Often, RNA is delivered at low efficiency and to only a small percentage of cells.

We have previously shown that proteins preloaded in red blood cells (RBCs) can be introduced into cells that express the influenza virus hemagglutinin (HA).[6-8] The technique exploits the low-pH-induced membrane

[1] V. C. Bond and B. Wold, *Mol. Cell. Biol.* **7**, 2286 (1987).
[2] J. E. Celis, *Biochem. J.* **223**, 281 (1984).
[3] C. A. Chen and H. Okayama, *BioTechniques* **6**, 632 (1988).
[4] P. L. Felgner, T. R. Gadek, M. Holm, R. Roman, H. W. Chan, M. Wenz, J. P. Northrop, G. M. Ringold, and M. Danielsen, *Proc. Natl. Acad. Sci. U.S.A.* **84**, 7413 (1987).
[5] E. Neumann, M. Schaefer-Ridder, Y. Wang, and P. H. Hofschneider, *EMBO J.* **1**, 841 (1982).
[6] S. Doxsey, J. Sambrook, A. Helenius, and J. White, *J. Cell Biol.* **101**, 19 (1985).
[7] S. J. Doxsey, F. M. Brodsky, G. S. Blank, and A. Helenius, *Cell (Cambridge, Mass.)* **50**, 453 (1987).
[8] H. Ellens, S. Doxsey, J. S. Glenn, and J. M. White, *Methods Cell Biol.* **31**, 155 (1989).

fusion activity of the HA.[9,10] The successful delivery of large amounts of protein to the majority of HA-expressing cells[6-8] encouraged us to consider applying HA-mediated fusion technology for the purpose of RNA delivery. We are particularly interested in delivering relatively large RNA molecules, for example, viral genomes and messenger RNAs (mRNAs) encoding specific proteins. Loading relatively large RNAs into RBCs such that they remain undegraded has, however, proved to be difficult.[11] We therefore sought an alternative carrier vehicle. In this chapter we describe a method that we have developed for delivering RNA encapsulated in liposomes to the cytoplasm of HA-expressing cells.

Our strategy for delivering liposome-encapsulated RNA into cells is similar to that used by influenza virus itself for infection of host cells (Fig. 1). In both cases, lipid vesicles with glycoproteins protruding from the bilayer and RNA within the lumen fuse with the plasma membrane of the target cell and fusion is promoted by the well-characterized acid-induced conformational change in HA.[9,10] In the case of the virus, the HA is present on the surface of the vesicle (i.e., the viral envelope). The virus binds to sialic acid-containing receptors on the target cell surface and fuses when it encounters low pH in the endosome. In this manner the influenza virus RNA genome gains access to the cell interior for replication (Fig. 1D). In our RNA delivery scheme, HA is expressed on the target cell surface and RNA-containing liposomes are prepared with the red blood cell sialoglyco-protein, glycophorin, which provides a specific attachment site for the HA.[12] After binding the liposomes to HA-expressing cells, fusion is induced (at the cell surface) by a brief drop in medium pH (Fig. 1F). The result is synchronous delivery of RNA into the cells.

Materials and Methods

Preparation of RNA

RNAs are synthesized in large quantities (several hundred micrograms) by *in vitro* transcription of linearized plasmid DNAs.[13] The DNA templates contain the gene of interest under the transcriptional control of

[9] T. Stegmann, D. Doms, and A. Helenius, *Annu. Rev. Biophys. Biophys. Chem.* **18**, 187 (1989).

[10] J. White, *Annu. Rev. Physiol.* **52**, 675 (1990).

[11] M. C. Rechsteiner and R. A. Schlegel, in "Microinjection and Organelle Transplantation Techniques: Methods and Applications." (J. E. Celis, A. Graessmann, and A. Loyter, eds.), p. 89. Academic Press, Orlando, FL, 1986.

[12] H. Ellens, J. Bentz, D. Mason, F. Zhang, and J. M. White, *Biochemistry* **29**, 9697 (1990).

[13] J. K. Yisraeli and D. A. Melton, this series, Vol. 180, p. 42.

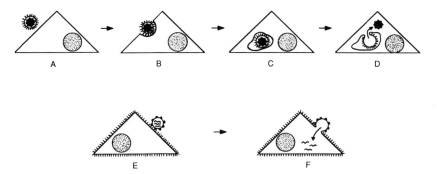

FIG. 1. Schematic of HA-mediated fusion events. Acid-induced HA-mediated fusion is depicted between a virus and a cell (A–D), and between a liposome and an HA-expressing cell (E and F). An influenza virus particle binds to a cell via HA spikes (A) and is endocytosed (B and C). When the virus encounters an endosomal pH of 5 (D), HA molecules are activated to promote fusion between the viral envelope and the endosomal membrane, thereby allowing the genome-containing nucleocapsid to enter the cytoplasmic compartment for subsequent replication. A liposome containing *in vitro*-transcribed RNA binds via glycophorin molecules in its membrane to HA molecules on the surface of the target cell (E). When the outside medium pH is lowered to 5 (F), the conformational change induced in the HA activates the protein to promote fusion of the two apposed membranes. The result is cytoplasmic delivery of the encapsulated RNA.

either an SP6 or T7 bacteriophage promoter. The transcription vector we have used also contains 5′ and 3′ untranslated sequences derived from *Xenopus* β-globin mRNA.[14] The genes encoding the enzymes chloramphenicol acetyltransferase (CAT)[15] and luciferase[16] are transcribed in the presence of 5′ m⁷GpppG 3′ (Pharmacia, Piscataway, NJ), a cap analog.[13] Following transcription, template DNA is digested with RNase-free DNase (Promega, Madison, WI). After two phenol:chloroform (1:1) extractions, the RNA solution is brought to 0.3 M sodium acetate and precipitated with 2.5 vol of ethanol. Following two washes with 70% ethanol, the final RNA pellet is either resuspended for immediate encapsulation into liposomes (see below) or stored in 70% ethanol at −70° for later use.

Preparation of Liposomes

The method for the preparation of liposomes is based on that of MacDonald and MacDonald.[17] A detailed characterization of the lipo-

[14] P. A. Krieg and D. A. Melton, *Nucleic Acids Res.* **12**, 7057 (1984).
[15] W. V. Shaw, this series, Vol. 43, p. 737.
[16] M. DeLuca and W. D. McElroy, this series, Vol. 57, p. 3.
[17] R. I. MacDonald and R. C. MacDonald, *J. Biol. Chem.* **250**, 9206 (1975).

somes is described elsewhere.[8,12] A lipid stock containing, per milliliter, 0.75 μmol each of cholesterol (Cat. No. C-8253; Sigma, St. Louis, MO) and the phospholipids egg phosphatidylcholine and plant phosphatidylethanolamine (Cat. Nos. 840051 and 840024, respectively; Avanti Polar Lipids, Pelham, AL) is prepared in chloroform (Cat. No. VW1430-3; VWR Scientific, San Francisco, CA). One-milliliter aliquots of this lipid stock are stored at $-70°$ (for up to several months) under highest purity argon in acid-cleaned glass screw-capped KIMAX tubes (Fisher Scientific, San Franscisco, CA) sealed with Teflon tape. For each preparation of liposomes, 0.5 ml of methanol, 3 μl of [14C]cholesterol oleate (NEC-638; NEN Research Products, Wilmington, DE) and 20 μl of 10-mg/ml glycophorin (Cat. No. G9511; Sigma) are added to a tube of aliquoted lipid stock on ice. The tube is immersed in a bath sonicator at room temperature (model G112SPIT; Laboratory Supplies Co., Inc., Hicksville, NY) three to five times for 5 sec each, and then placed on a Buechi RE 111 Rotovapor (Fisher Scientific, San Franscisco, CA) at room temperature until the contents appear dry (about 1 hr). The resulting lipid film is further dried using a lyophilizer or the Rotovapor for approximately 1–2 hr. Lipid films are used directly or after storing (for up to 2 weeks) under high-purity argon gas at $-20°$.

The RNA to be encapsulated is resuspended to a final concentration of 0.8 to 10 μg/μl in 300–400 μl of calcium- and magnesium-free phosphate-buffered saline (PBS-CMF) containing 10 mM dithiothreitol (DTT) and 1 unit RNasin/μl (Cat. No. N2111; Promega) and added to a lipid film. The solution is then flushed with argon, capped, sealed with Teflon tape, and rotated at 4° overnight on a rotary apparatus (Cat. No. 400-110; Labindustries, Inc., Berkeley, CA). The apparatus is tilted at a slight angle so that the RNA solution in the tube just covers the entire lipid surface on rotation. The tube contents are then vortexed for 10 min until they become homogenous and no patches of lipid remain adhering to the walls of the tube. The volume is recorded and a first aliquot is saved (about 3 μl). The rest of the solution is transferred to an 11×34 mm polyallomer tube (#347357; Beckman Instruments, Inc., Palo Alto, CA). A solution of 5% (w/v) sucrose in PBS-CMF containing 10 mM DTT is underlaid with a 23-gauge syringe until the tube is nearly full. The liposomes are pelleted at 4° for 90 min at 90,000 g in a Beckman TLS 55 swinging-bucket rotor. The supernatant is carefully removed and extracted with phenol : chloroform (1 : 1) to recover unencapsulated RNA. Typically only 1–2% of the total RNA is encapsulated. The unencapsulated recovered RNA can be used again for preparing new liposomes. The liposome pellet is washed twice, each time by resuspension with PBS-CMF and centrifugation at 4° for 30 min as above. The final liposome pellet is resuspended in about 300 μl PBS-CMF, the volume is recorded, and a second aliquot is removed. Both aliquots are analyzed in a scintillation counter. The percentage phospholipid recovered, and the phospholipid concentration of the final liposome solution, are calculated using the recorded volumes and assuming that all 1.5 μmol of phospholipid was present in the pregradient liposome preparation. Phospholipid recoveries range from 20 to 80%. The final liposomes are then stored at 4° until use. We have used RNA-containing liposomes stored for over 2 months with good success.

Target Cells

Several stable cell lines expressing large numbers of HA molecules on their surfaces ($> 3 \times 10^6$ HA trimers per cell) are available.[8,18] These lines were derived from NIH 3T3 (GP4f, HAb-2) and CHO (WTM) cells by transfection of the HA structural gene from the Japan and X : 31 strains of influenza virus, respectively. For the studies described here, GP4f cells are used. They are grown in complete medium [DME-H16, 1 g glucose/liter, 3.7 g $NaHCO_3$/liter, 100 U penicillin/ml, 100 μg streptomycin/ml, 10% (v/v) fetal calf serum] in a 5% CO_2 incubator. Cells are typically plated at 100,000 cells/well in 6-well cluster dishes (Falcon 3046; Becton Dickinson and Co., Lincoln Park, NJ) 2 days before use.

Fusion of Liposomes to Hemagglutinin-Expressing Cells

Processing one plate at a time, the cells are washed twice with DME-H16 containing no serum and incubated for 4 min at room temperature with 2 ml of a solution containing trypsin and neuraminidase [5 μg trypsin/ml, 1 mg neuraminidase/ml (Cat. Nos. T8642 and N2876, respectively; Sigma) in DME-H16 without serum]. Trypsin treatment cleaves the fusion-inactive HA precursor, HA_0, to the fusion-competent HA.[9,10] Neuraminidase treatment of the target cells enhances binding of RBCs and glycophorin-containing liposomes,[12] presumably by decreasing electrostatic repulsion. Studies on fusion of RBCs to GP4f cells indicate that the amount of neuraminidase can be reduced to 0.2 mg/ml.[19] Although we predict that lower concentrations of neuraminidase will also suffice for optimal binding of glycophorin-containing liposomes, this parameter has not yet been investigated. After incubation with the trypsin/neuraminidase solution, the cells are washed twice with complete medium containing 20 μg soybean trypsin inhibitor/ml (Cat. No. T9003; Sigma). They are then returned to the incubator in complete medium for 45–90 min to allow the cells to reflatten. After washing twice with binding buffer [RPMI 1640

[18] J. Sambrook, L. Rodgers, J. White, and M. J. Gething, *EMBO J.* **4**, 91 (1985).
[19] S. Morris, D. Sarkar, J. White, and R. Blumenthal, *J. Biol. Chem.* **264**, 3972 (1989).

(Cat. No. 430-1800; GIBCO-Bethesda Research Laboratories, Gaithersburg, MD), bovine serum albumin (BSA; 0.2%, w/v), 10 mM N-2-hydroxyethylpiperazine-N'-2-ethanesulfonic acid (HEPES) (pH 7.4), 35 mM NaCl], 2 ml of liposome solution (2.5–12.5 nmol phospholipid/ml in binding buffer) is added per well. The plates are centrifuged twice for 5 min each time at 4° and 500 g, with a 180° rotation of the plates between spins. This centrifugation augments liposome binding and fusion. Unattached liposomes are aspirated and the wells are washed once quickly but gently with fusion medium (binding buffer containing 10 mM succinate and brought to pH 4.75). Two milliliters of fusion medium is then added and the plate is held for 90 sec in a 37° water bath. To ensure good temperature equilibration, care is taken to prevent the trapping of air bubbles between the plate and the water. After this brief incubation, the fusion medium is aspirated, 4 ml of complete medium is added, and the plate is returned to the CO_2 incubator for the desired amount of time.

Harvesting of Cells and Analysis of RNA Expression

The cells are harvested by trypsinization, diluted in complete medium, pelleted, washed twice with PBS-CMF, and resuspended with 50 μl sucrose lysis buffer [250 mM sucrose, 10 mM Tris (pH 7.4), 10 mM ethylenediaminetetraacetic acid (EDTA)] in an Eppendorf centrifuge tube. After three freeze-thaw cycles, each involving successive incubations in liquid nitrogen and a 37° water bath, the lysates are centrifuged at 9000 g for 10 min at 4° in an Eppendorf centrifuge. The supernatants are assayed for protein concentration by the method of Bradford[20] and stored at −70° until use. Chloramphenicol acetyltransferase activity is assayed by thin-layer chromatography[21] or by phase extraction,[22] using equal amounts of protein from each sample. Luciferase assays are conducted as described.[23]

Results

Tightly Controlled Delivery

The narrow pH window in which HA is induced to undergo its fusion-activating conformational change should allow tight control over the delivery process. To document this we have monitored the delivery of liposome-encapsulated messenger RNAs encoding specific enzymes to HA-expressing target cells. Because translation of the RNAs into protein

[20] M. M. Bradford, *Anal. Biochem.* **72**, 248 (1976).
[21] C. M. Gorman, L. F. Moffat, and B. H. Howard, *Mol. Cell. Biol.* **2**, 1044 (1982).
[22] B. Seed and J.-Y. Sheen, *Gene* **67**, 271 (1988).
[23] J. R. De Wet, K. V. Wood, M. DeLuca, D. R. Helinski, and S. Subramani, *Mol. Cell. Biol.* **7**, 725 (1987).

can occur only if the mRNA reaches the cytoplasm (i.e., on intracellular ribosomes), measuring the activity of the protein products of the mRNAs indicates the extent of cytoplasmic delivery. Such an experiment is described in Fig. 2. Liposomes containing CAT mRNA are bound and fused to HA-expressing fibroblasts. Two days later the cells are harvested and assayed for CAT activity. Extensive expression of CAT mRNA is observed only in the cells that both received liposomes and were briefly treated at the HA fusion–activation pH of 4.8. A small amount of CAT enzyme activity is detected in cells receiving liposomes but maintained at pH 7.4, a pH that does not potentiate fusion activity of HA. This low level of expression may, however, also be HA mediated. It could result from the endocytosis of liposomes that are bound by surface HA molecules. When these HA-bound liposomes reach the endosomal compartment, acid-induced HA-mediated fusion would occur.

Majority of Cells Receive Liposomal Contents

To ascertain the percentage of cells in a population that receive liposomal contents, ricin A chain, a potent inhibitor of eukaryotic translation, is encapsulated in glycophorin-containing liposomes. The latter are bound and fused to HA-expressing cells. Two hours later, the protein synthetic activity of the cells is measured as described previously.[8,12] The results, shown in Fig. 3, suggest that ~90% of cells receive delivered contents at lipid concentrations ≥ 10 nmol phospholipid/ml. Further increasing the amount of liposomes added does not significantly increase the extent of delivery. Although this analysis applies to the delivery of liposome-encap-

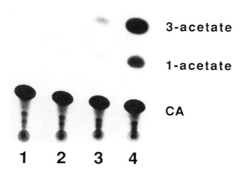

FIG. 2. Delivery of CAT mRNA-containing liposomes to GP4f cells. Cells receiving (lanes 3 and 4) and not receiving (lanes 1 and 2) CAT mRNA-loaded liposomes were treated with medium of either pH 7.4 (lanes 1 and 3) or pH 4.8 (lanes 2 and 4) for 90 sec. Cells were then grown in complete medium for 48 hr, harvested, and assayed for CAT activity (see Materials and Methods for details). CA, Unacetylated chloramphenicol; 3-acetate and 1-acetate, acetylated chloramphenicol products indicative of CAT activity.[21]

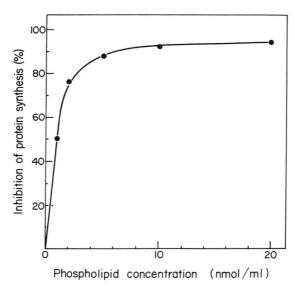

Fig. 3. Delivery of ricin A chain-containing liposomes to GP4f cells. Various concentrations of liposomes containing 0.2 mg ricin A chain/ml were bound and fused to GP4f cells as described in the Materials and Methods section. Two hours postdelivery, the incorporation of [^{35}S]methionine into protein was determined. The values represent the percentage of protein synthesis inhibition observed relative to cultures treated in the same manner but maintained at pH 7.4 to prevent fusion. [Reprinted with permission from H. Ellens, S. Doxsey, J. S. Glenn, and J. M. White, *Methods Cell Biol.* **31**, 155 (1989)].

sulated protein molecules, we expect that the same percentage of HA-expressing cells receive liposome-encapsulated mRNA molecules.

Translation of Delivered RNA and Comparison to Other Methods

The RNA that reaches the cytoplasm following liposome-mediated delivery is readily translatable. Translation products have been assayed as early as 1 hr postdelivery and increase with time, suggesting that the introduced RNA templates direct multiple rounds of translation (Fig. 4A). Chloramphenicol acetyltransferase activity is still readily detectable at 60 hr postdelivery (data not shown). Increasing the concentration of RNA in the liposomes over a fivefold range increases the amount of protein product produced (Fig. 4B).

The levels of desired protein products can also be increased by the use of Sindbis virus vectors.[24] The latter contain a cDNA copy of the Sindbis

[24] C. Xiong, R. Levis, P. Shen, S. Schlesinger, C. M. Rice, and H. V. Huang, *Science* **243**, 1188 (1989).

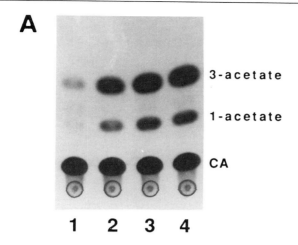

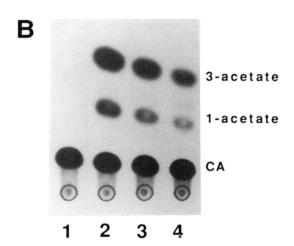

FIG. 4. Translation of delivered RNA. (A) Time course. Extracts of cells harvested at the following times postdelivery of CAT mRNA-containing liposomes to GP4f cells were assayed for CAT activity: 1 hr (lane 1), 3 hr (lane 2), 6 hr (lane 3), 12 hr (lane 4). (B) Effect of RNA concentration. Liposomes containing different concentrations of encapsulated CAT mRNA were delivered to GP4f cells. Six hours later, aliquots of cell extracts were assayed for CAT activity. Lane 1, no delivery; lane 2, 7.5 μg RNA/μl; lane 3, 3.8 μg/μl; lane 4, 1.5 μg/μl.

virus genome located downstream of an SP6 promoter such that *in vitro* transcription with SP6 RNA polymerase yields the infectious (+) strand of the virus. The viral structural genes in these vectors are replaced with DNA encoding the gene of interest. Thus no virus particles are produced. How-

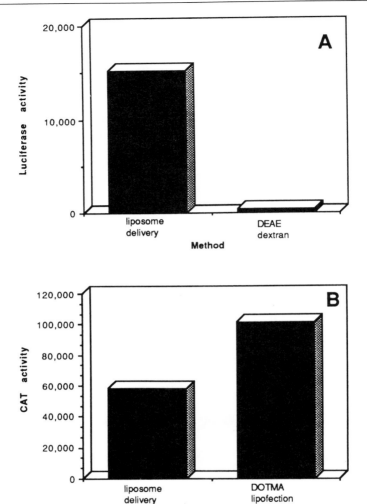

FIG. 5. Comparison to other methods. Experiments were conducted in triplicate in six-well dishes and cells (GP4f) were harvested about 8 hr posttreatment. DEAE-dextran transfection[25] and DOTMA lipofection[26] were performed as described, using 20 and 3.5 μg mRNA/well, respectively. Liposome delivery (0.9 μg mRNA/well) and extract preparation were as described under Materials and Methods. Extracts were assayed for luciferase (A) and CAT (B) activities following treatment with the corresponding mRNAs. Values are in 60-sec light emission counts (A)[23] and xylene-extractable counts per minute (B),[22] respectively. The corresponding activities normalized to the amount of mRNA used are shown in (C) and (D).

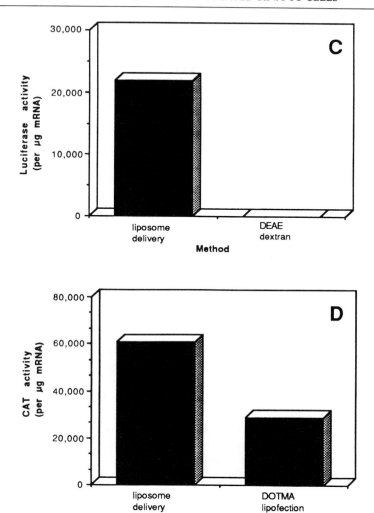

ever, because all of the genes encoding the proteins needed for replication and amplification of the Sindbis virus genome, as well as all of the cis-acting RNA regulatory sequences, are left intact, the gene of interest is amplified and expressed at high levels on introduction of the recombinant RNA genome into target cells.[24] The delivery of a Sindbis virus recombinant encoding CAT increases the production of CAT enzyme in target cells by two orders of magnitude over that obtained following delivery of standard SP6-transcribed CAT mRNAs (to between 10^5 and 10^6 molecules/cell). Our preliminary results also indicate higher expression of CAT activity when the liposome delivery protocol is conducted on cell lines (HAb-2 and WTM) that express HA at higher surface density than in the GP4f cells.

We have compared (Fig. 5) our delivery method to both DEAE-dextran transfection[25] and the N-[1-(2,3-dioleyloxy)propyl]-N,N,N-trimethyl-ammonium chloride (DOTMA) lipofection protocol.[26] The total amount of assayed protein product resulting from HA-mediated liposome delivery of encapsulated mRNA is considerably greater than that following the DEAE-dextran procedure (Fig. 5A and C) and roughly comparable to that resulting from the DOTMA lipofection technique (Fig. 5B and D).

Conclusions and Perspectives

In this chapter we have described a method for delivering liposome-encapsulated RNA into target cells that express a potent viral membrane fusion protein, the influenza virus HA, at their surface. To date we have used this method to deliver several different RNA species, including enzyme-encoding mRNAs and viral RNA genomes. The target cells remain viable after delivery and have been passaged several times. Perhaps the major advantage of the technique is the prospect of achieving synchronous cytoplasmic delivery to the majority of cells in a population. Because the same type of liposomes can be used for the delivery of both proteins[8,12] and RNA,[27] the method also offers the possibility of simultaneously delivering proteins and nucleic acids contained within the same liposome.

Several extensions of the technology are possible. The development of larger liposomes should increase the efficiency of delivery. The RNA-containing glycophorin liposomes that we use are $\sim 0.5\ \mu m$ in diameter. A twofold increase in diameter (to 1 μm) would increase the volume of the liposomes, and hence the expected delivery capacity, by eightfold. The use of gene amplification vectors[29] should allow the cloning of cell lines that express HA at higher surface density and that are thus able to fuse with greater numbers of liposomes.[12] Additional types of cell lines that express sufficient HA for use with the RNA delivery technology can also be established.

The HA-mediated RNA delivery technique should prove useful in

[25] G. Koch, *Curr. Top. Microbiol. Immunol.* **62,** 89 (1973).

[26] R. W. Malone, P. L. Felgner, and I. M. Verma, *Proc. Natl. Acad. Sci. U.S.A.* **86,** 6077 (1989).

[27] J. S. Glenn, J. M. Taylor, and J. M. White, *J. Virol.* **64,** 3104 (1990).

[28] J. S. Glenn and J. M. White, *J. Virol.* . **65,** 2357 (1991).

[29] R. J. Kaufman and P. A. Sharp, *J. Mol. Biol.* **159,** 601 (1982).

several areas of cellular and molecular biology. Using this methodology, questions concerning mRNA structure and function as well as message stability can be addressed. Simultaneously delivering a particular gene product to a large population of cells should permit the study of its effect on a biochemical, as opposed to a single-cell, scale. The method has been used in our laboratory for studying the replication of hepatitis delta virus,[27,28] a virus with an RNA genome. We also expect the method to be helpful in investigating other aspects of viral life cycles, such as packaging of the viral genome and budding of virus particles. Finally, the HA-mediated liposome delivery technique may be particularly well suited for studying the intracellular effects of antisense RNA[30] and ribozymes,[31] agents designed to destroy specific RNA molecules.

Acknowledgments

We would like to thank Bill Hansen for the SP6 vector encoding CAT, Charles Rice for the Sindbis virus vector, and Bob Malone and Inder Verma for help with experiments on the expression of mRNA encoding luciferase. We also thank Diane Mason for isolating the GP4f cell line and for excellent technical assistance, and Lori Hymowitz for manuscript preparation. J. Glenn was supported by the Medical Scientist Training Program (University of California, San Francisco) and H. Ellens was supported by an American Cancer Society Senior Postdoctoral Fellowship. The work was supported by a grant from the National Institutes of Health (AI22470) and an award from the Pew Scholars Program in the Biomedical Sciences to J. White.

[30] C. V. Cabrera, M. C. Alonso, P. Johnston, R. G. Phillips, and P. A. Lawrence, *Cell (Cambridge, Mass.)* **50,** 659 (1987).
[31] J. Haseloff and W. L. Gerlach, *Nature (London)* **334,** 585 (1988).

[27] Electroinjection

By G. A. NEIL and U. ZIMMERMANN

Introduction

For more than a century it has been recognized that high electric field pulses are capable of killing cells by lysis of the cell membrane.[1,2] Until 1972 it was believed that electric field-mediated release of intracellular proteins was possible only with associated destruction of the cells.[3] The

[1] A. J. H. Sale and W. A. Hamilton, *Biochim. Biophys. Acta* **148,** 781 (1967).
[2] A. J. H. Sale and W. A. Hamilton, *Biochim. Biophys. Acta* **163,** 37 (1968).
[3] E. Neumann and K. Rosenheck, *J. Membr. Biol.* **10,** 279 (1972).

discovery that short pulses of high electric field strength resulted in a reversible breakdown of the membrane without killing or significantly damaging the cell opened the field of electrical cell manipulation.[4,5] Zimmermann and co-workers showed for the first time that this reversible breakdown technique could be used not only for the release of low and high molecular weight substances from the cells, but also allowed the incorporation of normally membrane-impermeable substances into the cells.[6-19] These early observations were later confirmed by other groups[20-26] and electroinjection now enjoys wide use for the injection of nucleic acids and proteins into cells (see below).

Electric field-mediated injection (electroinjection) of substances into living cells is superior to osmotically mediated incorporation[27,28] in a variety of ways. It is not restricted to a few cell types because the phenomenon of reversible breakdown can be universally applied to all living cell membranes. Electroinjection can be readily controlled and the process can be observed microscopically. In addition, the precise control afforded by

[4] U. Zimmermann, J. Schultz, and G. Pilwat, *Biophys. J.* **13**, 1005 (1973).

[5] U. Zimmermann, G. Pilwat, and F. Riemann, *Biophys. J.* **14**, 881 (1974).

[6] U. Zimmermann, G. Pilwat, and F. Riemann, *Z. Naturforsch. C* **29C**, 304 (1974).

[7] U. Zimmermann, G. Pilwat, and F. Riemann, German Patent 2,405,119 (1974); U.S. Patent 4,081,340 (1974).

[8] F. Riemann, U. Zimmermann, and G. Pilwat, *Biochim. Biophys. Acta* **394**, 449 (1975).

[9] U. Zimmermann, G. Pilwat, C. Holzapfel, and K. Rosenheck, *J. Membr. Biol.* **30**, 135 (1976).

[10] U. Zimmermann, F. Riemann, and G. Pilwat, *Biochim. Biophys. Acta* **436**, 460 (1976).

[11] U. Zimmermann, G. Pilwat, F. Beckers, and F. Riemann, *Bioelectrochem. Bioenerg.* **3**, 58 (1976).

[12] U. Zimmermann and G. Pilwat, *Z. Naturforsch. C* **31C**, 732 (1976).

[13] U. Zimmermann, G. Pilwat, and B. Esser, *J. Clin. Chem. Clin. Biochem.* **16**, 135 (1978).

[14] J. Vienken, E. Jeltsch, and U. Zimmermann, *Cytobiologie* **17**, 182 (1978).

[15] U. Zimmermann, G. Pilwat, and J. Vienken, *Recent Results Cancer Res.* **75**, 252 (1980).

[16] U. Zimmermann, J. Vienken, and G. Pilwat, *Bioelectrochem. Bioenerg.* **7**, 553 (1980).

[17] U. Zimmermann, P. Scheurich, G. Pilwat, and R. Benz, *Angew. Chem., Int. Ed. Engl.* **20**, 325–344 (1981).

[18] U. Zimmermann, G. Kuppers, and N. Salhani, *Naturwissenschaften* **69**, 451 (1982).

[19] U. Zimmermann, *in* "Targeted Drugs" (E. P. Goldberg, ed.), p. 153. Wiley, New York, 1983.

[20] T. Y. Tsong, *Biosci. Rep.* **3**, 487 (1983).

[21] M. A. Yaseen, K. C. Pedley, and S. L. Howell, *Biochem. J.* **206**, 81 (1982).

[22] P. Lindner, E. Neumann, and K. J. Rosenbeck, *Membr. Biol.* **32**, 231 (1977).

[23] W. Schussler and G. Ruhenstroth-Bauer, *Blut* **49**, 213 (1984).

[24] K. Kinosita, Jr. and T. Y. Tsong, *Nature (London)* **272**, 258 (1978).

[25] D. E. Knight and M. C. Scrutton, *Biochem. J.* **234**, 497 (1986).

[26] D. Auer, G. Brandner, W. Bodemer, *Naturwissenschaften* **63**, 391 (1976).

[27] D. Auer and G. Brandner, *Z. Naturforsch. C* **31C**, 149 (1976).

[28] St. J. Updike, R. T. Wakamiya, and E. N. Lightfoot, *Science* **193**, 681 (1976).

widely available electroinjection devices provides a high degree of reproducibility. However, controversy remains with respect to the reversibility of electrically mediated cell membrane breakdown. Some authors have found that most of the cells subjected to electroinjection are killed, leaving only a few surviving cells for subsequent selection (see below). In contrast, with properly selected and optimized conditions, we have shown that this apparent disadvantage can be easily overcome.[29-34]

Reversibility of Membrane Breakdown

As is the case for electrofusion,[35] controlled membrane breakdown is the fundamental step in the incorporation of normally membrane-impermeable substances into living cells. Reversibility of the membrane breakdown requires that membrane and the cytosol perturbations caused by the application of electrical field pulses be sufficiently small to permit resealing without irreversibly damaging the cell. Postbreakdown resealing of the membrane occurs principally by the tendency of the lipid bilayer to regain its integrity, a rapid process.[36] Enzyme-mediated repair processes such as restoration of osmotic and electrolyte homeostasis by membrane "pumps," protein synthesis, and insertion into the cell membrane also come into play, but take longer to effect.[37] Because these processes are temperature dependent, the temperature profile during and after field application is important (see Temperature, below).

Electrical field-mediated membrane disruption must be of short duration if it is to be reversible, that is, lasting no more than a few microseconds.[11,30,37-45] With longer pulse duration, membrane breakdown

[29] U. Zimmermann, *Trends Biotechnol.* **1,** 149 (1983).
[30] H. Stopper, U. Zimmermann, and E. Wecker, *Z. Naturforsch. C* **40C,** 929 (1985).
[31] U. Zimmermann and H. Stopper, *in* "Biomembrane and Receptor Mechanisms" (E. Bertoli, D. Chapman, A. Cambria, and U. Scapagnini, eds.), Vol. 7, p. 371. Springer-Verlag, Berlin, 1987.
[32] H. Stopper, U. Zimmermann, and G. A. Neil, *J. Immunol. Methods* **109,** 145 (1988).
[33] R. Daumler and U. Zimmermann, *J. Immunol. Methods* **122,** 203 (1989).
[34] U. Zimmermann, P. Gessner, M. Wander, and S. K. H. Foung, *in* "Electromanipulation in Hybridoma Technology" (C. Borrebaeck and I. Hagen, eds.), p. 1. Stockton Press, New York, 1989.
[35] G. A. Neil and U. Zimmermann, this volume [27].
[36] R. Benz and U. Zimmermann, *Biochim. Biophys. Acta* **640,** 169 (1981).
[37] U. Zimmermann, *Biochim. Biophys. Acta* **694,** 227 (1982).
[38] U. Zimmermann, *Rev. Physiol. Biochem. Pharmacol.* **105,** 175 (1986).
[39] U. Zimmermann and H. Urnovitz, this series, Vol. 151, p. 194.
[40] C. A. Kruse, G. W. Mierau, and G. T. James, *Biotechnol. Appl. Biochem.* **11,** 571 (1989).
[41] K. Hashimoto, N. Tatsumi, and K. Okuda, *J. Biochem. Biophys. Methods* **19,** 143 (1989).
[42] K. Lindsey and M. G. K. Jones, *Plant Mol. Biol.* **10,** 43 (1987).

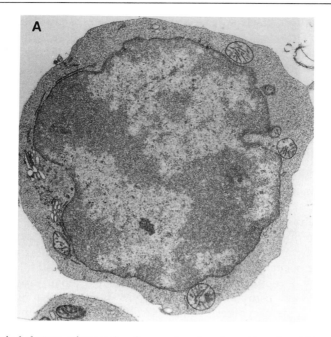

FIG. 1. Typical electron micrographs of mouse lymphocytes subjected to different pulse protocols. (A) Single field pulse of 14 kV cm^{-1} and 5-μsec duration; (B) single field pulse of 4 kV cm^{-1} and 40-μsec duration. The field pulse was applied in conductive, 30 mM KCl-containing pulse medium at 4°. The electron micrographs were made after the resealing process was completed at 37°. It is evident that the high-strength, short-duration field (A) did not lead to visible damage of the plasma and nuclear membrane or of the organelles. In contrast, application of longer pulses of low field strengths (B) resulted in the disruption of the membranes and in "vacuolization" of the cytosol. Magnification: × 8500.

usually becomes irreversible as illustrated in Fig. 1.[37–39,46–48] Breakdown of the membrane is, itself, rapid and can be demonstrated to occur within nanoseconds.[48] Long-duration applications of sufficiently high-magnitude field strength may cause prolonged loss of membrane integrity, resulting in large-scale electrolyte exchange with the environment of the cell. Such shifts may overwhelm the capacity of the cell to regain its homeostatic

[43] T. C. B. Schut, B. G. de Grooth, and J. Greve, *Cytometry* **11**, 659 (1990).
[44] J. G. Bliss, G. I. Harrison, J. R. Mourant, K. T. Powell, and J. C. Weaver, *Bioelectrochem. Bioenerg.* **20**, 57 (1988).
[45] H. Potter, *Anal. Biochem.* **174**, 361 (1988).
[46] R. Benz, F. Beckers, and U. Zimmermann, *J. Membr. Biol.* **48**, 181 (1979).
[47] U. Zimmermann and W. M. Arnold, *J. Electrostat.* **21**, 309 (1988).
[48] R. Benz and U. Zimmermann, *Biochim. Biophys. Acta* **597**, 637 (1980).

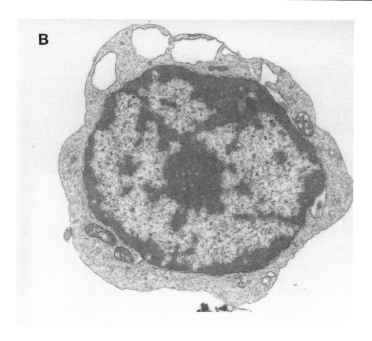

balance after restoration of membrane integrity. Because the membrane functions as a resistor, prolonged exposure to an electrical field may produce local heating that may, in turn, denature proteins and nuclear components.[49] This, too, can contribute to cell death.

To avoid irreversible destruction of the cells it is recommended that exponentially decaying, rather than square wave, pulses be administered to achieve membrane breakdown.[11,49] This circumvents the generation of lethal high current densities in the cell interior, which result from square-wave pulses.

Field Conditions

The strength of the field pulse is given by the Laplace equation.[35] Because of the radius dependence of the voltage required for membrane breakdown (critical field strength), up to 25 kV cm^{-1} is required for the permeabilization of bacterial cell membranes, whereas only about 1 kV cm^{-1} is required for the large plant protoplasts.[35] The angle dependence of the generated voltage dictates that electropermeabilization of the mem-

---

[49] G. L. Andreason and G. A. Evans, *BioTechniques* **6,** 650 (1988).

brane occurs first at membrane sites oriented parallel to the electrical field. Permeabilization in these membrane areas alone is sufficient to induce an exchange of ions and uncharged low molecular weight compounds. For release or uptake of macromolecules, larger areas of the membrane must be permeabilized and, thus, supracritical field strength must be applied. Empirically, it was found that, for the incorporation of bacterial plasmids, field strengths must be applied that are a factor of 5 to 8 higher in isosmolar media[32] and a factor of 2 to 3 higher in hypoosmolar media[33,34] than that calculated for permeabilization of membrane sites oriented in the electrical field direction, using the Laplace equation.[35] This is much higher than the field strengths used in electrofusion.[35]

Another important difference between electrofusion and electropermeabilization is the finding that the electrical breakdown of the cell membrane is optimally performed in a homogeneous electric field. In a homogeneous field, the cells are exposed to the same field strength at any given point. This has the consequence that uptake or release of substances, particularly macromolecules, is more nearly uniform for all cells, provided that the field strength applied is sufficiently high to exceed the breakdown voltage of smaller cells in the population.[50] Such a homogeneous field may be produced between two plate electrodes arranged parallel to one another. However, the introduction of cells into the field results in a field inhomogeneity that depends on the distance between the plate electrodes as well as on the density of the cell suspension.[50] For electroinjection, the distance between the electrode plates should, therefore, usually exceed 0.5 cm and the suspension density should not exceed 10^6 cells ml^{-1}. If these parameters are ignored, dielectrophoresis of the cells may occur during pulse application, which may result in unintentional cell fusion.[35,51] It follows that electroinjection applications demand higher voltage power supplies than those needed for electrofusion (wherein the distance between the chamber electrodes can be adjusted to a few hundred micrometers).

In an attempt to circumvent the voltage limitations of their power supplies, some authors have greatly lengthened the pulse duration (to the millisecond range) and/or used smaller electrode gap distances to perform electroinjection.[52-55] Such maneuvers are not recommended, however, for

[50] U. Zimmermann, in "Membrane Fusion" (J. Wilschut and D. Hoekstra, eds.), p. 665. Dekker, New York, 1990.
[51] U. Zimmermann and G. Pilwat, Abstr. 6th Int. Biophys. Congr., Kyoto, Japan, p. 140 (1978).
[52] J. A. Sokoloski, M. M. Jastreboff, J. R. Bertino, A. C. Sartorelli, and R. Narayanan, Anal. Biochem. 158, 272 (1986).
[53] Y. Watanabe, T. Meshi, and Y. Okada, FEBS Lett. 173, 247 (1984).
[54] J. Callis, M. Fromm, and V. Walbot, Nucleic Acids Res. 15, 5823 (1987).
[55] P. Christou, J. E. Murphy, and W. F. Swain, Proc. Natl. Acad. Sci. U.S.A. 84, 3962 (1987).

the reasons outlined above. In addition, one must be aware that the breakdown voltage is dependent on the pulse duration. As shown by Zimmermann and Benz[56,57] on artificial lipid bilayer membranes and giant algal cells, the breakdown voltage drops from about 1 V (at room temperature) to about 0.5 V when the duration time of the field pulse is extended from about 1 μsec to 10 μsec. The influence of pulse duration on the breakdown voltage of smaller (such as mammalian) cells is not known; however, measurements made using a particle analyzer suggest that pulse duration is likely to be important in these cells as well.[38]

Long-duration pulse-field application (in the millisecond range) may also produce secondary field effects, including membrane ionization.[58-60] This can lead to a "punch-through" effect of the membrane analogous to that seen in semiconductors. Punch-through results in a transient increase of membrane permeability and it is possible that some published methods based on long-duration pulse application produce uptake of macromolecules by this effect. A disadvantage of the punch-through effect is that it is difficult to control and, as a result, may result in the death of a large proportion of the cells so treated. Such methods are less reproducible and short-duration pulse application is, therefore, recommended.

There are a few exceptions in which pulse duration times longer than about 100 μsec must be applied. Prolonged pulse durations are needed in electroinjection under conditions favoring a long membrane charging time. The relaxation time of the exponential membrane charging process depends on both the radius of the cell and the external conductivity.[35] For the electroinjection of most cells in conductive solutions, charging of the membrane is rapid. Under these conditions, the steady state voltage is established in about 1 μsec, so that with pulse duration times of more than 5 μsec, the Laplace equation accurately predicts the required critical voltage for membrane breakdown of most cells. However, in the case of large cells, for example, oocytes with a diameter of more than 100 μm, the steady state voltage requires much more time to establish.[34] Hence the pulse duration time must be adjusted to about 100 to 500 μsec.

The second exception in which pulse duration times in the range of 100 μsec or longer must be applied is in the electroinjection of protein (Fig. 2).[61] To achieve maximum injection of proteins (e.g., albumins) into

[56] U. Zimmermann and R. Benz, *J. Membr. Biol.* **53**, 33 (1980).
[57] R. Benz and U. Zimmermann, *Planta* **152**, 314 (1981).
[58] H. G. L. Coster and U. Zimmermann, *Z. Naturforsch. C* **30C,** 77 (1975).
[59] H. G. L. Coster and U. Zimmermann, *Biochim. Biophys. Acta* **382,** 410 (1975).
[60] H. G. L. Coster and U. Zimmermann, *J. Membr. Biol.* **22**, 73 (1975).
[61] U. Zimmermann, R. Schnettler, G. Klock, H. Watzka, E. Donath, and R. Glaser, *Naturwissenschaften* **77**, 543 (1990).

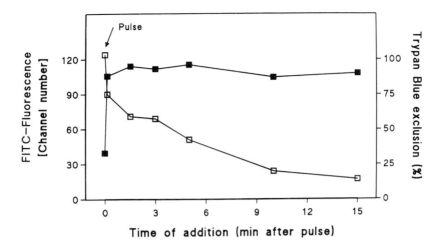

FIG. 2. Flow cytometric analysis of field-induced uptake of fluorescein isothiocyanate-labeled bovine serum albumin (4% FITC–BSA) into mouse L cells under hypoosmolar (90 mOsm) conditions. The protein was added either before or after the pulse at the time intervals indicated. Trypan blue (0.05%, v/v) was added after the pulse immediately or later, as indicated, in order to monitor the resealing process of the membrane by dye exclusion (determined by microscopy). Electroinjection was achieved by the application of a single field pulse of 4.55 kV cm^{-1} strength and 140-μsec duration. For further details, see Ref. 61. ($\square$) FITC fluorescence in the cells; ($\blacksquare$) trypan blue exclusion. It is evident that only 10% of the cells are stained, and are therefore dead, after a few seconds. In contrast, FITC–BSA was still taken up, when added after 10 to 15 min, although the uptake continuously decreased with time.

mammalian cells we found that pulse duration times on the order of 140 μsec are required if a field strength of about 4.5 kV cm^{-1} is used. At first glance, these results appear contradictory because this pulse duration leads to death of most of the cells in nucleic acid electroinjection protocols.[33,34] It is presumed that proteins in the external medium modulate cell membrane breakdown by an unknown mechanism. One explanation may lie in the observation that, compared with nucleic acids, much higher concentrations of protein must be present in the medium before significant uptake into the cytoplasm is evident after electroinjection (determined fluorometrically using fluorophore-labeled probes[61]). This high concentration of protein apparently blocks the membrane "pores" introduced by the field pulse and interferes with uptake as well as entry of the electrical field into the cell. Simultaneously, the release of essential ions and other low molecular weight compounds from the cell and the uptake of toxic compounds from the external medium are greatly reduced.

Medium Composition

From the foregoing considerations it is clear that the medium plays an important role in electroinjection. Because of the low heat dissipation during pulse application, conductive solutions can be used, including isosmolar electrolyte solutions. The temperature increases in the pulse medium after application of a single pulse or a pulse train is only 1 to 2°.[37,38] Thus many authors have used isosmolar phosphate-buffered NaCl solutions.[51,53,54,62-67] This is not, however, recommended, as shown by Zimmermann et al.,[16] because the uptake of Na^+ and release of K^+ through the electropermeabilized membrane along concentration gradients lead to an intracellular milieu of high toxicity. Sodium is an inhibitor of many enzymatic processes, whereas potassium is required as a cofactor of enzymatic processes.[16,68,69] The best alternative is, therefore, the use of electrolyte solutions in which sodium is completely replaced by potassium. Under these conditions an exchange of both ions after electropermeabilization is minimized. Although this is possible, Zimmermann et al.[16] showed that high concentrations of potassium in the external medium had adverse side effects on the membrane if the incubation time was longer than about 10 min. Experiments with different cell types showed that the elimination of NaCl and addition of 30 mM KCl to the pulse medium lead to optimum electroinjection of both low and high molecular weight substances.[30,32-34,53,61] Under these conditions only a small amount of the intracellular potassium is released, particularly if the resealing process was expedited by the appropriate temperature profile (see the next section). The amount of potassium is so small that the viability of the cells was not affected.[34] In addition, about 1 mM potassium phosphate buffer is added to the pulse medium and, in order to maintain isosmolarity, the appropriate amount of inositol is added.

We have found that, as in the case of electrofusion,[35] the performance of electroinjection in hypoosmolar solutions yielded a much higher uptake of DNA and proteins.[32-34,67] Experiments using a variety of mammalian

[62] E. Neumann, M. Schaefer-Ridder, Y. Wang, and P. H. Hofschneider, *EMBO J.* 1, 841 (1982).

[63] F. Toneguzzo, A. Keating, S. Glynn, and K. McDonald, *Nucleic Acids Res.* 16, 5515 (1988).

[64] R. A. Winegar, J. W. Phillips, J. H. Youngblom, and W. F. Morgan, *Mutat. Res.* 225, 49 (1989).

[65] H. Potter, L. Weir, and P. Leder, *Proc. Natl. Acad. Sci. U.S.A.* 81, 7161 (1984).

[66] R. Chakrabarti, D. E. Wylie, and S. M. Schuster, *J. Biol. Chem.* 264, 15494 (1989).

[67] J. C. Knutson and D. Yee, *Anal. Biochem.* 164, 44 (1987).

[68] U. Zimmermann, G. Pilwat, and T. Gunther, *Biochim. Biophys. Acta* 311, 442 (1973).

[69] M. Lubin and D. Kessel, *Biochem. Biophys. Res. Commun.* 2, 249 (1960).

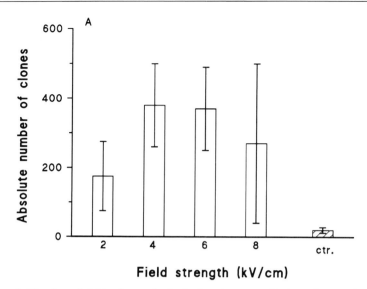

Field strength (kV/cm)

FIG. 3. Number of stable clones obtained after electrotransfection of mouse L cells and J774.A1 macrophages in 75 mOsm, 30 m*M* KCl-containing pulse solutions. The concentration of the linearized plasmid, pSV-2*neo* (conferring resistance to the antibiotic G-418) was 1 μg ml^{-1} for L cells (A and B) and 5 μg ml^{-1} for macrophages (C). The plasmid was added to the pulse medium before field pulse application. A single field pulse of 5-μsec duration at 4° was used. The cell suspension density was varied between 10^5 and 2×10^6 cells/ml. If not otherwise stated, the postincubation time in the pulse medium was 2 to 10 min at 4°. The temperature was raised to 37°. After appropriate dilution with culture medium, the cells were kept for a further 30 to 60 min at this temperature before they were transferred into complete growth medium. After 48 hr the culture medium was replaced by selection medium containing the antibiotic G-418. Resistant clones were counted after about 10 days. (A) Mouse L cells: Clone number as a function of the field strength. The clone number is given per 10^5 cells ml^{-1} because no dependence of the clone number on suspension density was observed. Data are the mean of three independent experiments ± standard deviation. Hatched column: Clones obtained in isosmolar pulse medium, using normal field treatment.[30,32,33] (B) Mouse L cells: Clone number as a function of the postincubation time at 4° per 2×10^5 cells ml^{-1} after a 5-μsec duration field pulse of 5 kV cm^{-1} strength under hypoosmolar conditions. It can be seen that the postincubation time is critical and should not exceed 10 min. Hatched column: Control experiments in isosmolar pulse medium. The data are the means of six experiments ± standard deviations. (C) Macrophages: Clone number as a function of the strength of the field pulse per 2×10^6 cells ml^{-1}. Hatched columns: Control experiments in isosmolar medium. In this case the yield of clones is lower compared to mouse L cells because of the poor expression of the plasmid in this cell type. Data are the mean of three independent experiments ± standard deviation.

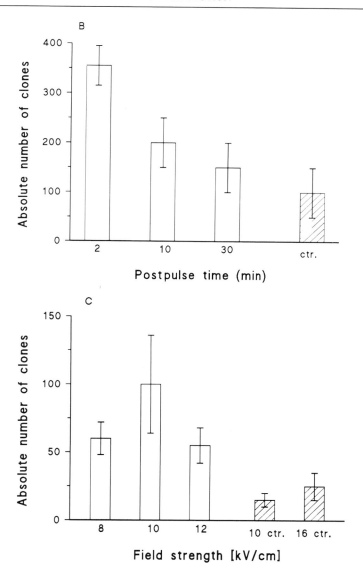

cell types showed that osmolarities between 75 and 90 mOsm gave the best results (Fig. 3).[33,34] Electroinjection of plant protoplasts is also superior under hypoosmolar conditions (U. Zimmermann *et al.*, unpublished). An interesting exception is the electroinjection of yeast cells. With these cells,

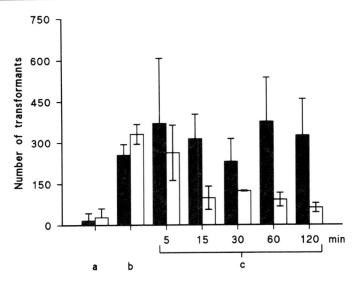

FIG. 4. Electrotransfection of protoplasts of the yeast strain AH 215 of *Saccharomyces cerevisiae* with supercoiled plasmid DNA pADH 040-2. The suspension density was adjusted to 10^9 cells ml^{-1}. The composition of the isosmolar medium was as described in Fig. 3. Three field pulses of 10 kV cm^{-1} strength and 10-μsec duration were applied. The time interval between consecutive pulses was adjusted to 2 min. The DNA (10 μg ml^{-1}) was added either before (b) or in time intervals of 5 to 120 min after (c) field application to the pulse medium. (a) Control experiments in which DNA was added without pulse application. Clones were counted 2 to 3 weeks after transfer of the pulsed cells to regeneration and selection agar. Solid columns: Number of clones when pulse application was performed at 4°; open columns: Number of clones when pulse application was performed at room temperature. Note that low temperature during pulse application resulted in significantly higher yields, in particular when the DNA was added after the pulse, indicating that the resealing process of the membranes is slow. The yield of clones is less than in electrotransfection experiments on mammalian cells (see Fig. 2) despite higher cell numbers and higher concentrations of DNA. The main reasons are the poor cell wall regeneration capacity of the protoplasts (about 5%) and the use of supercoiled DNA. If linearized DNA can be used (as in Fig. 2), the yield of clones is at least an order of magnitude higher.

electroinjection of DNA under isosmolar solutions of 1.2 Osm gives the highest yield of transformants (Figs. 4 and 5[70]). It should be mentioned that hypoosmolar electroinjection of bacteria has thus far not been performed. Whereas hypoosmolar electroinjection requires the application of a single breakdown pulse for optimal results, isosmolar electroinjection requires several successive pulses.[30,32,71] One must be aware that the interval between consecutive pulses must be carefully chosen. In contrast to the

[70] A. Salek, R. Schnettler, and U. Zimmermann, *FEMS Lett.* **70**, 67 (1990).
[71] K. Lindsey and M. G. K. Jones, *Planta* **172**, 346 (1987).

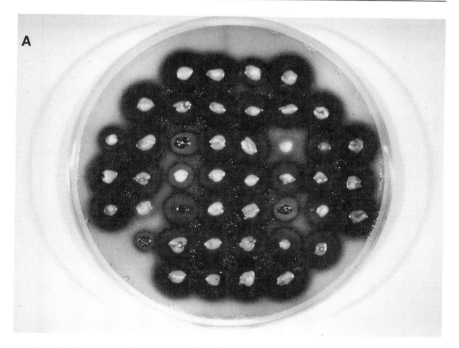

FIG. 5. Typical results of assays for killer activity in (A) transformed killer-negative variants of the *Saccharomyces cerevisiae* superkiller mutant strain T 158C and (B, see page 352) in the industrial (killer-sensitive) *S. cerevisiae* strain AS-4/H2 (rho⁻). *Note:* Killer strains of yeasts secrete a polypeptide toxin to which they are immune but which is lethal to sensitive strains of their own species and frequently to those of other species and genera of yeast. For electroinjection 10 μg circular dsRNA for K_1 toxin strain of the superkiller mutant strain T 158C was used per milliliter (for isolation, see Salek *et al.*[70]). For electroinjection the protoplasts suspended in isosmolar solution (1.2 M sorbitol plus 30 mM KCl, 1 mM CaCl$_2$ and small amounts of phosphate buffer) were exposed to three field pulses of 18.2 kV cm^{-1} strength and 40-μsec duration at 4°. The pulses were applied with a time interval of 2 min. Ten to 20 min after pulsing, the protoplasts were transferred to cell wall regeneration agar and, later on, to selection agar. The selection agar contained cells of the supersensitive *S. cerevisiae* strain S 6-1 (ade^{-1}). Transformed cells are indicated by a zone of growth inhibition of the supersensitive cells bounded by a ring of dead cells (visualized by using 0.03% methylene blue). The diameter of the ring of the inhibition zone of the dead cells indicates the strength of killer activity. It is seen that the number of transformants in (B) is much less than in (A). The reason is that the expression of the electroinjected dsRNA in the industrial yeast strain is strongly diminished. The high yield of stable transformants of the killer-negative variant of T 158C shows the efficiency of the electroinjection method.

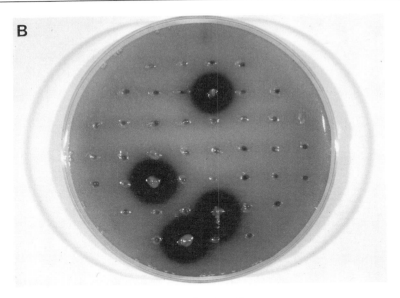

FIG. 5. *(continued)*

procedure used in electrofusion, the time interval between pulses should be at least 30 sec and as long as 120 sec. This is done to allow resealing of the membrane between pulses and thus to prevent buildup of a membrane potential. In electrofusion, the rapid application of several pulses under isosmolar conditions leads to an enlargement of the membrane "pores" in the contact zone due to secondary processes, which is desirable for intermingling of the apposed membranes. Prolonged delay between pulses allows the cells to change their orientation to the field lines by Brownian motion. New areas are thus exposed to the subsequent breakdown pulses, facilitating the entry of molecules located close to the electropermeabilized area.

A concentration of DNA of about $1-5$ μg ml^{-1} in the medium is generally sufficient to generate maximal numbers of stable transformants, provided optimal field conditions (outlined above) are employed. The cell suspension density should not exceed 10^6 cells ml^{-1} for mammalian cells and large plant protoplasts. We have found that 10^4-10^5 cells ml^{-1} is optimal. Larger suspension densities may be used with yeast protoplasts and bacteria owing to their smaller size and therefore their lower field disturbance.

An important question is whether Ca^{2+} and Mg^{2+} should be added to the electroinjection medium. For injection of DNA and RNA we have found that the addition of 1 mM Ca^{2+} and Mg^{2+} to the external medium enhances the efficiency of electroinjection in yeast protoplasts.[70] The addition of 0.1 mM Ca^{2+} is also optimal for plant protoplasts. Conversely, the addition of divalent cations reduces the number of stable transformants generated by the electroinjection of DNA into mammalian cells.[33,34] It is possible that Ca^{2+} precipitation of DNA may increase the concentration of DNA on the surface of the yeast protoplasts, thus facilitating entry into these smaller cells. Karube et al.[72] observed that Ca^{2+} concentrations in the pulse medium as high as 10 mM enhance electroinjection yields in yeast cells. It should be noted, however, that Ca^{2+} concentrations in this range can result in DNA transfection in the absence of pulse application. Despite the fact that divalent cations reduce electroinjection efficiencies in mammalian cells, we do not recommend the use of chelating agents such as ethylenediaminetetraacetic acid (EDTA).[35] Such agents are readily introduced into the cytoplasm after breakdown pulse application and may be toxic under these circumstances.

Temperature

The temperature at which electroinjection is performed is critical. Some authors have suggested that the procedure should be performed at room temperature[62,64,72,73] whereas others have found 4° to be optimal, particularly when whole plasmids are to be introduced into the cell.[44,45,55,67,71] Temperature comparisons can be made only after consideration of field and medium parameters. This is because the field strength required for membrane breakdown increases significantly below about 10°. At 4° the breakdown voltage is approximately twofold higher than at room temperature, that is, about 2 V.[17] Correspondingly, the field strength required for uptake must be doubled when electroinjection is performed at 4°. Using these calculations, Zimmermann et al.[10-15,17,30,31,36-39] showed that electroinjection at 4° is usually superior to room temperature when membrane-impermeable substances are taken up by diffusion along concentration gradients after membrane breakdown. The reason underlying this is that the membrane resealing process is strongly temperature dependent (Fig. 4). Studies with dye uptake showed that the membrane reseals in 1 to 5 min at 37°,[16] whereas at 4° resealing proceeds slowly (after 30 min substances are still exchanged through the membrane by diffusion both under hypo- and isosmolar conditions).

Thus, to maximize uptake of substances across the permeabilized

[72] I. Karube, E. Tamiya, and H. Matsuoka, FEBS Lett. 182, 90 (1985).
[73] D. A. Spandidos, Gene Anal. Technol. 4, 50 (1987).

membrane, electroinjection at 4° is recommended for most cells. However, because intracellular contents are simultaneously released while the membrane is permeable, and because homeostatic enzymes are inhibited at low temperatures, many mammalian cells are compromised by low-temperature electroinjection. For mammalian cells we have shown[33] that maintaining the medium temperature at 4° for about 2 min, then increasing it to 37°, gives optimal results (Fig. 3B).

We have discovered that the resealing process of the membrane is much more complicated than previously suspected. This is shown in Fig. 2 for uptake of the dye, trypan blue, and fluorescein-labeled protein into mouse L cells. Trypan blue staining of electric field-treated cells in protein-containing solutions showed that the membranes of these cells were virtually impermeable to this low molecular weight substance after a few seconds at 37°. In contrast, flow cytometric analysis showed that under the same field and temperature conditions, these cells were permeable to fluorescein-labeled albumin (Fig. 2). This result indicates that, as at 4°, protein can be taken up by diffusion across the permeabilized membrane provided that the concentration is high. However, in contrast to the results of the trypan blue experiment, significant uptake of the protein molecules was still observed in about 60 to 70% of the mouse L cells when the protein was added a few minutes after field application in both hypoosmolar solution and nearly isosmolar (260 mOsm) solutions (Fig. 6). It is interesting to note that under these circumstances the proteins were trapped in discrete vesicles (Fig. 6). Similar results were obtained with SP2/0 hybridoma cells by using fluorescein-labeled albumin. Varying the Ca^{2+} concentration in the external medium showed that the vesicle formation after pulse application was enhanced when 0.1 mM calcium acetate was present (U. Zimmermann et al., unpublished).

Similarly, although yeast protoplasts are likewise impermeable to dye a few minutes after exposure to reversible electrical membrane breakdown, plasmid DNA is taken up by these cells for up to 120 min after breakdown pulse application. The yield of yeast transformants under these conditions is still reasonably high at 4°, but is diminished at room temperature (Fig. 4). Microscopic analysis of the yeast protoplasts also indicated the formation of vesicular structures.

These results suggest that two mechanisms of macromolecule uptake operate in field-treated cells: uptake by passive diffusion and by field-induced "endocytotic" processes.[61] The latter process can apparently proceed for some time (depending on the cell type and on the macromolecule to be incorporated) even though membrane impermeability to low molecular weight compounds is restored. Formation of vesicular structures during osomotically induced expansion and subsequent shrinkage of plant

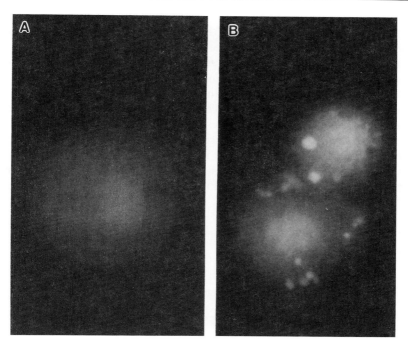

FIG. 6. Fluorescence photomicrographs of mouse L cells containing electroinjected FITC–BSA. For experimental details, see Fig. 2. (A) FITC–BSA (4%) distribution in the cells if the protein was added before field application. (B) FITC–BSA distribution if the protein was added 3 min after the breakdown pulse. The photographs were made 100 min after transfer to isosmolar medium.

protoplasts has also been reported.[74] These vesicles may be excess membrane material formed when the cells swell under hypoosmolar conditions and later internalized when the cells are returned to isosmotic conditions.

By analogy, the breakdown pulse creates a "hypoosmolar state" within the permeabilized cells because of ion equilibration and subsequent water uptake along its concentration gradient to maintain internal colloid osmotic pressure. This is supported by the observed increase in cell volume after breakdown pulse application and clearly depends on the osmolarity of the external pulse medium.[9] Such volume expansion is maximal in isosmolar solutions. In hypoosmolar solutions the cells are "preswollen," but electrical field-induced water uptake will still occur to some extent. In both cases the original cell volume and the internal osmotic pressure are

[74] W. J. Gordon-Kamm and P. L. Steponkus, *Protoplasma* **123**, 83 (1984).

reestablished in isosmolar solutions after the resealing process has been completed. It should also be noted that water uptake after electropermeabilization is partly dependent on the concentration of macromolecules in the medium. This may be demonstrated by the field treatment of red blood cells in varying concentrations of inulin in electrolyte solutions (Fig. 7). The striking similarities between the osmotic phenomena reported for plant protoplasts and the field-induced osmotic processes that accompany membrane breakdown and subsequent resealing suggest that the formation of vesicular structures results primarily from osmotically created incorporation and subsequent removal of membrane-like material rather than being a direct effect of the electrical field. This "field-induced" endocytosis or pinocytosis is of great interest for basic research and may also have practical applications in the future. However, the researcher who is pri-

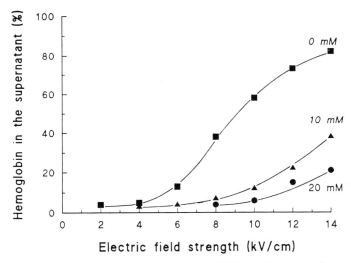

FIG. 7. Demonstration of the role of the internal colloid osmotic pressure for water uptake and cell swelling after electropermeabilization of the membrane for the human erythrocyte system (single field pulse of 40 μsec). Hemolysis of the erythrocytes was used as an indicator. Pulse application in isosmolar electrolyte solutions resulted in increasing hemolysis beyond a critical field strength of 4 kV cm^{-1} (■) because the colloid osmotic pressure of the hemoglobin is not compensated (after electropermeabilization of an increasing portion of cells of variable size) by an equivalent osmotic pressure of molecules of larger size in the pulse medium. Addition of increasing concentrations of inulin [10 mM (▲) and 20 mM (●)] resulted in a corresponding shift of the threshold field strength for the onset of hemolysis to higher field strengths. Apparently a higher degree of permeabilization of the membrane is needed and, therefore, higher field strengths (because of the angle dependance of the breakdown voltage) in order to achieve unhindered diffusion of the inulin molecules through the membrane.

marily interested in high-efficiency uptake of membrane-impermeable substances will find electroinjection at 4° and addition of the substances to be incorporated before or just after field application most suitable.

Protocols

Bacteria

Electroinjection of bacteria requires the application of large field strengths for membrane breakdown, owing to the small size of the cells. Using the Laplace equation, field strengths of approximately 20–25 kV cm^{-1} would appear to be needed. However, most authors have reported success using field strengths on the order of 6–10 kV cm^{-1}.[75-85] In these studies small electrode gap distances, long pulse applications (on the order of milliseconds), and high cell suspension densities (10^{10} cells ml^{-1}) were used. Under these circumstances one would expect that the field generated must be nonhomogeneous and that irreversible membrane breakdown would occur in the majority of the cells. Many investigators have not been able to transfect bacteria successfully using such protocols (personal communications to the authors). Nevertheless, the empirical application of these methods has apparently resulted in large numbers of both gram-negative and gram-positive transformants.[75-85]

Possible explanations for the success of such empirically derived protocols in bacteria include (1) the likelihood that killing the vast majority of cells has little consequence for the success of bacterial electroinjection, (2) the need for transfer of smaller numbers of plasmids for successful bacterial transformation (only one copy may be required, because the DNA does not need to be integrated into the bacterial chromosome), and (3) the possibility that electrical breakdown of the cell membrane is not required for the transfection of some bacteria. With regard to the latter explanation,

[75] T. Aukrust and I. F. Nes, *FEMS Microbiol. Lett.* **52**, 127 (1988).
[76] J. F. Miller, W. J. Dower, and L. S. Tompkins, *Proc. Natl. Acad. Sci. U.S.A.* **85**, 856 (1988).
[77] A. Taketo, *Biochim. Biophys. Acta* **949**, 318 (1988).
[78] W. J. Dower, J. F. Miller, and C. W. Ragsdale, *Nucleic Acids Res.* **16**, 6127 (1988).
[79] N. M. Calvin and P. C. Hanawalt, *J. Bacteriol.* **170**, 2796 (1988).
[80] E. S. Cymbalyuk, L. V. Chernomordik, N. E. Broude, and Yu. A. Chizmadzhev, *FEBS Lett.* **234**, 203 (1988).
[81] G. Zealey, M. Dion, Sh. Loosmore, R. Yacoob, and M. Klein, *FEMS Microbiol. Lett.* **56**, 123 (1988).
[82] A. Suvorov, J. Kok, and G. Venema, *FEMS Microbiol. Lett.* **56**, 95 (1988).
[83] J. D. Oultram, M. Loughlin, T. J. Swinfield, J. K. Brehm, D. E. Thompson, and N. P. Minton, *FEMS Microbiol. Lett.* **56**, 83 (1988).
[84] R. Wirth, *Forum Mikrobiol.* **12**, 507 (1989).
[85] K. Ito, T. Nishida, and K. Izaki, *Agric. Biol. Chem.* **52**, 293 (1988).

Benz[86] has shown that porins abundantly present in the cell walls of gram-negative bacteria are highly conductive. Gram-positive bacteria are surrounded by a murein "sac" that is also highly conductive.[87] It is unlikely that electrical breakdown of these structures occurs, given the conditions described above. During prolonged electrical pulse application, ionic fluxes occur at the gram-negative outer membrane and gram-positive murein sac. This phenomenon may be associated with local heating and osmotic processes that may, in turn, lead to rupture of the outer membrane or murein sac (in a fashion similar to the events occurring in Ca^{2+} coprecipitation of DNA[87]). This would allow plasmid DNA to pass the outer membrane (or murein sac) and enter the periplasmic space, from which point it is rapidly taken up into the bacterial cytoplasm. Such methods may be less controllable, reproducible, and applicable to electroinjection of other macromolecules, such as proteins.

When we performed bacterial electroinjection in preliminary experiments, using sound biophysical principles, that is, using high voltage fields with larger electrode gaps, short-duration pulse application, and lower cell densities, we found that the guidelines outlined above for larger cells also hold true for bacteria. Further experiments are underway in our laboratories to refine this technology. The following is a method devised by Dower et al.[78]

Electrotransformation of Bacteria

1. Harvest the cells in log growth phase. Wash and resuspend at a density of $2-4 \times 10^{10}$ ml^{-1} in chilled, nonconductive [10% (v/v) glycerol] medium.

2. Add the DNA (dissolved in low ionic strength buffer) to the cell suspension at a concentration of $0.1-2.5$ μg ml^{-1}.

3. Transfer 100 μl of the cell suspension to a chilled 0.2-cm electrode chamber (capacity about 2 ml). Care must be taken to remove all bubbles from sealed chambers to avoid explosion. If the electroinjection device is not fitted with cooling elements, the chamber must be maintained at 4° by other means. Caution must be exercised to avoid electrical contact with ice or condensation on the chamber surface if ice packs are used. Connect the chamber to the power supply (e.g., Biojet CF, Biomed, Theres, Germany).

4. Apply a single pulse of 12.5 kV cm^{-1} strength and 5-msec duration. Immediately after pulse application add 1 ml of room temperature nutrient medium, for example, SOC, and resuspend the cells with a Pasteur pipette. Place the cell suspension in a 37° incubator or water bath for 1 hr.

[86] R. Benz, *Annu. Rev. Microbiol.* **42,** 359 (1988).
[87] A. G. Sabelnikov, *Bioelectrochem. Bioenerg.* **22,** 271 (1989).

5. Plate the cell suspension on appropriate selection agar.

A straightforward protocol for direct electrotransfer of plasmid or chromosomal DNA from "donor" to "recipient" bacteria has been described by Kilbane and Bielaga.[88] DNA purification steps are completely obviated by using this technique.

Yeast Protoplasts

Molecular-grade reagents should be used wherever possible to avoid contamination with potentially harmful trace metals. This is particularly true of the water used to make solutions, for which high-performance liquid chromatography (HPLC)-grade water is optimal. The field strength should be adjusted according to the diameter of the protoplasts used, using the integrated Laplace equation.

1. Prepare the protoplasts, and wash and resuspend them at a density of 10^9 ml^{-1} in a chilled solution of 30 mM KCl, 1 mM CaCl$_2$, 0.3 mM KH$_2$PO$_4$, 0.85 mM K$_2$HPO$_4$, and 1.2 M sorbitol.
2. Add the material to be electroinjected to the solution. For nucleic acids, $5-10$ μg ml^{-1} has proved optimal in our hands.
3. Add an appropriate amount of the cell suspension to a precooled (4°) electroinjection chamber. Connect the chamber to the power supply.
4. Apply three exponentially decaying field pulses of strength 18 kV cm^{-1} and 40-μsec duration (or three pulses of 10 kV cm^{-1} and 10-μsec duration) at an interval of 2 min. After pulse application, allow the temperature of the chamber to increase to room temperature (or set the power supply to increase the temperature to 25° over $10-20$ min).
5. Gently flush the cells from the chamber into 5-ml centrifuge tubes containing 3 ml of complete growth medium (CGM) at room temperature. Allow the cells to remain undisturbed for 30 min.
6. After this incubation, the cells may be gently washed and resuspended in regeneration/selection agar and poured into petri dishes for colony selection.

Plant Protoplasts

The following protocol is suitable for both vacuolated and evacuolated plant protoplasts (G. Klöck and U. Zimmermann, unpublished). The technique requires a slightly hypoosmotic pulse medium.

1. Prepare the protoplasts and wash and resuspend them at a density of

---

[88] J. J. Kilbane and B. A. Bielaga, *BioTechniques* (in press).

10^5 ml^{-1} in a chilled solution of 0.45 mM mannitol, 0.1% (w/v) bovine serum albumin (BSA), and 0.1 mM CaCl$_2$.

2. Add the material to be electroinjected to the solution. For nucleic acids, $5 - 10$ μg ml^{-1} is optimal.

3. Add an appropriate amount of the cell suspension to a precooled (4°) electroinjection chamber. Connect the chamber to the power supply.

4. Apply three exponentially decaying field pulses of strength 4 kV cm^{-1} and 40-μsec duration or three pulses of 10 kV cm^{-1} and 15-μsec duration at an interval of $1-2$ min. After pulse application, allow the temperature of the chamber to increase to 20° (or set the power supply to increase the temperature to 20°) and maintain for about 1 hr.

5. Gently flush the cells from the chamber into 5-ml centrifuge tubes containing 3 ml of cell wall regeneration medium at room temperature. Allow the cells to remain undisturbed for 30 min.

6. After completing their incubation, the cells may be washed gently and resuspended in selection medium.

Mammalian Cells

Hypoosmolar Electroinjection. This protocol has been tested with both adherent cells (mouse L cells and macrophages) as well as cells growing in suspension (mouse hybridoma cells) and yields higher numbers of transformants than the isosmolar protocol given below.[34] It is, therefore, the method of choice for most mammalian cells. Cells in logarithmic growth phase are most suitable for electroinjection. Where RNA is to be electroinjected, the water should first be treated with diethyl pyrocarbonate and autoclaved to inhibit RNase.[89]

1. Treat the cells with 0.1 mg ml^{-1} dispase (6 U ml^{-1}, grade 1; Boehringer GmbH, Mannheim, Germany) in CGM for 1 hr or 1 mg ml^{-1} pronase (Boehringer GmbH). Wash the cells twice in RPMI 1640 medium.

2. Resuspend the cells in chilled 30 mM KCl, 1 mM sodium phosphate buffer (pH 7.0) and an appropriate amount of inositol to adjust the osmolarity to 75 mOsm [an osmometer such as the Osmomat 030 osmometer (Gonotec, Berlin, Germany) may be used]. The cell suspension density should be adjusted to 2×10^5 ml^{-1} to 1.0×10^6 ml^{-1}.

3. Add the material to be injected, that is, linearized plasmid DNA, RNA, or protein. For nucleic acids, a concentration of $1-5$ μg ml^{-1} is optimal. For proteins, a concentration of $10-50$ mg ml^{-1} should be used.

4. Add the cell suspension to a precooled electroinjection chamber

[89] P. Chomczynski and N. Sacchi, *Anal. Biochem.* **162**, 156 (1987).

with an electrode gap of 0.5–1 cm. Connect the chamber to the power supply.

5. Apply a single exponentially decaying field pulse of 4–6 kV cm^{-1} strength and 5-μsec duration. After pulse application, allow the temperature of the chamber to increase to room temperature (or set the power supply to increase the temperature to 25° over 10 min).

6. Gently flush the cells from the chamber into 5-ml centrifuge tubes containing 3 ml of 37° CGM. Place the cells in a 37° incubator or water bath and allow them to remain undisturbed for 30 min.

7. After completing their incubation, the cells may be washed gently and resuspended in appropriate culture vessels containing CGM supplemented with selection medium (e.g., geneticin for plasmids containing the neomycin resistance gene). If fluorophore-labeled proteins are injected, the cells may now be selected by flow cytometry.

Isosmolar Electroinjection. For those cells that cannot survive under hypoosmolar conditions, the use of isosmolar electroinjection may be a useful alternative. The protocol is identical to the hypoosmolar protocol except that (1) the pulse medium contains sufficient inositol to adjust the osmolarity to 300 mOsm, and (2) three or more breakdown pulses of strength 10 kV cm^{-1} and 5-μsec duration are administered at intervals of 1 min.

Acknowledgments

Supported by the Deutsche Forschungsgemeinschaft (SFB 176, B5), and the Federal Ministry for Research and Technology (DARA50WB9212-b) to U.Z. and the National Aeronautics and Space Administration (NASA-18433) to G.A.N. The expert assistance of Drs. Gerd Klöck and Reine Schnettler is gratefully acknowledged.

[28] pH-Sensitive Liposomes for Delivery of Macromolecules into Cytoplasm of Cultured Cells

By ROBERT M. STRAUBINGER

Studies to advance our understanding of membrane structure and function through the use of model systems have spawned a companion objective: to exploit lipid vesicle technology in devising vehicles for cellular delivery of a wide range of compounds and macromolecules. In particular, work to elucidate the controls and mechanisms of membrane fusion have inspired attempts to engineer the fusion of lipid vesicles with the membranes of target cells. Early recognition of the role of calcium as an apparent regulatory agent in a wide range of membrane fusion events led to the development of pure lipid model systems with a high selectivity for calcium as a modulator of fusion.[1,1a] Attempts to utilize calcium-dependent fusion for intracellular delivery have had some success; unfortunately, it is probable that this success obtains more from the fortuitous interaction of the constituent acidic phospholipids with cells than from the susceptibility of acidic phospholipids to undergo Ca^{2+}-dependent membrane fusion with the plasma membrane.

Intense efforts to elucidate the mechanisms by which liposomes interact with cells occurred in parallel with rapid progress in understanding the mechanism by which certain lipid-enveloped viruses entered cells: acid-dependent fusion of the viral envelope with cellular membranes, with subsequent intracellular liberation of the viral contents.[2] These lines of investigation inspired the development of pure lipid systems that undergo fusion at mildly acidic pH, along with mechanistic studies to understand the molecular events occurring during acid-dependent fusion of lipid vesicles, and morphological and biochemical studies to determine the mechanism by which such lipid vesicles deliver impermeant compounds and macromolecules to the cell interior.

The purpose of the present chapter[3] is to review aspects of liposome–cell interaction that inspired the development of acid-destabilized lipid vesicles, the strategies used in creating pH-sensitive liposomes, their characteristics *in vitro,* and their potential as cellular delivery vehicles. Methodology will be presented to allow the preparation and evaluation of pH-sensitive liposomes. Finally, hypotheses will be advanced to explain critical steps that determine the efficiency of cytoplasmic delivery by pH-sensitive liposomes.

Liposome–Cell Interaction

Charge-Mediated Binding to Cell Surface

As the techniques and probes for detecting liposome-mediated delivery to cells have improved, it has become apparent that liposomes bearing net negative electrostatic charge have a higher efficiency of delivery to many

[1] J. Wilschut, *in* "Membrane Fusion" (J. Wilschut and D. Hoekstra, eds.), p. 89. Dekker, New York, 1991.
[1a] D. Papahadjopoulos, S. Nir, and N. Düzgüneş, *J. Bioenerg. Biomembr.* **22,** 157 (1990).
[2] J. M. White, M. Kielian, and A. Helenius, *Q. Rev. Biophys.* **16,** 151 (1983).
[3] A previous volume in this series contains contributions on many aspects of the properties and uses of liposomes, and it is highly recommended as background to the present work: "Methods in Enzymology," Vol. 149 (1987).

cell types *in vitro* than do neutral liposomes composed of zwitterionic phospholipids. Such a phenomenon is rationalized by studies at the ultra-structural level[4] that demonstrated liposome entry into the endocytic apparatus of cells, principally through the cell surface "coated pits" commonly associated with the process of receptor-mediated endocytosis[5] (Fig. 1). Receptor-mediated endocytosis is a rapid, high-efficiency, and constitutive property of many cells that is responsible for concentration and internalization of nutrient macromolecules and other ligands that are present in the extracellular medium, often in trace quantities. Although it is not clear that liposomes enter cells exclusively through the coated pit/coated vesicle pathway, as opposed to morphologically distinct endocytic mechanisms (e.g., the smooth "coatless" vesicle pinocytic pathway[6]), it is clear that endocytosis is the principal means by which most liposome formulations interact with cells.

Presently there are no satisfactory explanations as to why negatively charged liposomes localize in clathrin-coated pits, the cell surface regions that function in an apparently specific manner as the first event in endocytosis. Hypotheses to explain liposome localization include the accumulation of (complementary) positive charge[7] in the coated pits, or charge-dependent adsorption of liposomes to "prosthetic" macromolecules, either in the extracellular medium or on the cell surface, that confer the ability to bind to specific cell surface receptors. In specialized cells such as macrophages, liposomes may bind to a nonspecific "scavenger receptor" that recognizes arrayed negative charge,[8] the natural function of which may be to clear oxidized lipoproteins and senescent cells from the circulation. As hypothesized below, elucidating the mechanism by which liposomes bind to the cell surface is critical to improving the efficiency by which pH-sensitive liposomes deliver their contents to cells.

Internalization into Endocytic Pathway

Subsequent to the binding of liposomes to the cell surface and localization in clathrin-coated pits, liposomes enter the cell within endocytic vesicles. Figure 1A depicts the likely early events in liposome–cell interaction. It is widely held that vacuolar H^+-translocating "pumps" rapidly acidify the lumen of endocytic vesicles[9]; it is clear from diverse experimen-

---

[4] R. M. Straubinger, K. Hong, D. S. Friend, and D. Papahadjopoulos, *Cell (Cambridge, Mass.)* **32**, 1069 (1983).

[5] M. S. Brown and J. L. Goldstein, *Proc. Natl. Acad. Sci. U.S.A.* **75**, 3330 (1979).

[6] S. C. Silverstein, R. M. Steinman, and Z. A. Cohn, *Annu. Rev. Biochem.* **46**, 669 (1977).

[7] N. Ghinea and N. Simionescu, *J. Cell Biol.* **100**, 606 (1985).

[8] K. Nishikawa, H. Arai, and K. Inoue, *J. Biol. Chem.* **265**, 5226 (1990).

[9] R. G. W. Anderson and L. Orci, *J. Cell Biol.* **106**, 539 (1988).

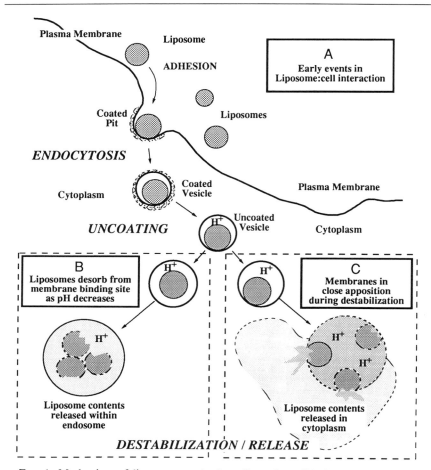

FIG. 1. Mechanism of liposome uptake by cells and possible intracellular fates. (A) Liposomes adsorb to the cell surface by a mechanism that depends on liposome negative charge. Liposomes enter the endocytic apparatus of cells through a classic pathway: localization in clathrin-coated pits and internalization into endocytic vesicles. The vesicle lumen acidifies rapidly, providing a potential trigger for inducing fusion of pH-sensitive liposomes with endocytic vesicle membranes. (B) and (C) represent alternative intracellular fates of pH-sensitive liposomes: (B) liposomes desorb from the endosome membrane as weakly acidic, liposome-stabilizing constituents become protonated and electrostatically neutral. Liposomes collapse, leaking their contents into the endocytic vesicle lumen; (C) liposomes are maintained in close apposition to the endocytic vesicle membrane during the drop in vesicular pH, leading to liposome-mediated destabilization of the endocytic vesicle membrane.

tal systems that internalized liposomes rapidly encounter a continuously acidifying compartment.[10-12] The rate and extent of endocytic vesicle acidification could have considerable impact on the efficiency of delivery by liposomes that use pH as a trigger to undergo membrane destabilization. In studies using specific ligands to carry probes into the endocytic pathway, the pH has been observed to fall to pH 6.5 within 5 min of formation of the endocytic vesicles.[13] Maximal acidification as low as pH 4.6 has been reported as the intravesicular pH in macrophages,[14,15] but the values may be higher in many other cells; pH 5.5 was observed as the lowest pH achieved in endocytic vesicles of fibroblasts.[13] Fluorescence studies using methods detailed below showed that the maximal acidity to which liposome contents are exposed in epithelioid cells (CV-1) is close to the value cited for fibroblasts, pH 5.6.[10] The maximal acidity reported by liposomes endocytosed by several macrophage-like cell lines (J774, RAW264.7, P388D1) was not stated explicitly, but is approximately pH 5.8–6.5.[11]

Method for Observing Binding and Intracellular Processing of Liposomes

Although many methods have been devised to observe the behavior of liposomes in the presence of cells, one general-purpose and one special-purpose fluorescence method will be detailed here. The general method is most useful for developing liposome formulations and optimizing incubation conditions for cytoplasmic delivery. The specific, more recent method allows continuous and rapid detection of liposome binding, endocytosis, and intracellular fate in living cells. pH-sensitive liposomes as delivery vehicles continue to evolve, and it is likely that optimization of the vehicle will depend heavily on observing the performance of liposomes during the critical early steps in endocytosis and intracellular processing.

Fluorescent dextrans are available commercially from a number of suppliers, and can be captured in liposomes at relatively high concentration. Because of the mass of these macromolecules (≥ 10 kDa), it is unlikely that liposome-encapsulated dextrans delivered to the endocytic apparatus of cells could gain access to the cytoplasm unless the integrity of endosomal or lysosomal membranes were compromised. Therefore fluo-

[10] R. M. Straubinger, D. Papahadjopoulos, and K. Hong, *Biochemistry* **29**, 4929 (1990).
[11] D. Daleke, K. Hong, and D. Papahadjopoulos, *Biochim. Biophys. Acta* **1024**, 352 (1990).
[12] C.-J. Chu, J. Dijkstra, M.-Z. Lai, K. Hong, and F. C. Szoka, *Pharm. Res.* **7**, 824 (1990).
[13] B. Tycko and F. R. Maxfield, *Cell (Cambridge, Mass.)* **26**, 643 (1982).
[14] C. T. DeDuve, T. DeBarsy, B. Poole, A. Trouet, P. Tulkens, and F. VanHoof, *Biochem. Pharmacol.* **23**, 2495 (1974).
[15] S. Ohkuma and B. Poole, *Proc. Natl. Acad. Sci. U.S.A.* **79**, 2578 (1978).

rescent dextrans, labeled with fluorescein, rhodamine, or other fluorophores, are excellent probes to detect the transfer of liposome contents to the cytoplasm. Cytoplasmic delivery is observed in the microscope as fluorescence diffusely filling the cytoplasm.[16] Smaller dextran (e.g., 10 kDa) distributes into the nucleus, whereas larger dextran (e.g., ≥ 40 kDa) is excluded from the nucleus, leaving a characteristic void in the diffusely labeled cytoplasm.[16] Highly confined "punctate" fluorescence is interpreted as evidence of liposomes bound to the cell surface or confined within vesicles of the endocytic pathway. A second, more recent assay allows the discrimination between these two possibilities.

HPTS[17] is a membrane-impermeant water-soluble fluorescent compound whose excitation spectrum (350–450 nm) is highly sensitive to pH.[18-21] The emission spectrum, centered at 510 nm, is invariant with changes in pH. Because of the magnitude of the pH-dependent excitation shift, standard fluorescence microscope filter sets are well suited for an observer to detect relatively small changes in pH. Previous work[10] with the particular "standard" filter sets of one manufacturer showed a 10-fold change in observed fluorescence over the range of pH 5.5 to 7.5, the range over which destabilization of most current pH-sensitive liposomes occurs. Changes of lesser magnitude are detected reliably without the aid of a photometer or image processing equipment.

Reagents

Fluorescent dextran, 10–40 kDa, labeled with rhodamine, fluorescein, or other probe appropriate for the fluorescence filters on hand: An alternative to HPTS if only the "cytoplasmic delivery" end point is of interest

HPTS (5 mM) in isotonic buffer,[22] pH 7.4 (or as compatible with pH-sensitive liposome formulation)

DPX[23] (50 mM) fluorescence quencher optionally included in HPTS solution: Adjust buffer tonicity accordingly if used

---

[16] R. M. Straubinger, N. Düzgüneş, and D. Papahadjopoulos, *FEBS Lett.* **179**, 148 (1985).

[17] HPTS, Hydroxypyrene-1,3,6-trisulfonate; known also as pyranine.

[18] K. Kano and J. H. Fendler, *Biochim. Biophys. Acta* **509**, 289 (1978).

[19] C. M. Biegel and J. M. Gould, *Biochemistry* **20**, 3474 (1981).

[20] K. A. Giuliano and R. J. Gillies, *Anal. Biochem.* **167**, 362 (1987).

[21] One supplier of HPTS is Molecular Probes, Inc. (Eugene, OR).

[22] For example, 145 mM NaCl, 10 mM TES [N-tris(hydroxymethyl)methyl-2-aminoethanesulfonic acid]; pH 7.4.

[23] DPX, p-Xylenebispyridinium bromide.

Equipment

Fluorometer and/or fluorescence microscope, the latter equipped with
the following:
  "Wide-band fluorescein" or other blue-band fluorescence filter set
    with excitation in the range of 440–490 nm (450 nm optimal) and
    emission in the range of ≥510 nm (510 nm optimal)
  Wide-band violet fluorescence filter set with excitation in the range of
    360–410 nm (400 nm optimal) and emission in the range of
    ≥450 nm (510 nm optimal)
  Photometer (optional; many photomicroscope camera controllers are
    suitably accurate for use as a substitute) or video camera and video
    image digitizer

Procedure. In the case of fluorescent dextran, the material is dissolved
at high concentration in a buffer that is isotonic with the medium in which
incubation with cells will take place. Typically, dextran is encapsulated at
1–10 mg/ml. Similarly. HPTS would be dissolved in a buffer that is
isotonic with the medium in which incubation with cells will take place.
The HPTS may be captured in liposomes at concentrations higher than
5 mM; the concentration of other osmotically active constituents such as
NaCl must be reduced accordingly. At >5–20 mM, HPTS undergoes
increasingly significant fluorescence self-quenching; depending on the application (e.g., maximal cellular delivery of HPTS required for high signal)
some quenching of HPTS in the intact liposomes may be tolerable. DPX, a
collisional quencher of HPTS, also may be included in the HPTS solution.[11,12] The fluorescence of intact HPTS/DPX-containing liposomes is
low, and HPTS/DPX liposomes have been used to observe cell-mediated
(or pH-mediated) changes in liposome integrity: leakage of liposome contents relieves DPX-dependent fluorescence quenching, and the resulting
increase in HPTS fluorescence signal is related to the fraction of liposomes
that have released their contents (leaked).

Fluorescent dextran, HPTS, or DPX/HPTS solutions are captured in
liposomes prepared by the desired procedure (see Ref. 24 and below), and
unencapsulated dye is removed, usually be gel chromatography. The exact
procedure used in incubating liposomes with cells depends on the objective
to be accomplished. To obtain an overview of the quantity of liposomes
taken up by a particular cell type, as well as the intracellular fate of
liposome contents, a small quantity of liposomes is added to cells in a

---

[24] For background information on general methods for preparing and characterizing liposomes, the reader is directed to a previous chapter on the subject: R. M. Straubinger and D.
Papahadjopoulos, this series, Vol. 101, p. 512.

serum-free buffer. Serum reduces the stability of many liposome formulations. After a defined interval, free liposomes and leaked liposome contents are washed away and cells are examined under the fluorescence microscope or in the fluorometer. Generally, the quantity of liposome phospholipid incubated with cells is in the range of $10-200$ nmol/10^6 cells. The rate of liposome binding to cells is such that most interaction generally occurs within 20 to 30 min of addition of liposomes to cells. Buffers such as Dulbecco's phosphate-buffered saline,[25] containing $0.5-4$ mM divalent cations (Ca^{2+} and Mg^{2+}) are used commonly during liposome–cell interaction. For cells that attach to a substratum, acid-cleaned or polylysine-coated[26] glass coverslips may be used, so that cells need not be detached from the substrate for observation in the fluorescence microscope.

By fluorescence microscopy, the determination of whether cytoplasmic delivery has occurred is based on the observation of fluorescence diffusely filling the cytoplasm; highly confined "punctate" fluorescence is interpreted as evidence for extracellular liposomes or intracellular vesicles containing liposome contents. HPTS, fluorescent dextran, or many other membrane-impermeant markers such as phycobiliproteins (e.g., phycoerythrin) would be satisfactory for demonstrating these end points. In many cases, the determination of whether punctate fluorescence corresponds to cell surface or intracellular liposome contents is important; one would want to know whether liposomes were bound but not internalized by cells, or whether liposomes collapsed in endocytic vesicles without transferring their contents to the cytoplasm. Such a determination is made most easily using HPTS as a liposome-encapsulated probe.

The fluorescence of HPTS excited at 450 nm is highly pH dependent; fluorescence decreases drastically as pH decreases. In contrast, fluorescence excited at 413 nm is invariant with pH, and fluorescence is inversely related to pH at ≤ 400 nm. Thus blue-band fluorescence filters excite HPTS at wavelengths where fluorescence is quenched under acidic conditions, but highly visible under neutral or basic conditions. Violet-band fluorescence filters excite HPTS at wavelengths where fluorescence is intense under all conditions. Therefore, comparison of fluorescence using blue vs violet excitation reveals whether liposome-delivered HPTS is in neutral or acidic compartments; HPTS in acidic compartments is poorly visible under blue excitation, but highly visible under violet excitation. In

---

[25] D-PBS: 137 mM NaCl, 3 mM KCl, 17 mM Na$_2$HPO$_4$, 1 mM KH$_2$PO$_4$, pH 7.4.

[26] Conditions may be determined empirically for optimal cell adhesion and minimal cytotoxicity; 1500-kDa polylysine diluted in distilled water to 10 μg/ml has been used as a starting point; coverslips are rinsed in the polylysine solution and allowed to air dry. For sterility, the polylysine may be filter sterilized or the coated coverslips may be exposed to ultraviolet light.

contrast, HPTS in neutral-pH compartments is intensely fluorescent when excited at either wavelength.

An example of blue-band filters would be standard fluorescein filter sets; although these filter sets tend to have an excitation maximum (490 nm) that is not well matched to the ("pH-sensitive") 450-nm HPTS peak, the high intensity of HPTS at neutral pH makes such suboptimal filter sets useful for many applications.

An example of violet-band filters would be those used to visualize Hoechst DNA fluorescent stains, and these filters tend to be well matched to HPTS fluorescence maxima. More detailed discussions of filters and the pH-dependent spectra of liposome-encapsulated HPTS may be found in Straubinger et al.[10] and Daleke et al.[11]

When the HPTS-based assay is applied to cells, it can be observed that liposomes report exclusively neutral pH immediately after addition to cells. Within 5 min of binding, clear evidence of acidification of liposome contents can be found (i.e., liposomes visible under violet, but not blue, excitation). With time, an increasing proportion of liposome-delivered dye reports acidification. Under conditions in which HPTS is known to be in the cytoplasm (e.g., after direct needle injection into cells[27]), HPTS gives diffuse labeling of the entire cell, and reports neutral pH. One potential problem we have found with the use of HPTS as a probe for intracellular delivery is that cytoplasmically localized HPTS is extruded from cells by an anion transport mechanism[28] that is probenecid-sensitive.[27] Under certain conditions (e.g., cells in complete nutrient medium and incubated at 37°), the cytoplasm could be cleared of HPTS within 1 hr.[27] Therefore observations should be made as befits the incubation conditions, so that cellular export of HPTS does not give a false-negative report of cytoplasmic delivery. Alternatively, fluorescent dextran could be used in parallel experiments, as cytoplasmic dextran is stable for long periods. Although never tested, it is likely that fluorescent dextrans having spectra complementary to the HPTS spectrum (e.g., Texas Red dextran) could be coencapsulated with HPTS to provide both short- and long-term assessment of cytoplasmic delivery.

In cases in which DPX has been included in HPTS liposomes to render intact liposomes relatively nonfluorescent, cell surface liposomes are much less apparent. By observing cells as they interact with HPTS/DPX-containing liposomes, it has been possible to monitor the stage at which cell-induced liposome leakage occurs, because leakage and dilution relieve

[27] R. M. Straubinger and D. J. Chin, unpublished observations (1988).
[28] T. H. Steinberg, A. S. Newman, J. A. Swanson, and S. C. Silverstein, J. Cell Biol. 105, 2695 (1987).

the DPX-mediated quenching of HPTS fluorescence. It has been found that certain liposome formulations leak their contents into the surrounding endocytic vesicle early in the process of endocytosis.[11]

Quantitative data also may be derived from experiments using HPTS-containing liposomes. Photometers attached to fluorescence microscopes can be used to quantify the fluorescence of cells excited alternately with violet- and blue-band filters. The ratio of these intensities is related directly to pH; a calibration curve is constructed by adjusting the pH of a known concentration of HPTS to different pH values, placing these solutions in a small chamber attached to a microscope slide, and observing the violet-to-blue fluorescence ratio. In cases in which a digital camera controller is attached to the microscope, one may be able to use the controller's photometer,[10] provided either the field intensity or "photographic exposure length" is displayed.

Spectrofluorometers may be used to give highly reliable data on the progression of HPTS-containing liposomes into acidic compartments, but cannot discriminate cytoplasmic (diffuse) fluorescence from cell surface liposomes because HPTS in both compartments reports approximately neutral pH. Therefore the method will not be discussed. The reader is referred to Straubinger et al.[10] and Daleke et al.[11] for further details.

Formulation of pH-Sensitive Liposomes

pH-dependent membrane fusion has been reported under a wide range of conditions, both in pure lipid systems and mediated by proteins, peptides, or amino acids. However, the number of systems that have been exploited for cytoplasmic delivery is limited; in many model systems, membrane fusion or destabilization occurs at pH values well below those thought to be present in any cellular compartment. A comprehensive review of background work is beyond the scope of the present chapter. The reader is directed to Düzgüneş et al.[29] for a more detailed perspective on pH-sensitive liposomes in the general context of membrane fusion.

With few exceptions, the fundamental approach for the formulation of pH-sensitive pure lipid systems relies on a single straightforward strategy: liposomes are composed predominantly of an amphipathic molecule that is unable to form stable bilayer vesicles under physiological conditions, and liposomes are caused to form by the inclusion of a second amphiphile that conditionally stabilizes the bilayer, dependent on pH.[29a] In virtually all

[29] N. Düzgüneş, R. M. Straubinger, P. A. Baldwin, and D. Papahadjopoulos, in "Membrane Fusion" (J. Wilschut and D. Hoekstra, eds.), p. 713. Dekker, New York, 1991.
[29a] N. Düzgüneş, R. M. Straubinger, P. A. Baldwin, D. S. Friend, and D. Papahadjopoulos, *Biochemistry* **24**, 3091 (1985).

studies of cellular delivery to date, phosphatidylethanolamine (PE) has been used as the predominant "unstable" amphipath. Pure PE can be induced to form bilayer vesicles at high pH (approximately pH$\geq$9) or in media of low ionic strength,[30-32] but PE liposomes collapse into alternate lipid bilayer conformations under physiological ionic strength and neutral pH. A variety of phosphatidylethanolamines, either from natural sources or synthesized with defined acyl chains, have been included in formulations of pH-sensitive liposomes.

There is much greater diversity in examples of the second, "stabilizing" amphipath, which imparts conditional (pH-dependent) stability to PE-rich bilayers. It is likely that the choice of the stabilizing amphipath has considerable impact on the performance of pH-sensitive liposome formulations, determining, among other properties, the pH at which membrane destabilization is triggered,[33] or the permeability of pH-sensitive liposomes in the presence of serum components or target cells. In many strategies, the stabilizing amphipath carries a weakly acidic functional group. Examples include oleic acid (OA),[16,29a,34] succinyl-PE,[35] cholesteryl hemisuccinate (CHEMS),[12,36] palmitoylhomocysteine,[37,38] and synthetic diacyl amphiphiles with head groups such as serine.[39] At pH $>$ pK_a, the amphiphile is charged and stabilizes the bilayer; at pH $<$ pK_a, protonation of the amphipath results in an uncharged or reduced-charge species that is unable to stabilize PE-rich bilayers in a configuration compatible with an extended bilayer membrane.

Upon acid-triggered transitions to nonbilayer or other conformations that are more stable thermodynamically, PE-rich liposomes have been observed to leak their aqueous contents, and to form larger structures with the coalescence of membrane components. Coincidentally, many pH-sensitive, PE-rich liposome formulations have been shown to deliver a variety of membrane-impermeant compounds to various cell types. Evidence for cytoplasmic delivery by pH-sensitive liposomes has been garnered from *in vitro* and *in vivo* systems, but the mechanism by which cytoplasmic delivery occurs has not been demonstrated definitively. It is not clear whether

[30] D. Papahadjopoulos, *Biochim. Biophys. Acta* **163**, 240 (1968).
[31] D. Papahadjopoulos and J. C. Watkins, *Biochim. Biophys. Acta* **135**, 639 (1967).
[32] D. A. Kolber and D. H. Haynes, *J. Membr. Biol.* **48**, 195 (1979).
[33] D. Collins, F. Maxfield, and L. Huang, *Biochim. Biophys. Acta* **987**, 47 (1989).
[34] J. Connor and L. Huang, *J. Cell Biol.* **101**, 582, (1985).
[35] A. J. Schroit, J. Madsen, and R. Nayar, *Chem. Phys. Lipids* **40**, 373 (1986).
[36] H. Ellens, J. Bentz, and F. C. Szoka, *Biochemistry* **23**, 1532 (1984).
[37] M. B. Yatvin, W. Kreutz, B. A. Horwitz, and M. Shinitzky, *Science* **210**, 1253 (1980).
[38] J. Connor, M. B. Yatvin, and L. Huang, *Proc. Natl. Acad. Sci. U.S.A.* **81**, 1715 (1984).
[39] R. Leventis, T. Diacovo, and J. R. Silvius, *Biochemistry* **26**, 3267 (1987).

pH-sensitive liposomes undergo acid-triggered fusion with the lumenal side of the endocytic vesicle membrane, depicted in Fig. 1C, or whether pH-dependent collapse of large numbers of PE-rich liposomes within endocytic vesicles exerts a general detergent-like effect that leads to gross defects in the integrity of the cell-derived membrane. Additional mechanisms for cytoplasmic delivery may be possible, but an incomplete understanding of the intracellular processing of liposomes hampers the proposal of alternative hypotheses.

Procedures for Preparation of pH-Sensitive Liposomes

Several methods have been used in preparing pH-sensitive liposomes, reflecting the robust diversity of liposome technology, as well as the special requirements arising during the preparation of liposomes that are conditionally stable. In the procedure that follows, two liposome compositions and two methods of preparation are mentioned. The reader is referred to a previous volume of this series[24] for a more explicit discussion of the liposome preparation methods outlined below.

The simplest procedure for forming liposomes is to mix the desired lipid components in an organic solvent such as chloroform, deposit the lipids on the walls of a glass tube by evaporating the organic solvent, and hydrate the dried lipids with a solution containing the material to be encapsulated. Generally $10-50 \mu$mol of phospholipid is used with a 0.5- to 1.0-ml aqueous phase. Vigorous vortex mixing usually aids in resuspending the dried lipids. The method may be referred to generally as the MLV (multilamellar vesicle) method, as the liposomes formed tend to have multiple membrane bilayers.[40] Subsequent processing steps may modify liposome properties considerably; extensive sonication of MLV results in particles of $30-50$ nm, and extrusion of MLV through polycarbonate membranes with defined pore diameters[41] produces liposomes with diameters approximating that of the membrane pore.

A second useful method for preparing liposomes is the REV (reversed-phase evaporation) method,[42] which generally has a higher efficiency of capture, more homogeneous particle size distribution, and theoretically a greater capacity to capture solutes of large Stokes radius, compared to the MLV method. In the REV method, lipid components are mixed in the desired stoichiometry, usually in organic solvents. The solvent is evaporated and the lipids are resuspended in diethyl or isopropyl ether. A defined

[40] A. D. Bangham, M. Standish, and J. Watkins, *J. Mol. Biol.* **13**, 238 (1964).
[41] F. C. Szoka, F. Olson, T. Heath, W. Vail, E. Mayhew, and D. Papahadjopoulos, *Biochim. Biophys. Acta* **601**, 559 (1980).
[42] F. C. Szoka and D. Papahadjopoulos, *Proc. Natl. Acad. Sci. U.S.A.* **75**, 4194 (1978).

volume of aqueous buffer is added to the ether–lipid mixture; typically 1 ml of ether contains 10 μmol of phospholipid and a 0.4-ml aqueous phase. The mixture is sonicated briefly (5–15 sec) in a bath-type sonicator to form a stable emulsion. Subsequently the ether is removed under reduced pressure and constant agitation; when all traces of ether have been removed, the liposome preparation is complete.

Formulations for pH-Sensitive Liposomes

The predominant constituent of most pH-sensitive liposomes studied as vehicles for cellular delivery is PE. Sources of PE used include soybeans, enzymatically trans-esterified from egg phosphatidylcholine, or synthesized with defined acyl chains such as dioleoyl or dipalmitoyl. Because the method is still evolving, it is not clear whether there exists a single best choice for intracellular delivery. One important point, regardless of the source of the PE, is that the material must be pure; it is abundantly clear that the property of pH sensitivity is abolished with relatively small quantities of other amphiphiles.[43]

The choice of the best secondary "stabilizing" amphiphile likewise continues to evolve. Much work has examined the delivery characteristics of liposomes containing oleic acid (OA); a common formulation is OA : PE in a 3 : 7 mol : mol ratio. For reasons yet to be determined, it is difficult to form liposomes of OA : PE (3 : 7) by the REV method; however, liposomes form readily by the MLV method. Increasing the pH to $\geq 8.5–9.0$ improves the behavior of the preparation during the REV procedure, but the chemical stability of lipids decreases at alkaline pH.

A second formulation that may have advantages over OA : PE (3 : 7) consists of cholesteryl hemisuccinate (CHEMS) and dioleoyl-PE (DOPE).[12,36] Liposomes of CHEMS : DOPE (4 : 6) form readily using the REV method and have considerably greater stability than OA : PE formulations. In addition, the biophysics, liposome–cell interaction, and the efficiency of cellular delivery has been studied in a more comprehensive way for CHEMS : PE liposomes than for perhaps any other pH-sensitive liposome formulation.[12,36]

Cellular Delivery by pH-Sensitive Liposomes

Although a rich literature exists reporting on cellular delivery by pH-sensitive liposomes, the sheer variety of formulations, incubation conditions, assay end points, and target cell types makes a comparison of

[43] D. Liu and L. Huang, *Biochim. Biophys. Acta* **1022**, 348 (1990).

performance difficult. One previous, exceptionally comprehensive work[12] estimated that $> 0.01 - 10\%$ of the cell-associated CHEMS : DOPE liposomes delivered their contents to the cytoplasm, depending on the probe used to detect cytoplasmic delivery. Cytoplasmic delivery by most other formulations has been studied in a less systematic manner. In the space remaining, we will use previously published work and considerable conjecture to support hypotheses regarding the mechanisms determining the efficiency of cytoplasmic delivery.

Although experiments demonstrate clearly that OA : PE liposomes mediate the delivery of large and impermeant molecules to the cytoplasm, such liposomes are exceedingly unstable in the presence of serum, and particularly in the presence of cells. Thus a major fraction of liposome contents is lost, perhaps before internalization of liposomes. It is likely that leakage is mediated by the presence of fatty acid "sinks" such as cell membranes and serum proteins. These may extract OA readily from liposomes, leading to collapse of the bilayer vesicle. In contrast, CHEMS : DOPE liposomes retain a higher fraction of their contents, perhaps the single determinant most responsible for the apparent superiority of CHEMS : DOPE in mediating cytoplasmic delivery. A second determinant of the efficiency of intracellular delivery may relate to the electrostatic charge of the liposomes and the specific target cell type. It was mentioned above that cellular uptake of liposomes may result from fortuitous binding to a "receptor" for negative charge. It has become clear that different cell types vary in the amount of negative charge required for cellular delivery. Whereas liposome uptake and delivery is maximal with just 30% negative charge for cells such as CV-1, other cell types, such as J774 (murine macrophage tumor), show little delivery by liposomes having less than 50% charge. These conclusions are based on experiments with non-pH-sensitive liposomes, utilizing either the effect of a cytostatic drug[44,45] or the rate of pH change as reported by HPTS[45,46] as assays for internalization of liposomes. It is clear that liposome electrostatic charge has a major impact on cellular binding and internalization, but it is unknown whether there are further charge-dependent differences in intracellular processing events. Differential intracellular fates or intracellular targeting could lead to significantly different efficiencies of triggering the acid-sensitive destabilization of pH-sensitive liposomes, and ultimately the efficiency of cytoplasmic delivery. What is clear from the data available is that OA : PE $(3 : 7)$ or

[44] T. D. Heath and C. S. Brown, *J. Liposome Res.* **1**, 303 (1990).
[45] A. Sharma, N. Lopez Straubinger, and R. M. Straubinger, *Pharm. Res.,* in press.
[46] K. D. Lee, K. Hong, and D. Papahadjopoulos, *Biochim. Biophys. Acta* **1103**, 185 (1991).

CHEMS:DOPE (4:6) liposomes carry a fraction of negative charge that could make them maximally efficient for delivery to cells that have a low requirement for negative charge, such as CV-1,[16] and, almost paradoxically, minimally efficient for delivery to cells that have a requirement for high negative charge, such as J774 cells.[12]

One means to circumvent the variable efficacy of negative charge in mediating liposome–cell interaction is the attachment of ligands such as antibodies to the liposome surface, either as a general method to promote liposome binding to receptors of cells such as macrophages (e.g., Fc receptors), or as a method to target liposomes to specific cell surface components for which antibodies or other ligands exist. An approach that has been used with OA:PE liposomes is to acylate antibodies to provide a hydrophobic anchor with which to bind antibodies to liposomes.[47] This methodology has been reviewed previously.[48] Other methods for coupling antibodies to liposomes may also work,[49] but are not mentioned explicitly in the literature on pH-sensitive liposomes.

Attachment of antibodies or other ligands to mediate binding of pH-sensitive liposomes to cells may have important, but less obvious, impact on the efficiency of delivery. As mentioned above, it has been difficult to compare the performance of pH-sensitive liposome formulations, owing to the dissimilar test systems used. In some cases, investigators working with apparently similar formulations have achieved apparently different results. The resolution may be that liposome-bound antibodies may provide a second function in addition to mediating liposome binding to the cell surface. Antibodies may also maintain the liposome and cellular membrane in close apposition during the process of pH-dependent destabilization, thus improving the likelihood that destabilization of the liposome will destabilize the enclosing endocytic vesicle membrane as well. In cases in which weakly acidic amphiphiles such as oleic acid are used to formulate pH-sensitive liposomes, the OA performs two functions: (1) it stabilizes the PE-rich bilayer; and (2) it mediates binding of liposomes to the cell surface. During acidification of the endocytic vesicle enclosing internalized liposomes, protonation of the fatty acid and reduction in liposome net negative charge may allow both the destabilization of the liposome and the desorption of the liposome from the endosome membrane. Thus, at the moment of peak liposome instability and tendency to undergo membrane fusion, liposomes may not reside in sufficiently close apposition to cellular membranes to permit fusion and cytoplasmic delivery. Figures 1B and 1C

[47] J. Connor and L. Huang, *Cancer Res.* **46**, 3431 (1986).
[48] L. Huang, J. Connor, and C.-Y. Wang, this series, Vol. 149, p. 88.
[49] F. Martin, W. Hubbell, and D. Papahadjopoulos, *Biochemistry* **20**, 4429 (1981).

suggest plausible late events following endocytosis of pH-sensitive liposomes. In Fig. 1B protonation of the acidic stabilizing amphiphile reduces the force originally mediating binding of liposomes to the cell surface. Liposomes could aggregate, fuse with each other, and release their contents within the lumen of the endocytic vesicle. Figure 1C depicts liposomes maintained in apposition to the endocytic vesicle membrane. Destabilization of pH-sensitive liposomes in close proximity to cellular membranes could result either in liposome–endosome fusion, or in a gross destabilization of the endosome membrane. Either event likely would lead to cytoplasmic delivery of liposome contents. Clearly, testing these hypotheses will provide the basis for further optimization of pH-sensitive liposomes for cytoplasmic delivery.

Section VII

Protoplast Fusion

[29] Plant Protoplast Fusion and Somatic Hybridization

By P. T. LYNCH, M. R. DAVEY, and J. B. POWER

Introduction

Conventional plant breeding is often limited by pre- and/or postzygotic incompatibility barriers, and fusion of somatic cells to generate somatic hybrid plants has been considered as a method of overcoming such limitations.[1] The lack of constraints to interspecific or intergeneric protoplast fusion permits hitherto reproducibly isolated plant genomes to be combined at the protoplast (heterokaryon) level, thus providing the basis for the generation of novel hybrids. Protoplast fusion also enables the genetic manipulation of vegetatively propagated crops, such as sterile or subfertile individuals, and those plants, including woody species, with naturally long life cycles.[2] Somatic hybridization of highly heterozygous species also provides an element of predictability in relation to the hybrid, because meiotic recombination is avoided. Cytoplasmic factors, such as mitochondrial-based cytoplasmic male sterility, may also be transferred from one species to another by protoplast fusion.[3]

Somatic hybridization involves four discrete, yet interrelated, stages: (1) protoplast isolation and culture with efficient plant regeneration in at least one of the fusion partners; (2) induced protoplast fusion, preferably at high frequency, without loss of viability; (3) the development of a selection strategy incorporating somatic hybrid plant regeneration; and (4) the confirmation of hybridity or cybridity. Such confirmation utilizes cytological and morphological markers, and a range of biochemical-based techniques.

Protoplast Isolation and Culture

The ability to isolate protoplasts that, when cultured under defined conditions, divide mitotically and regenerate plants has now been established for many species,[4-7] including woody plants.[8]

[1] G. Pelletier and Y. Chupeau, *Physiol. Veg.* **22**, 377 (1984).
[2] Y. Y. Gleba and K. M. Sytnik, "Protoplast Fusion—Genetic Engineering in Higher Plants," Monogr. Theor. Appl. Genet., No. 8. Springer-Verlag, Berlin, 1984.
[3] A. Kumar and E. C. Cocking, *Am. J. Bot.* **74**, 1289 (1987).
[4] S. C. Maheshwari, R. Gill, N. Maheshwari, and P. K. Gharyal, *Results Probl. Cell Differ.* **12**, 3 (1986).
[5] M. R. Davey and J. B. Power, *in* "Progress in Plant Protoplast Research" (K. J. Puite, J. J. M. Dons, H. J. Huizing, A. J. Kool, M. Koornneef, and F. A. Krens, eds.), p. 15. Kluwer Academic Publishers, Dordrecht, Boston, London, 1987.

Protoplast Fusion

Induced protoplast fusion can be achieved using chemical[9] and electrical treatments.[10] In both cases, fusion is a two-stage process. First, protoplasts are brought into close membrane contact, the degree of plasma membrane adhesion depending on the parental protoplasts. Tight contact may occur only in localized regions between adhering protoplasts.[11] Subsequently, the plasma membranes are stimulated to interact, for example, by modification of the electrical charges on the membranes,[12] resulting in protoplast fusion.

Fusion generates products (heterokaryons) with two or more nuclei within a mixed cytoplasm containing organelles from the parental protoplasts. The cytoplasms derived from the respective parental protoplasts mix at different rates within the heterokaryons, according to the protoplast types.[9] Cell wall formation and nuclear fusion to produce hybrid cells occur early in culture. Nuclear fusion takes place either during interphase by the formation of nuclear bridges, or at the first mitosis.[11] The fate of plastids in hybrid cells varies, and includes loss of one parental type or recombination between plastids of the two parents.[13] Vacuoles in heterokaryons may fuse,[14] and microtubules integrate.[15] However, the fate of other cell organelles is unclear.

The extent of protoplast fusion, heterokaryon formation, and survival of fusion products can be monitored using naturally occurring visual markers. Thus heterokaryons can be readily identified following the fusion of chlorophyll-containing leaf mesophyll protoplasts with suspension cell

[6] Y. P. S. Bajaj, in "Biotechnology in Agriculture and Forestry. 8. Plant Protoplasts and Genetic Engineering I" (Y. P. S. Bajaj, ed.), p. 3. Springer-Verlag, Berlin, 1989.

[7] R. P. Finch, P. T. Lynch, J. P. Jotham, and E. C. Cocking, in "Biotechnology in Agriculture and Forestry. 14. Rice" (Y. P. S. Bajaj, ed.), p. 251. Springer-Verlag, Berlin, 1991.

[8] S. J. Ochatt and J. B. Power, in "Comprehensive Biotechnology 2" (M. Moo-Young, G. S. Warren, and M. W. Fowler, eds.), p. 99. Pergamon, New York, 1992.

[9] J. A. Saunders and G. W. Bates, in "Cell Fusion" (A. E. Sowers, ed.), p. 497. Plenum, New York, 1987.

[10] J. A. Saunders, B. F. Matthews, and P. D. Miller, in "Electroporation and Electrofusion in Cell Biology" (E. Neumann, A. E. Sowers, and C. A. Jordan, eds.), p. 343. Plenum, New York, 1989.

[11] L. C. Fowke, in "Biotechnology in Agriculture and Forestry. 18. Plant Protoplasts and Genetic Engineering I" (Y. P. S. Bajaj, ed.), p. 289. Springer-Verlag, Berlin, 1989.

[12] B. Hahn-Häqerdal, K. Hosono, A. Zachrisson, and C. H. Bornman, Physiol. Plant. 67, 359 (1986).

[13] H. Lörz, in "Plant Genetic Engineering" (J. H. Dodds, ed.), p. 27. Cambridge Univ. Press, Cambridge, 1985.

[14] F. Constabel, H. Koblitz, J. W. Kirkpatrick, and S. Rambold, Can. J. Bot. 58, 1032 (1980).

[15] B. Hahne and F. Hoffmann, Plant Sci. 47, 199 (1986).

protoplasts lacking this pigment.[16] Fluorescent dyes have also been used as visual markers to label protoplasts.[17]

Chemically Induced Protoplast Fusion

The plasma membranes of isolated plant protoplasts have a net negative electrical charge of approximately $10-35$ mV,[18] as a consequence of which adjacent protoplasts naturally repel each other. To induce the close membrane contact required for membrane fusion, the charges on the surfaces of protoplasts must be neutralized by exposure, for example, to polycations such as polyethylene glycol (PEG), or by the use of a high-pH solution. A number of protocols have been described for chemically induced protoplast fusion.[9,19] The use of PEG coupled with solutions buffered at high pH in the presence of Ca^{2+} (high pH/Ca^{2+}) is the most commonly used method to induce protoplast fusion. Carbonyl-free PEG has been shown to improve protoplast fusion, to diminish the formation of large protoplast aggregates, and to retain protoplast viability.[20]

General Protocols for Chemically Induced Fusion of Plant Protoplasts

Polyethylene Glycol Treatment. The following steps are required.

1. Protoplasts are suspended in CPW13M solution (Table I), typically at a density of 2.0×10^5 ml^{-1}, and 4.0-ml aliquots of each of the respective protoplast suspensions are mixed in 16-ml capacity screw-capped centrifuge tubes (Corning, Ltd., Stone, Staffordshire, England). The protoplasts are pelleted by centrifugation (100 g; 10 min, 22°C) and the supernatant removed.

2. Aliquots (2.0 ml) of PEG solution (Table I) are added to the pellets and the protoplasts gently resuspended prior to incubation at 22° for 10 min.

3. The PEG solution is diluted, at 5-min intervals, by the addition of 0.5, 1.0-, 2.0-, 2.0-, 3.0-, and 4.0-ml aliquots of CPW9M solution (Table I). Protoplasts are gently resuspended after each dilution.

4. Protoplasts are centrifuged (100 g; 10 min, 22°) and the supernatant removed. Subsequently, they are resuspended in an appropriate culture

[16] R. P. Finch, I. H. Slamet, and E. C. Cocking, *J. Plant Physiol.* **136**, 592 (1990).
[17] K. P. Pauls and P. V. Chuong, *Can. J. Bot.* **65**, 834 (1987).
[18] T. Nagata and G. Melchers, *Planta* **142**, 235 (1978).
[19] J. B. Power, M. R. Davey, M. McLellan, and D. Wilson, "Laboratory Manual: Plant Tissue Culture." University of Nottingham, 1989.
[20] P. K. Chand, M. R. Davey, J. B. Power, and E. C. Cocking, *J. Plant Physiol.* **133**, 480 (1988).

TABLE I
COMPOSITION OF FUSION AND WASHING SOLUTIONS

Solution	Composition[a]
CPW13M	KH_2PO_4 (27.2), KNO_3 (101.0), $Ca_2Cl_2 \cdot 2H_2O$ (1480.0), KI (0.16), $MgSO_4 \cdot 7H_2O$ (246.0), $CuSO_4 \cdot 5H_2O$ (0.025), 13% (w/v) mannitol, pH 5.8, autoclaved
CPW13M/ Ca^{2+}	As above, but supplemented with 7.4 g $CaCl_2 \cdot 2H_2O$ per liter
CPW9M	As CPW13M, but with 9% (w/v) mannitol
PEG	30% (w/v) Polyethylene glycol 6000 (Koch-Light, Ltd., Haverhill, England), 4% (w/v) sucrose, 0.01 M $CaCl_2 \cdot 2H_2O$, autoclaved
High pH/ Ca^{2+}	0.05 M Glycine–NaOH buffer, 1.1% (w/v) $CaCl_2 \cdot 2H_2O$, 10% (w/v) mannitol, pH 10.4, filter sterilized
Purified PEG	PEG 1540 (Boehringer-Mannheim, Indianapolis, IN) in N-2-hydroxyethylpiperazine-N'-2-ethanesulfonic acid (HEPES) buffer, pH 8.0, filter sterilized
Electrofusion solution	0.05 mM $CaCl_2 \cdot 2H_2O$, 11% (w/v) mannitol, filter sterilized

[a] Data in milligrams per liter unless indicated otherwise.

medium before plating at a density of 5.0×10^4 ml^{-1} (this plating density depends on the protoplast partners used for fusion).

High-pH/Ca^{2+} Treatment. Three steps are required.

1. Protoplasts are suspended in CPW13M solution, spun down as in step 1 of the previous section, and 8.0 ml of a high-pH/Ca^{2+} fusion solution (Table I) added. The protoplasts are gently resuspended, immediately centrifuged (60 g; 3 min, 22°) and maintained at 30° for 15 min.

2. Sterile distilled water (2.0 ml/tube) is added and gently mixed with the fusion solution, leaving the protoplast pellet intact. Incubation is continued for a further 10 min (30°).

3. The supernatant is removed, the protoplasts washed once in CPW13M/Ca^{2+} solution (Table I), and resuspended in the appropriate culture medium.

Polyethylene Glycol with High-pH/Ca^{2+} Treatment. Fusion frequencies have been enhanced by the use of PEG in combination with high pH/Ca^{2+}.[21] The success of this latter modification probably relates to a combined effect of the two fusogens, which have separate modes of action. Primarily, PEG acts as a protoplast agglutinator, whereas high pH/Ca^{2+} modifies the surface charges of the plasma membrane.[9]

[21] K. N. Kao and M. Saleem, *J. Plant Physiol.* **122**, 217 (1986).

1. Protoplasts are treated with PEG as described in steps 1 and 2 for the section on polyethylene glycol treatment (above), but are diluted with 8.0-ml volumes of high-pH/Ca^{2+} solution per tube. Protoplasts are incubated at 22° for 10 min.

2. The protoplasts are centrifuged (60 g; 3 min) and treated as in step 3 of the previous section.

Purified Polyethylene Glycol Fusion Treatment. Polyethylene glycol is known to reduce the viability of fusion products and this cytotoxic effect has been attributed to membrane dehydration[12] and impurities in the polymer, such as α-tocopherol and phenolic derivatives.[22,23] An improved procedure, using PEG preparations (MW 1540) with a low carbonyl content,[20] has been developed for plant protoplasts. This method is applicable to a wide range of plant protoplast systems and results in a high frequency of heterokaryon survival compared with treatments using unpurified PEG.

1. Protoplasts of the species to be fused are suspended separately in 13% (w/v) mannitol solution at a density of 1.0×10^5 ml^{-1} and are allowed to stand for 5–10 min.

2. Equal volumes of the protoplasts suspensions are mixed and 1.0- to 1.5-ml aliquots dispensed into the wells of a 25-compartment 120-mm^2 grid dish (Sterilin, Ltd., Hounslow, Middlesex, England).

3. An aliquot (0.5 ml) of the purified, low-carbonyl PEG solution (Table I) is added and the mixture left for 15–20 min at 22°.

4. One milliliter of 5% (w/v) mannitol solution is added and the fused protoplasts are left for approximately 5 min to become spherical.

5. The mixture of PEG and mannitol solution is removed and the protoplasts are washed in 13% (w/v) mannitol solution. Protoplasts are left in this concentration of mannitol solution for 30 min before a final wash in 13% (w/v) mannitol solution and transfer to culture medium.

Other Methods of Chemical Fusion. Several other methods have been developed for the chemical fusion of plant protoplasts, but none has been used as extensively as the four procedures already described. Three examples of other compounds used to fuse protoplasts are given in the following three sections.

Dextran and dextran sulfate: Both high molecular weight dextran and dextran sulfate have been used to induce protoplast aggregation, although dextran sulfate was found to be toxic to protoplasts. Protoplast fusion has

---

[22] K. Honda, Y. Maeda, S. Sasakawa, H. Ohno, and E. Tsuchida, *Biochem. Biophys. Res. Commun.* **100,** 442 (1981).

[23] K. Honda, Y. Maeda, S. Sasakawa, H. Ohno, and E. Tsuchida, *Biochem. Biophys. Res. Commun.* **101,** 165 (1981).

also been achieved by using dextran in the presence of organic salts but, to date, somatic hybrid plants have not been reported using this method.[24]

Polyvinyl alcohol: Protoplast adhesion and fusion have been induced by polyvinyl alcohol (PVA) in the presence of $CaCl_2$ and mannitol, with little loss of protoplast viability.[25] Again, this technique has not resulted in the production of somatic hybrid plants.

Agarose and calcium nitrate: Protoplasts plated at high density (1.0×10^5 ml^{-1}) in 2.0% (w/v) agarose in $0.2\ M\ Ca(NO_3)_2$ can be kept in close contact during subsequent treatment with a high-pH/$Ca(NO_3)_2$ solution, which induces the protoplasts to fuse. After a 20-min incubation period the fusion solution is replaced by culture medium. Somatic hybrid and cybrid plants have been produced between *Solanum tuberosum* and *Solanum nigrum* by using this method.[26]

Electrofusion of Protoplasts

Using electrofusion to fuse plant protoplasts can have several advantages when compared with chemically induced fusion.[27] For example, electrofusion eliminates the need for toxic chemical fusogens and extensive washing procedures. The areas of membrane disturbance are restricted to zones of membrane contact, thus maintaining protoplast viability. Most important, electrofusion usually results in a higher frequency of heterokaryon formation.[28]

Electrofusion has several inherent shortcomings. Having to suspend protoplasts in an essentially electrolyte-free solution may adversely affect protoplast viability due to a loss of membrane stability and leakage of cellular electrolytes. Generally, the electronics required for electrofusion are sophisticated and, as a consequence, expensive. Additionally, only relatively small volumes of material can be fused at one time because of the restricted volume of the fusion chamber. Thus, flat chambers of 7-μl capacity, and helical fusion chambers of 200-μl volume, have been explored.[29,30] A convenient electrode system, consisting of a series of parallel

[24] I. Kishinami and J. M. Widholm, *Plant Cell Physiol.* **28**, 211 (1987).

[25] T. Nagata, *Naturwissenschaften* **65**, 263 (1978).

[26] H. Binding, M. Zuba, J. Rudnick, and G. Mordhorst, *J. Plant Physiol.* **133**, 409 (1988).

[27] G. W. Bates, J. A. Saunders, and A. E. Sowers, *in* "Cell Fusion" (A. E. Sowers, ed.), p. 367. Plenum, New York, 1987.

[28] A. Zachrisson and C. H. Bornman, *Physiol. Plant.* **67**, 507 (1986).

[29] G. Pilwat, U. Zimmermann, and H. P. Richter, *FEBS Lett.* **133**, 169 (1981).

[30] U. Zimmermann and J. Vienken, *in* "Hybridoma Technology in Agricultural and Veterinary Research" (N. J. Stern and H. R. Gamble, eds.), p. 173. Rowman & Allanheld, Totowa, NJ, 1984.

brass plates that fit the square wells of a Sterilin 25-compartment dish, enabling volumes of 1.0 ml or more to be handled, has been constructed.[31]

When subjected to a nonuniform alternating electric (AC) field, protoplasts suspended in an electrolyte-free solution move together to form "pearl chains" in which point-to-point membrane contact develops between adjacent protoplasts. Such pearl chains form because the polarized protoplasts move toward the region of higher field strength (dielectrophoresis), and become attracted to each other (mutual dielectrophoresis). Fusion is stimulated by short pulses of direct current (DC), which causes breakdown of the closely aligned membranes. During this process, membrane lipids become randomly oriented and pores develop in the plasma membranes of the protoplasts. Membrane bridges result, leading to the actual fusion process,[32] with the cytoplasms of adjacent protoplasts becoming continuous. Protoplasts can also be brought together and fused by using microelectrodes, avoiding the necessity for a potentially damaging AC field.[33] This technique, if combined with single-cell culture, can permit hybrid cell formation from a defined pair of protoplasts.[34]

Chemical treatments of protoplasts prior to electrofusion have been reported to improve both protoplast stability and fusion frequency. Proteases, polyamines, and dimethyl sulfoxide have been used. These compounds probably decrease membrane fluidity and increase membrane lipid domains.[35-37]

To maximize heterokaryon formation, it is important to optimize the conditions under which short pearl chains (preferably consisting of pairs of protoplasts) are formed. This can be achieved by minimizing the alignment time and AC field strength. Generally, smaller protoplasts less than 25 μm in diameter must be maintained at high densities, usually in excess of 5.0×10 ml^{-1}, in order to maximize heterokaryon formation.

Example of An Electrofusion Protocol. The electrofusion of protoplasts of *Rudbeckia hirta* and *Rudbeckia laciniata* illustrates a typical electrofusion protocol[38] using the plate electrode system.[31]

[31] J. W. Watts and J. M. King, *Biosci. Rep.* **4**, 335 (1984).
[32] U. Zimmermann and H. B. Urnovitz, this series, Vol. 151, p. 194.
[33] H. Morikawa, Y. Hayashi, Y. Hirabayashi, M. Asada, and Y. Yamada, *Plant Cell Physiol.* **29**, 189 (1988).
[34] H. G. Schweiger, J. Dirk, H.-U. Koop, E. Kranz, G. Neuhaus, G. Spangenberg, and D. Wolff, *Theor. Appl. Genet.* **73**, 769 (1987).
[35] P. T. Lynch, S. Isaac, and H. A. Collin, *Planta* **178**, 207 (1989).
[36] L. J. Nea, G. W. Bates, and P. J. Gilmer, *Biochim. Biophys. Acta* **897**, 293 (1987).
[37] L. J. Nea and G. W. Bates, *Plant Cell Rep.* **6**, 337 (1987).
[38] J. S. Al-Atabee, B. J. Mulligan, and J. B. Power, *Plant Cell Rep.* **8**, 517 (1990).

1. Leaf mesophyll protoplasts of *R. hirta* and callus protoplasts of *R. laciniata* are isolated as described.[39] Protoplasts of *R. laciniata* are stained with fluorescein diacetate (25 μg ml^{-1}) during enzyme incubation.

2. Protoplasts of the two species are washed twice in electrofusion medium (Table I) and mixed (1:1) to give a final protoplast density at 2.0×10 ml^{-1}. Aliquots (1.0 ml) of the protoplast mixture are transferred into the 9 central wells of a 25-compartment square-grid dish. The dish is gently agitated to distribute the protoplasts evenly throughout each well, and the protoplasts are allowed to settle for 5 min.

3. The dish is placed on the stage of an inverted microscope, which is situated in a laminar air-flow hood. The parallel-plate electrode assembly is sterilized by immersion in ethanol (for 30 sec) and allowed to dry in the sterile air flow.

4. The electrode is inserted into one of the wells of the dish and the protoplasts are subjected to an AC field of 0.5 MHz, 54 V cm^{-1} for approximately 45 sec. Subsequently, two DC pulses of 810 V cm^{-1}, and 2000 μsec duration, are applied to induce protoplast fusion. The AC field is reduced to zero over a 15-sec period.

5. Twenty-microliter aliquots of CPW13M solution (Table I) are added to each well of the dish. After 15 min the protoplasts are gently transferred to 8.0-ml capacity screw-capped centrifuge tubes. The protoplasts are allowed to settle for 15–20 min, after which the electrofusion-CPW13M solution is withdrawn. The protoplasts are resuspended in culture medium. Examples of the electrical parameters used to fuse plant protoplasts are given in Table II.[35,38,39a-d]

Selection of Somatic Hybrids

Despite efforts to increase protoplast fusion frequencies, the formation of viable, binucleate heterokaryons is typically restricted to less than 5% of the protoplast population. Therefore, it is necessary to select these fusion products against a background of homokaryons, unfused parental protoplasts, and/or multiple fusion bodies. Several selection methods have been described, but a universally applicable system has not yet been developed. Some commonly used selection systems are described below.

Genetic Complementation

Complementation methods depend on fusion of two protoplast systems, each of which carries different recessive selectable markers. The resulting somatic hybrid cells are functionally restored. A range of comple-

[39] J. S. Al-Atabee and J. B. Power, *Plant Cell Rep.* **6,** 414 (1987).

TABLE II
ELECTROFUSION PARAMETERS

Plant species	Alignment		Fusion		Ref.
	Field strength (V cm^{-1})	Frequency (MHz)	Field strength (V cm^{-1})	Pulse period (μsec)	
Apium graveolens	200	1.5	150	99	35
Nicotiana tabacum, N. plumbaginifolia	300	0.5	500	2000	39a
Oryza sativa	100	2	1000	50	39b
Picea abies, Pinus sylvestris	150	0.35	2000–4000	50–1000	39c
Rudbeckia hirta, R. laciniata	54	0.5	810	2000	38
Solanum tuberosum, S. brevidens	100	1	1250–1500	10	39d

mentation systems has been used to recover somatic hybrid tissues and somatic hybrid plants.

Use of Chlorophyll-Deficient Mutants (Albinos). Fusion of protoplasts from two nonallelic chlorophyll-deficient lines results in somatic hybrid cells that are chlorophyll proficient, as in the case of the fusion of protoplasts from albino cell lines of *Medicago sativa* and *Medicago borealis.*[40] Selection can also be based on complementation between wild-type and albino lines. Thus somatic hybrid cells between wild-type mesophyll protoplasts of *Petunia parodii* and protoplasts from an albino line of *Petunia inflata* exhibit chlorophyll synthesis and sustained growth.[41]

Use of Light-Sensitive Mutants. The fusion of mesophyll protoplasts from a light-sensitive mutant of *Nicotiana plumbaginifolia* with wild-type *mesophyll* protoplasts of *Nicotiana gossei,* irradiated with 200 J Kg^{-1} of ^{60}Co γ rays (0.066 J kg^{-1} sec^{-1} dose rate) prior to fusion, has been used to select heterokaryon-derived green hybrid cell colonies.[42] Regenerated plants have the morphology of *N. plumbaginifolia,* and normal green coloration.

[39a] J. D. Hamill, J. W. Watts, and J. M. King, *J. Plant Physiol.* **129,** 111 (1987).
[39b] K. Toriyama and K. Hinata, *Theor. Appl. Genet.* **76,** 665 (1988).
[39c] U. Kirsten, H. E. Jacob, M. Tesche, and S. Kluge, *Stud. Biophys.* **119,** 85 (1987).
[39d] N. Fish, A. Karp, and M. G. K. Jones, *Theor. Appl. Genet.* **76,** 260 (1988).
[40] D. M. Gilmour, M. R. Davey, and E. C. Cocking, *Plant Cell Rep.* **8,** 29 (1989).
[41] L. S. Schnabelrauch, F. Kloc-Bauchan, and K. C. Sink, *Theor. Appl. Genet.* **70,** 57 (1985).
[42] P. Medgyesy, R. Golling, and F. Nagy, *Theor. Appl. Genet.* **70,** 590 (1985).

Use of Nitrate Reductase-Deficient Lines. Following fusion, parental protoplasts from nitrate reductase-deficient (NR⁻) cell lines are eliminated by their inability to utilize nitrate in the culture medium. This deficiency can be overcome in hybrid tissues through complementation by the other fusion partner. Thus, for hybrid cell/tissue selection, two selectable markers are required, with the result that protoplasts from NR⁻ lines are combined with those carrying other selectable markers. For example, protoplasts from NR⁻ *Nicotiana tabacum* fused with wild-type *Nicotiana glutinosa* pollen tetrad protoplasts (which do not undergo sustained cell division) produce hybrid cells that utilize nitrate. Such cells regenerate to form green plants.[43] Other types of autotrophic plant mutants can be employed in somatic hybridization selection schemes, including amino acid autotrophic lines for the intraspecific fusion of *Datura innoxia* protoplasts.[44]

Use of Resistance Markers. Dominant characteristics for traits such as resistance to herbicides[45] and amino acid analogs[46] are employed in selection. When protoplasts from two separate and mutually exclusive resistant lines are fused, the tolerance of each parental species is acquired by the somatic hybrid cells and the latter exhibit dual resistance. Unfused parental protoplasts and homokaryons are eliminated during selection. Intraspecific somatic hybrids are produced between parental lines of *S. tuberosum* that have resistance to different amino acid analogs, including *S*-aminoethylcysteine and *S*-methyltryptophan.[47]

Use of Double Mutants. Protoplasts of many potential fusion partners are of the wild type and, as a result, do not possess any markers suitable for selection. One method of overcoming this limitation is to construct a parental line carrying both negative and positive selectable markers, that is, an auxotrophic trait and a resistant trait. Only the heterologous fusion products with complemented auxotrophic–resistant traits will survive selection. Somatic hybrids between *Sinapis turgida* and *Brassica oleracea,* using protoplasts from a double mutant (NR⁻ and an *S*-methyltryptophan resistant) of *S. turgida,* have been produced by this approach.[48]

Use of Transformed Cell Lines. Resistance markers used in somatic hybrid selection schemes can be introduced by transformation. Protoplasts from transformed lines of *S. tuberosum,* carrying kanamycin or hygromy-

[43] A. Pirrie and J. B. Power, *Theor. Appl. Genet.* **72,** 48 (1986).
[44] P. K. Saxena and J. King, *Plant Cell, Tissue Organ Cult.* **9,** 61 (1987).
[45] J. Gressel, N. Cohen, and H. Binding, *Theor. Appl. Genet.* **67,** 131 (1984).
[46] M. E. Horn, T. Kameya, J. E. Brotherton, and J. M. Widholm, *Mol. Gen. Genet.* **192,** 235 (1983).
[47] S. E. de Vries, E. Jacobsen, M. G. K. Jones, A. E. H. M. Loonen, M. J. Tempelaar, J. Wijbrondi, and W. J. Feenstra, *Theor. Appl. Genet.* **73,** 451 (1987).
[48] K. Toriyama, T. Kameya, and K. Hinata, *Planta* **170,** 308 (1987).

cin B resistance genes, are fused, resulting in hybrid tissue that is resistant to both antibiotics. In forage legumes, kanamycin resistance combined with the use of the metabolic inhibitor sodium iodoacetate (see the next section), are used to select somatic hybrids between *Lotus corniculatus* and *Lotus tenuis.*[49]

Use of Antimetabolites. Complementation selection systems can also be based on the use of irreversible biochemical inhibition, which blocks metabolic pathways when the parental protoplasts are treated prior to fusion.[50] Inactivated parental lines cannot undergo cell division in their own right, but hybrid cells exhibit metabolic complementation and undergo sustained growth. The metabolic inhibitor sodium iodoacetate is used in combination with other markers, including lack of sustained cell division in one of the parental protoplast lines, to select somatic hybrid plants.[51] An example of this selection system involves the fusion of sodium iodoacetate-inactivated *Oryza sativa* protoplasts with protoplasts from a range of wild *Oryza* species. Protoplasts of the wild species fail to divide in culture. Iodoacetate usage requires a careful determination of treatment levels, so as to minimize cross-toxicity from parental protoplasts.[50]

Use of Tumorous Growth of F_1 Hybrids

To permit continued development of regenerated shoots from calli derived from the fusion of protoplasts of *Nicotiana langsdorffii* and *Nicotiana glauca,* the tissues are grafted onto plants of *Nicotiana glauca.*[52] Tumor formation, a characteristic of the sexual F_1 hybrid between these two *Nicotiana* species, is observed on the scion, thus providing a method for somatic hybrid selection.

Use of Differential Growth and Plant Regeneration

The differential response of parental protoplasts to culture conditions provides a method for selecting somatic hybrid tissues. Following the fusion of iodoacetamide-inactivated *O. sativa* protoplasts with those of *Echinochloa oryzicola,* the treated protoplasts are cultured in a medium that supports the growth of rice protoplasts and somatic hybrid cells, but not protoplasts of *E. oryzicola.*[53]

[49] M. A. Aziz, P. K. Chand, M. R. Davey, and J. B. Power, *J. Exp. Bot.* **41,** 471 (1991).
[50] C. T. Harms, *in* "Plant Protoplasts" (L. C. Fowke and F. Constabel, eds.), p. 169. CRC Press, Boca Raton, FL, 1985.
[51] R. Nehls, *Mol. Gen. Genet.* **166,** 117 (1978).
[52] P. S. Carlson, H. Smith, and R. D. Dearing, *Proc. Natl. Acad. Sci., U.S.A.* **69,** 2292 (1972).
[53] R. Terada, J. Kyozaka, S. Nishibayashi, and K. Shimamoto, *Mol. Gen. Genet.* **210,** 39 (1987).

The different mechanisms of plant regeneration also provide a method for somatic hybrid selection. Thus plant regeneration in *R. hirta* occurs through shoot formation, whereas shoot production in *R. laciniata* is via rhizogenesis. Somatic hybrids and plants of *R. laciniata* are regenerated through rhizogenesis. The somatic hybrids are identified by the presence of pigmented roots, a feature of *R. hirta*.[38]

Use of Electrical Stimulation

Electrical pulse treatments have been shown to enhance the division of plant protoplast-derived cells,[54] and to stimulate shoot formation from protoplast-derived cells of several plants, including woody species such as *Prunus avium × pseudocerasus*.[55] This technology is applied successfully in somatic hybridization. Thus electroporation of parental protoplasts prior to electrofusion promotes the division of heterokaryons and facilitates the recovery of somatic hybrids between the two woody species *Pyrus communis* var. *pyraster* and *Prunus avium × pseudocerasus*.[56] Somatic hybrid tissues are not produced when parental protoplasts are not electro-stimulated prior to fusion. Electrostimulation of protoplast division and plant regeneration may prove particularly useful in cases in which parental protoplasts respond to this treatment with increased growth and plant regeneration, especially if used in combination with other selection techniques.

Physical Isolation of Heterokaryons

Biochemical complementation/selection systems usually lead to preferential recovery of amphidiploid somatic hybrids.[57] Asymmetric hybrids, such as those possessing one complete genome but only a few chromosomes of the other parent, are likely to be lost during selection due to an inability of the cells to survive the strong selection pressure, through incomplete complementation to growth proficiency.[57] This, combined with the lack of suitable selectable markers for many parental species, makes physical identification, isolation, and culture of fusion products an important alternative. Heterokaryons can be identified by a dual-labeling system, such as red chlorophyll autofluorescence used in combination with the

[54] E. L. Rech, S. J. Ochatt, P. K. Chand, J. B. Power, and M. R. Davey, *Protoplasma* 141, 169 (1987).
[55] S. J. Ochatt, P. K. Chand, E. L. Rech, M. R. Davey, and J. B. Power, *Plant Sci.* 54, 165 (1988).
[56] S. J. Ochatt, E. M. Patat-Ochatt, E. L. Rech, M. R. Davey, and J. B. Power, *Theor. Appl. Genet.* 78, 35 (1989).
[57] E. C. Cocking, M. R. Davey, D. Pental, and J. B. Power, *Nature (London)* 293, 265 (1981).

yellow-green fluorescence of fluorescein diacetate.[58] Fluorescein diacetate labeling combined with the use of red fluorochromes such as rhodamine isothiocyanate has also been employed.[59]

Somatic hybrid tissues of *Medicago* species[60] and somatic hybrid plants of *Solanum* species[61] have been recovered from dual-labeled heterokaryons by using micromanipulation. However, micromanipulation is a laborious technique and the number of heterokaryons that can be selected with ease is limited.

Flow cytometry is another procedure that permits the selection of larger numbers (usually several thousand) of labeled heterokaryons.[62] Until recently, the range of somatic hybrid plants recovered from flow-sorted heterokaryons was limited to the genera *Nicotiana* and *Brassica*.[63,64] However, sorting has been extended to fused protoplasts from a wide combination of plant species, in some cases with somatic hybrid plant production.[65]

Confirmation of Hybridity

The first indication of the hybridity of cell lines/callus is their ability to survive the selection procedure. To eliminate potential problems such as reversion, cross-feeding, and residual leakiness from the selection system, additional confirmation is required at both the callus and plant levels.[66] Verification of hybridity requires demonstration of the presence and expression of genetic traits from both parents.

Morphological Characteristics of Regenerated Plants

Intermediate morphologies can be used to identify somatic hybrid material. Leaf shape and size[56] and floral characteristics, including flower size, color, and number of ray florets, can be evaluated.[38] Ideally, several independent characteristics should be considered. The more distant the

[58] G. Patnaik, E. C. Cocking, J. Hamill, and D. Pental, *Plant Sci. Lett.* **24,** 105 (1982).
[59] T. L. Barsby, J. F. Shepard, R. J. Kemble, and R. Wong, *Plant Cell Rep.* **3,** 165 (1984).
[60] D. M. Gilmour, M. R. Davey, and E. C. Cocking, *Plant Sci.* **53,** 267 (1987).
[61] K. J. Puite, S. Roest, and L. P. Pijnacker, *Plant Cell Rep.* **5,** 262 (1986).
[62] D. W. Galbraith, *in* "Cell Culture and Somatic Cell Genetics of Plants" (I. K. Vasil, ed.), Vol. 1, p. 433. Academic Press, London, 1984.
[63] C. L. Afonso, K. R. Harkins, M. A. Thomas-Compton, A. E. Krejci, and D. W. Galbraith, *Bio/Technology* **3,** 811 (1985).
[64] N. Hammatt, A. Lister, N. W. Blackhall, J. Gartland, T. K. Ghose, D. M. Gilmour, J. B. Power, M. R. Davey, and E. C. Cocking, *Protoplasma* **194,** 34 (1990).
[65] C. Sjodin and K. Glimelius, *Theor. Appl. Genet.* **77,** 651 (1989).
[66] R. Nehls, G. Krumbiegel-Schroeren, and H. Binding, *Results Prob. Cell Differ.* **12,** 67 (1986).

taxonomic relationship between the parental species, the greater the number of morphological characteristics that are available for assessment. Morphological features, such as pigmentation[67] and relative growth rates,[68] can be used to identify hybridity, even in protoplast-derived callus. In some cases, morphological analysis may be complicated by abnormalities arising from aneuploidy, somatic incompatibility, or somaclonal variation from the effects of the tissue culture procedure.[50]

Chromosomal Complement of Hybrids

The chromosome complements from actively dividing somatic cells, such as those from root tips, provide further evidence of hybridity and of ploidy levels. Hybrid plants are identified by their chromosome numbers,[43] and the structure and size of somatic cell chromosomes when compared with the karyotypes of parental species.[69] In some cases, chromosome counts may be inaccurate due to doubling or elimination of chromosomes.[70]

Isoenzyme Analysis

The different electrophoretic mobilities of isoenzymes that catalyze basic cell functions can be used to identify hybrid tissues/plants, as is the case of somatic hybrids between wild pear and colt cherry,[56] *Rudbeckia* species,[38] and *Oryza* species.[71] Hybrid tissue may possess isoenzyme band profiles characteristic of each parent, as well as additional bands. These additional bands may be regarded as possible artifacts, or as hybrid molecules or genes present in parent cells that are expressed within the new genetic background.[72]

Molecular Analysis

The development of molecular techniques, such as restriction fragment analysis and DNA hybridization of nuclear and organelle DNAs,[1] has permitted detailed analysis of the genetic constitution of somatic hybrids. Specific patterns of restricted DNA of both mitochondria and chloroplasts

---

[67] K. Klimaszewska and W. A. Keller, *Plant Sci.* **58**, 211 (1988).

[68] M. Niizeki, *in* "Biotechnology in Agriculture and Forestry. 8. Plant Protoplasts and Genetic Engineering" (Y. P. S. Bajaj, ed.), p. 410. Springer-Verlag, Berlin, 1989.

[69] L. R. Wetter and K. N. Kao, *Theor. Appl. Genet.* **576**, 272 (1980).

[70] F. D'Amato, *CRC Crit. Rev. Plant Sci.* **3**, 73 (1985).

[71] Y. Hayashi, J. Kyozuka, and K. Shimamoto, *Mol. Gen. Genet.* **214**, 6 (1988).

[72] H. Binding, G. Krumbiegel-Schroeren, and R. Nehls, *Results Probl. Cell Differ.* **12**, 37 (1986).

confirm hybridity, and elucidate organelle segregation and DNA recombination patterns.[73] Species-specific DNA fragments are used to determine the relative parental contributions to somatic hybrids.[74] Restriction fragment length polymorphism (RFLP) mapping[75] permits a more detailed examination of the inheritance of nuclear and organelle genomes in somatic hybrids. Thus a variety of established methods are available that permit accurate determination of the presence of genetic material from both parents in somatic hybrids.

Future Prospects for Plant Protoplast Fusion and Somatic Hybridization

Although plant protoplast fusion is now a routine procedure, methods are still being refined and new techniques developed, including radio-frequency electric field-induced fusion (electroacoustic fusion)[76] and laser-induced cell fusion.[77] Electroacoustic fusion may prove particularly useful for small protoplasts, which often require extreme treatment, such as high fusogen concentrations or longer DC pulses. However, in general, the culture of protoplasts postfusion and hybrid cell selection present more problems than the actual process of fusion.

Although conventional methods of plant breeding will continue to play a major role in crop improvement, somatic hybridization will offer a unique opportunity for achieving gene flow in plants, particularly for the transfer of reproductively isolated multigenic traits. The application of protoplast fusion to plant breeding depends on continued extension of the range of crop plants that can be regenerated from protoplasts, together with refinement of the procedures for the selection of somatic hybrid tissues and plants.

[73] A. Morgan and P. Maliga, *Mol. Gen. Genet.* **209,** 240 (1987).

[74] M. W. Saul and I. Potrykus, *Plant Cell Rep.* **3,** 65 (1984).

[75] G. Kochert, "Introduction to RFLP Mapping and Plant Breeding Applications," *in* Rockefeller Found. Int. Program Rice Biotechnol., New York, 1989.

[76] D. C. Chang, *in* "Electroporation and Electrofusion in Cell Biology" (E. Neumann, A. E. Sowers, and C. A. Jordan, eds.), p. 215. Plenum, New York, 1989.

[77] E. Schierenberg, *in* "Cell Fusion" (A. E. Sowers, ed.), p. 409. Plenum, New York, 1987.

[30] Insertion of Lipids and Proteins into Bacterial Membranes by Fusion with Liposomes

By Arnold J. M. Driessen and Wil N. Konings

Introduction

Cytoplasmic membrane vesicles derived from *Escherichia coli* and many other bacteria provide a well-defined model system to study membrane-associated energy-transducing processes.[1] Functional and structural properties of these membrane vesicles have been presented elsewhere.[2,3] An important feature of these membranes is their ability to generate a proton-motive force (Δp) when supplied with a source of energy (oxidizable substrates, light, etc.) that can be utilized by redox enzymes (i.e., electron transfer chains) embedded in the membrane. Studies performed with these membrane vesicles have been essential for our understanding of the critical role of the Δp as a driving force and/or regulator of energy-transducing functions of membranes.[4] Such a functional membrane vesicle system is, however, not generally available for bacteria. For instance, fermentative bacteria lack electron carriers such as cytochromes and are therefore unable to generate a Δp by electron flow.[5] These bacteria generate a Δp by the hydrolysis of ATP catalyzed by a membrane-bound H^+-translocating F_0F_1-ATPase. The possibility of generating a Δp by ATP hydrolysis in membrane vesicles with the *in vivo* polarity of the cytoplasmic membrane is prevented by the inaccessibility of the catalytic site of the F_0F_1-ATPase to externally added nucleotides. Because of the limited Δp-generating abilities of these vesicles, a detailed analysis of energy-transducing processes taking place in these bacteria seemed not to be possible. A method has been developed in our laboratory that allows the fusion-mediated insertion of a variety of Δp-generating systems into bacterial membranes while energy-conserving properties of the membrane are retained.[6–8] This procedure can

[1] H. R. Kaback, *Annu. Rev. Biophys. Chem.* **15**, 279 (1986).
[2] H. R. Kaback, this series, Vol. 22, p. 99.
[3] W. N. Konings, this series, Vol. 56, p. 378.
[4] K. J. Hellingwerf and W. N. Konings, *Adv. Microb. Physiol.* **26**, 125 (1985).
[5] W. N. Konings, B. Poolman, and A. J. M. Driessen, *Crit. Rev. Microbiol.* **16**, 419 (1989).
[6] A. J. M. Driessen, K. J. Hellingwerf, and W. N. Konings, *Microbiol. Sci.* **4**, 173 (1987).
[7] A. J. M. Driessen, *Antonie van Leeuwenhoek* **55**, 139 (1989).
[8] A. J. M. Driessen and W. N. Konings, *in* "Membrane Fusion: Molecular Mechanisms, Significance in Cell Biology and Biotechnological Applications" (J. Wilschut and D. Hoekstra, eds.), p. 777. Dekker, New York, 1991.

also be used for the bulk enrichment of the bacterial membrane with exogenous lipids. Principles of this method and specific details of its applications in studies on the mechanism of solute translocation and lipid–protein interactions in lactic acid bacteria are described.

Growth of Bacteria and Isolation of Membrane Vesicles

Lactococcus lactis ML_3 (or *L. lactis* ssp. *cremoris* Wg_2) is grown at 30° on MRS broth[9] with 0.3% (w/v) lactose at a controlled pH of 6.4 in a 5-liter fermenter.[10] An exponential culture (A_{660} 1.0–1.2) is harvested by centrifugation (10 min, 10,000 g, 4°), washed with 0.1 M potassium phosphate, pH 7.0, and suspended in 100 ml of 0.1 M potassium phosphate, pH 7.0, containing 10 mM $MgSO_4$ and 500 mg egg lysozyme (E. Merck AG, Darmstadt, Germany). The solution is incubated for 30 min at 30°. Subsequently, saturated K_2SO_4 is added to a final concentration of 0.15 M, which results in lysis of the cells. After 5 min, lysed cells are diluted with 180 ml of 0.1 M potassium phosphate, pH 7.0, containing 50 μg RNase/ml (Miles Laboratories, Ltd., Slough, England) and 50 μg DNase/ml (Miles). Incubation is continued for 20 min, followed by the addition of 15 mM ethylenediaminetetraacetic acid (EDTA)–KOH (pH 7.2). After 10 min the $MgSO_4$ concentration is increased to 20 mM, and particulate material is collected by centrifugation (30 min, 48,200 g, 4°). Pellets are suspended in 210 ml of 50 mM potassium phosphate, pH 7.0, containing 10 mM $MgSO_4$ and centrifuged at low speed (1 hr, 750 g, 4°) to remove whole cells and cell debris. Membranes are collected from the supernatant by centrifugation (30 min, 48,200 g, 4°), resuspended in the same buffer at 10–15 mg of protein/ml, and stored in liquid N_2. This procedure yields closed membrane vesicles with an average diameter of 100 nm.[10] Their orientation is the same as in intact cells.[10,11]

Reconstitution of Δp-Generating Systems into Liposomes

Of the many Δp-generating systems described, only a few are suitable to function in fused membrane systems. The protein should have the following desirable properties: (1) it should be easily isolated and purified in large quantities by relatively simple procedures; (2) it should be stable against denaturation, with retention of high activity; and (3) when reconstituted it should be able to generate a Δp of considerable magnitude and defined

[9] J. C. Man, M. Rogosa, and M. E. Sharpe, *J. Appl. Bacteriol.* **23,** 130 (1960).
[10] R. Otto, R. G. Lageveen, H. Veldkamp, and W. N. Konings, *J. Bacteriol.* **149,** 733 (1982).
[11] A. J. M. Driessen and W. N. Konings, *Biochim. Biophys. Acta* **1015,** 87 (1990).

polarity. Reconstitution procedures for cytochrome-c oxidase, photosynthetic reaction centers, and bacteriorhodopsin are described in this section.

Preparation of Liposomes. Phospholipids dissolved in chloroform/methanol (9:1, v/v) are dried under a stream of N_2 gas. Traces of solvent are removed under vacuum for 1 hr by a rotary evaporator. The dried lipid film is hydrated for 10 min in buffer A [20 mM N-2-hydroxyethylpiperazine-N'-2-ethanesulfonic acid (HEPES)–KOH (pH 7.0), 50 mM KCl] at 20 mg of lipid per milliliter, and dispersed by the use of a sonic bath (Sonicor, Sonicor Instruments, New York, NY). The suspension is then sonicated in 1-ml portions, using a tip sonicator (MSE Scientific Instruments, West Sussex, England) at an intensity of 4 μm (peak to peak) for 10 min with intervals of 15 sec sonication and 45 sec rest. N_2 is flushed over the suspension to reduce lipid oxidation. Sonication is performed at temperatures that ensure the liquid crystalline state of the lipid bilayer.

Cytochrome-c Oxidase. Cytochrome-c oxidase is a redox-linked H^+ pump that interacts asymmetrically with the electron donor cytochrome c (Fig. 1A). Reduced cytochrome c will only be able to donate electrons to oxidase molecules with their cytochrome c-binding site exposed to the outer surface of the liposomal membrane. This guarantees the exclusive generation of a Δp, inside negative and alkaline. Cytochrome-c oxidase can be isolated in large quantities from bovine heart mitochondria[12] by the method of Yu *et al.*[13] The heme a content of the purified oxidase is typically 10.2–10.7 nmol/mg of protein as estimated from $\Delta A_{605-630}$ after dithionite reduction, using an extinction coefficient of 13.5 mM^{-1}cm^{-1}.[14] A chloroform/methanol solution containing 40 mg of acetone/ether-washed *E. coli* phospholipid (Sigma Chemical Co., St. Louis, MO)[15] is first dried under N_2, lyophilized and hydrated in 2 ml of buffer A containing 30 mM n-octyl-β-D-glucopyranoside (Sigma), and then sonicated on ice for 5 min (cycles of 15 sec sonication and rest) with a probe at an output of 4 μm (peak to peak) under N_2. Cytochrome-c oxidase (9 nmol of heme a) is added and detergent is removed by dialysis at 4° for 20 hr against a 500-fold volume of buffer A with 2 changes. Proteoliposomes are stored in liquid N_2, and before use thawed at 20° and then sonicated at 4° for 8 sec with a microtip at an output of 2 μm (peak to peak). Proteoliposomes have a diameter of 60 to 140 nm. About 65–70% of the cytochrome-c oxidase molecules have their cytochrome c-binding site exposed to the external

---

[12] T. E. King, this series, Vol. 10, p. 202.
[13] C. A. Yu, L. Yu, and T. E. King, *J. Biol. Chem.* **250**, 1383 (1975).
[14] T. Yonetani, *J. Biol. Chem.* **236**, 1680 (1965).
[15] P. Viitanen, M. J. Newman, D. L. Foster, T. H. Wilson, and H. R. Kaback, this series, Vol. 125, p. 429.

A

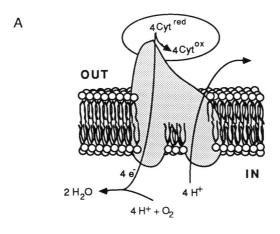

OUT

4 Cyt ^{red}
4 Cyt ^{ox}

IN

4 e$^-$

2 H$_2$O

4 H$^+$

4 H$^+$ + O$_2$

B

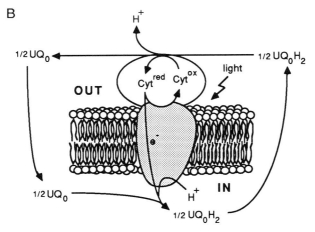

H$^+$

1/2 UQ$_0$

1/2 UQ$_0$H$_2$

Cyt ^{red} Cyt ^{ox}

light

OUT

⊕$^-$

IN

1/2 UQ$_0$

H$^+$

1/2 UQ$_0$H$_2$

C

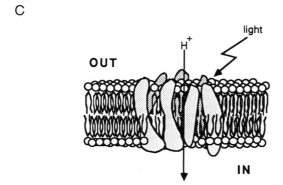

light

H$^+$

OUT

IN

Fig. 1. Protonmotive force-generating systems used in membrane fusion experiments: (A) cytochrome-c oxidase, (B) photosynthetic reaction center, and (C) bacteriorhodopsin.

aqueous phase. The respiratory control index is between 2.5 and 3.0.[16] The intrinsic maximal capacity of cytochrome-c oxidase to generate a Δp (the electromotive force) at pH 6.0 has been estimated to be -170 mV for the transmembrane electrical potential ($\Delta\psi$) and 2.2 pH units for the transmembrane pH gradient (ΔpH).[17]

Reaction Centers. Like cytochrome-c oxidase, the light-dependent Δp-generating reaction center (RC) interacts specifically with cytochrome c at the outer surface of the membrane (Fig. 1B). Cyclic electron transfer in this system requires the presence of membrane-permeable quinones. Illumination results in the generation of a Δp, inside negative and alkaline. The magnitude of Δp can be adjusted by varying the light intensity. Because oxygen is not required, anaerobic conditions can be employed. Reaction centers with light-harvesting complex I still attached (RCLH$_I$) are extracted from chromatophores of *Rhodopseudomonas palustris* by a modified procedure of Varga and Staehelin,[18] using octylglucoside and deoxycholate.[19] The RC concentration is estimated from ΔA_{880} of the reduced (sodium dithionite) and oxidized (potassium ferricyanide) form, using an extinction coefficient of 113 mM^{-1}cm^{-1}.[20] The RCLH$_I$ complexes are reconstituted into liposomes at a ratio of 1.4 nmol RC/mg of lipid by detergent dialysis (see the previous section).[19] The RCLH$_I$ liposomes range from 50 to 150 nm in diameter. More than 95% of the RCs have their cytochrome c-binding site accessible at the external aqueous phase. The electromotive force at pH 8.0 is -210 mV for the $\Delta\psi$.[19]

Bacteriorhodopsin. Bacteriorhodopsin (bR) is a light-driven H$^+$ pump (Fig. 1C) organized in a two-dimensional crystalline lattice (purple membranes) in the cytoplasmic membrane of halophilic bacteria. Most reconstitution procedures result in the formation of proteoliposomes containing bR, in which on illumination an everted polarity of the Δp, that is, inside positive and acid, is generated.[21] Purple membranes are prepared from *Halobacterium halobium* by extensive washing and sucrose density gradient centrifugation.[22] The amount of bR is estimated from A_{560}, using an extinction coefficient of 63 mM^{-1}cm^{-1}.[23] Purple membranes suspended in

[16] W. de Vrij, A. J. M. Driessen, K. J. Hellingwerf, and W. N. Konings, *Eur. J. Biochem.* **156,** 431 (1986).

[17] A. J. M. Driessen, unpublished results (1989).

[18] A. R. Varga and L. A. Staehelin, *J. Bacteriol.* **161,** 921 (1985).

[19] D. Molenaar, W. Crielaard, and K. J. Hellingwerf, *Biochemistry* **27,** 2014 (1988).

[20] G. Feher and M. Y. Okamura, *in* "The Photosynthetic Bacteria" (R. K. Clayton and W. R. Sistrom, eds.), p. 349. Plenum, New York, 1978.

[21] P. W. M. van Dijck and K. van Dam, this series, Vol. 88, p. 17.

[22] D. Oesterhelt and W. Stoeckenius, this series, Vol. 30, p. 667.

[23] D. Oesterhelt and B. Hess, *Eur. J. Biochem.* **37,** 316 (1973).

distilled water are stored in liquid N_2 at a concentration of 450 μM bR. Reconstitution of bR into liposomes is accomplished by sonication.[21] A chloroform/methanol solution containing 2 mg of egg yolk L-α-phosphatidylcholine and 18 mg of acetone/ether-washed *E. coli* phospholipid is dried under N_2, lyophilized, and dispersed into 0.5 ml of 40 mM HEPES–KOH, pH 7.0, and 100 mM KCl. The lipid suspension is then diluted with an equal volume of distilled water containing 76.8 nmol of bR and sonicated (cycles of 15 sec sonication and 45 sec rest) at 4° for 20 min with a tip at an output of 4 μm (peak to peak) under N_2. Liposomes containing bR are heterogeneous in size and number of bR molecules per liposome.[24] The bR is predominantly incorporated in an inside-out direction, that is, H^+ pumping is from outside to inside.[24,25] The electromotive force at pH 7.0 equals +143 mV for the $\Delta\psi$ and 2.7 pH units for the ΔpH.[26]

Commercial preparations of bacteriorhodopsin and cytochrome-*c* oxidase are available from Sigma.

Freeze-Thaw Sonication-Induced Fusion

Proteoliposomes are fused with membrane vesicles of *L. lactis* by a freeze-thaw sonication (FTS) procedure (Fig. 2). Although the mechanism by which fusion takes place is not fully understood, membranes are subjected to stresses during the freeze-thaw step.[27] Freeze-induced dehydration of the hydrophilic surface of the bilayer results in lateral phase separations and the formation of hexagonal II-phase structures.[28] Lipid domains may be formed from which the intrinsic membrane proteins are excluded. The spatial distance between the bilayers decreases on extensive dehydration. These may be the conditions that overcome the strong repulsive hydration forces that normally confer stability on the membranes by preventing their contact and fusion.[29,30]

Fusion Procedure. *Lactococcus lactis* membrane vesicles (1 mg of protein) and (proteo)liposomes (10 mg of lipid) mixed in a final volume of 0.6 ml of buffer A are rapidly frozen into liquid N_2 and then slowly thawed

[24] P. W. M. van Dijck, K. Nicolay, J. Leunissen-Bijvelt, K. van Dam, and R. Kaptein, *Eur. J. Biochem.* **117**, 639 (1981).

[25] J. C. Arents, K. J. Hellingwerf, K. van Dam, and H. V. Westerhoff, *J. Membr. Biol.* **60**, 95 (1981).

[26] A. J. M. Driessen, K. J. Hellingwerf, and W. N. Konings, *Biochim. Biophys. Acta* **891**, 165 (1987).

[27] M. J. Taylor, *in* "The Effects of Low Temperatures on Biological Systems" (B. W. M. Grout and G. J. Morris, eds.), p. 3. Edward Arnold, London.

[28] P. J. Quinn, *J. Bioenerg. Biomembr.* **21**, 3 (1989).

[29] R. P. Rand, *Annu. Rev. Biophys. Bioeng.* **10**, 277 (1981).

[30] R. P. Rand and V. A. Parsegian, *Annu. Rev. Physiol.* **48**, 201 (1986).

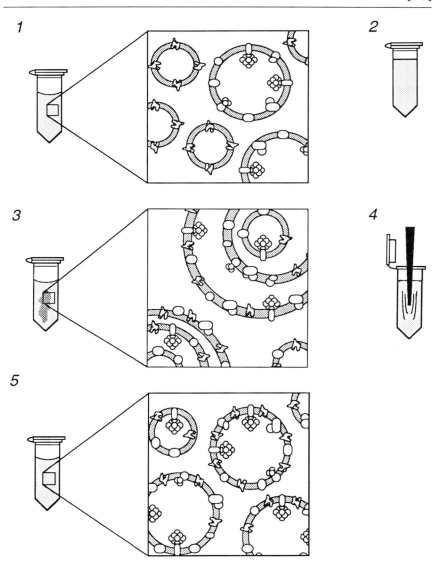

FIG. 2. Schematic representation of the freeze-thaw sonication procedure to fuse bacterial membrane vesicles with proteoliposomes containing a protonmotive force-generating system. A mixture of membrane vesicles and proteoliposomes (1) is rapidly frozen into liquid N_2 (2), slowly thawed at room temperature (3), and then sonicated (4) to form unilamellar hybrid membranes (5).

at room temperature.[31,32] Fusion results in the formation of multilamellar vesicles that are nonhomogeneous in both size and the number of lamellae (Fig. 2). Aggregates are dispersed by brief vortexing to form a turbid, but homogeneous, suspension. To form unilamellar vesicles, the suspension is sonicated at 4° for 8 sec, using a microtip at an output of 2 μm (peak to peak). Fused membranes are collected by centrifugation (1 hr, 280,000 g_{max}, 4°) and suspended at 10 mg of protein/ml in buffer A (or any other buffer solution).

Factors Affecting Fusion. With different biological membranes it may be necessary to optimize the convenience and reproducibility of the fusion method. A number of general principles are formulated below. The efficiency of membrane fusion is determined primarily by the method of freezing and of thawing. Optimal fusion requires rapid freezing and slow thawing. Freezing in either liquid N_2 ($-196°$) or solid CO_2/ethanol ($-80°$) is equally effective. The use of a freezer ($-20°$) is not recommended. Membranes can be stored in liquid N_2 for at least 1 year without detectable loss in activity. With respect to the method of thawing, good results are obtained with either slow thawing on ice (0°), in cold air (4°) or at room temperature (20°). The suspension should be left undisturbed while thawing. The buffer composition may affect the efficiency of membrane fusion.[33-35] The method of fusion is compatible with most low ionic strength buffers. Cryopreservatives such as glycerol and sucrose are inhibitory. The intensity and duration of sonication depends on the type of equipment used. Sonication should result in a marked decrease in turbidity of the suspension. Sonic baths and tip sonicators are equally effective, although the reproducibility is usually better with the latter. A problem associated with tip sonication is the release of traces of titanium in the sample. These may be removed by a 1-min spin in a microfuge. Prolonged sonication is undesirable because it generates small vesicles that are, in most studies, less useful than large vesicles. Instead of sonication, frozen-thawed membranes can be extruded through 100-nm pore size polycarbonate filters at high pressure,[36] using an extrusion device (Lipex Biomembranes, Vancouver, British Columbia, Canada). Vesicles produced by extrusion have a defined size distribution and large entrapment volume.

[31] A. J. M. Driessen, W. de Vrij, and W. N. Konings, *Proc. Natl. Acad. Sci. U.S.A.* **82,** 7555 (1985).

[32] A. J. M. Driessen, W. de Vrij, and W. N. Konings, *Eur. J. Biochem.* **154,** 617 (1986).

[33] U. Pick, *Arch. Biochem. Biophys.* **212,** 186 (1981).

[34] N. Oku and R. C. MacDonald, *Biochemistry* **22,** 855 (1983).

[35] K. Anzai, M. Yoshida, and Y. Kirino, *Biochim. Biophys. Acta* **1021,** 21 (1990).

[36] F. Olson, C. A. Hunt, F. C. Szoka, W. J. Vail, and D. Papahadjopoulos, *Biochim. Biophys. Acta* **557,** 9 (1979).

The FTS procedure may have some adverse effects on the orientation of intrinsic membrane proteins. With respect to the inserted proteins, bR appears to be most stable in the inside-out orientation. Proteoliposomes with bR reconstituted in the *in vivo* orientation change their net direction of H^+ pumping on fusion, suggesting that a significant fraction of the bR molecules undergo a reversal in orientation.[17] Cytochrome-*c* oxidase is partially scrambled by the fusion procedure,[32] whereas the orientation of the $RCLH_I$ complex remains unchanged.[37] Less information is available with respect to the orientation of the endogenous proteins of *L. lactis* membranes. Partial inversion of the orientation of the H^+-translocating ATPase has been observed.[8]

Physical Characterization of Fused Membranes

Fusion between bacterial membranes and (proteo)liposomes can be tested by sucrose density gradient centrifugation, freeze-fracture electron microscopy, and quantitative membrane fusion assays. The buoyant density of fused membranes is between the buoyant densities of the starting membrane preparations.[31,32] Electron microscope images show a low intramembranous particle density as a result of the immense dilution of the membrane proteins by the insertion of exogenous phospholipid.[31,38] Independent of the lipid composition of the liposomes, native *L. lactis* membrane vesicles (i.e., membrane structures with a high intramembranous particle density) are virtually undetectable after fusion.[38]

Quantitative Fusion Assay. Lactococcus lactis membrane vesicles are labeled with 4 mol% (total phospholipid phosphorus) octadecylrhodamine B chloride (R_{18}) according to the procedure of Hoekstra *et al.*[39] At this R_{18} concentration, a linear relationship exists between the efficiency of self-quenching and the concentration of R_{18} in the membrane. To 2 ml of membrane vesicle suspension containing 50 μmol of phospholipid phosphorus, 20 μl of 100 mM R_{18} (Molecular Probes, Eugene, OR) dissolved in ethanol is added under vigorous vortexing.[32,40] After incubation for 1 hr in the dark at 20°, nonincorporated R_{18} is removed by chromatography on Sephadex G-75 (1-cm diameter × 20-cm height). Labeled membranes eluting in the void volume are washed twice with buffer A and fused with

[37] W. Crielaard, A. J. M. Driessen, D. Molenaar, K. J. Hellingwerf, and W. N. Konings, *J. Bacteriol.* **170**, 1820 (1988).
[38] G. In 't Veld, A. J. M. Driessen, J. A. F. Op den Kamp, and W. N. Konings, *Biochim. Biophys. Acta* **1065**, 203 (1991).
[39] D. Hoekstra, T. de Boer, K. Klappe, and J. Wilschut, *Biochemistry* **23**, 5675 (1984).
[40] A. J. M. Driessen, T. Zheng, G. In 't Veld, J. A. F. Op den Kamp, and W. N. Konings, *Biochemistry* **27**, 865 (1988).

(proteo)liposomes by FTS. R_{18} fluorescence of the fused membranes is determined prior to (F_0^{fused}) and after the addition of Triton X-100 (1%, v/v) (F_∞^{fused}) and corrected for sample dilution. Excitation and emission of R_{18} are at 560 and 590 nm, respectively. Because intrinsic membrane proteins contribute to the surface area of the lipid bilayer, the fusion efficiency will be overestimated when calculated on the basis of the linearity of the assay and amount of lipid present. Therefore the maximal level of fluorophore dilution is determined experimentally. Fused membranes are solubilized in the presence of 30 mM octylglucoside, reconstituted by slow dialysis at room temperature,[41] and assayed for R_{18} fluorescence in the absence (F_0^{rec}) and presence (F_∞^{rec}) of detergent. The relative level of fluorophore dilution of the labeled membrane vesicles (F_0^{mem}/F_∞^{mem}) is taken as the zero level. The fusion efficiency is calculated using Eq. (1).

$$\text{Fusion efficiency} = \left(\frac{F_0^{fused}}{F_\infty^{fused}} - \frac{F_0^{mem}}{F_\infty^{mem}} \right) \left(\frac{F_0^{rec}}{F_\infty^{rec}} - \frac{F_0^{mem}}{F_\infty^{mem}} \right)^{-1} 100\% \quad (1)$$

The validity of this approach relies on the assumption that detergent solubilization and reconstitution results in complete mixing of the phospholipids. Because of spectral overlap, the R_{18} fusion assay cannot be used with $RCLH_I$ proteoliposomes. The assay is compatible with proteoliposomes containing bR provided that bR is first bleached by light in the presence of hydroxylamine.[42] The standard protocol for the FTS-induced fusion between *L. lactis* membrane vesicles and (proteo)liposomes containing *E. coli* phospholipids yields a fusion efficiency of 85–90%. Fusion exhibits only a moderate dependency on the phospholipid (polar head groups[40] and fatty acyl chain[38]) composition and cholesterol content[43] of the (proteo)liposomes.

Determination of Internal Volume. The internal volume of the fused membranes is estimated from the trapped amount of the fluorophore calcein present during FTS treatment.[44] Bacterial membranes are fused with (proteo)liposomes in the presence of 100 μM calcein, and 20–50 μl of this suspension is diluted into 2 ml of buffer A. The stability of the signal is improved by the presence of 10 μM EDTA. Excitation and emission of calcein are at 480 and 520 nm, respectively. Fluorescence is measured before (F_0) and after (F_C) the addition of 100 μM $CoCl_2$. Entrapped calcein is quenched by Co^{2+} after the disruption of the membranes with 1%

[41] A. J. M. Driessen, D. Hoekstra, G. Scherphof, and J. Wilschut, *J. Biol. Chem.* **260**, 10880 (1985).
[42] A. J. M. Driessen, K. J. Hellingwerf, and W. N. Konings, *Biochim. Biophys. Acta* **808**, 1 (1985).
[43] T. Zhen, A. J. M. Driessen, and W. N. Konings, *J. Bacteriol.* **170**, 3194 (1988).
[44] N. Oku, D. A. Kendall, and R. C. MacDonald, *Biochim. Biophys. Acta* **691**, 332 (1982).

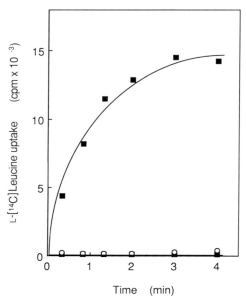

FIG. 3. Leucine uptake by membrane vesicles of *L. lactis* fused with cytochrome-*c* oxidase proteoliposomes by freeze-thaw sonication (■). Uptake of leucine was assayed in the presence of reduced cytochrome *c*. No uptake of leucine is observed with unfused *L. lactis* membrane vesicles (○) or cytochrome-*c* oxidase proteoliposomes (□). [From A. J. M. Driessen, W. de Vrij, and W. N. Konings, *Proc. Natl. Acad. Sci. U.S.A.* **82**, 7555 (1985).]

(v/v) Triton X-100 (F_R). The specific internal volume is calculated using Eq. (2).

$$\text{Internal volume} = \left(\frac{F_C - F_R}{F_0 - F_R}\right)C^{-1} \tag{2}$$

C is the protein (or lipid) concentration of the stock solution, in milligrams per milliliter.

Functional Characterization of Fused Membranes

A major application of the fusion procedure is shown in Fig. 3. Membrane vesicles of *L. lactis* were fused with cytochrome-*c* oxidase proteoliposomes by the FTS procedure and then assayed for their capacity to accumulate [^{14}C]leucine in the presence of reduced cytochrome *c*. Unlike the individual membrane preparations, fused membranes rapidly accumulate this amino acid, thereby providing functional evidence for fusion.

Fused membranes have been used as a model system in studies on the mechanism of energy coupling of Δp to uptake and efflux of amino acids[37,45–49] and peptides,[50] and the role of the internal pH[45,46] and phospholipids[38,40,43] on solute transport. Similar studies have been performed with fused membrane vesicles of the obligate anaerobes *Clostridium acetobutylicum* and *Clostridium fervidus*.[51,52] The system awaits application in energizing plasma membrane vesicles derived from eukaryotic cells. This section describes experimental details of the measurement of Δp and solute transport in fused membranes.

Measurements of Protonmotive Force and Uptake of Solutes

Cytochrome-c Oxidase. Δp-driven amino acid uptake and the generation of $\Delta\psi$ (inside negative) are measured simultaneously in a temperature-controlled vessel at 25°.[46,48] The vessel is equipped with a tetraphenylphosphonium (TPP$^+$)-selective electrode (constructed by the method of Shinbo *et al.*[53]) to monitor the distribution of the lipophilic cation TPP$^+$. The incubation mixture contains buffer A, 2 μM TPP$^+$, 1 mM MgCl$_2$, 20 μM horse heart cytochrome c, and fused membranes (0.5 mg of protein) in a total volume of 2 ml. The suspension is aerated by a constant flow of water-saturated oxygen over the surface. Cytochrome c is reduced by 10 mM ascorbate–KOH (pH 7.0) in the presence of 200 μM N,N,N',N'-tetramethyl-p-phenylenediamine. For uptake studies, L-[^{14}C]leucine (12.4 TBq/mol; final concentration, 4.5 μM; Amersham, Buckinghamshire, England) or another radiolabeled solute at the desired concentration is added, and samples of 50 μl are taken at timed intervals from the incubation mixture. Samples are diluted in 2 ml of 100 mM LiCl and filtered over nitrocellulose filters (0.45-μm pore size; Schleicher & Schuell, Dassel, Germany). More than 98% of the membranes are retained on these filters. Filters are washed with 2 ml of 100 mM LiCl, dried, and transferred to scintillation vials. Radioactivity is then measured, with a liquid scintillation counter, after the addition of scintillation fluid. $\Delta\psi$ is

[45] A. J. M. Driessen, S. de Jong, and W. N. Konings, *J. Bacteriol.* **169**, 5193 (1987).

[46] A. J. M. Driessen, J. Kodde, S. de Jong, and W. N. Konings, *J. Bacteriol.* **169**, 2748 (1987).

[47] A. J. M. Driessen, B. Poolman, R. Kiewiet, and W. N. Konings, *Proc. Natl. Acad. Sci. U.S.A.* **84**, 6093 (1987).

[48] A. J. M. Driessen, K. J. Hellingwerf, and W. N. Konings, *J. Biol. Chem.* **262**, 12438 (1987).

[49] A. J. M. Driessen, C. van Leeuwen, and W. N. Konings, *J. Bacteriol.* **171**, 1453 (1989).

[50] E. J. Smid, A. J. M. Driessen, and W. N. Konings, *J. Bacteriol.* **171**, 292 (1989).

[51] A. J. M. Driessen, T. Ubbink-Kok, and W. N. Konings, *J. Bacteriol.* **170**, 817 (1988).

[52] G. Speelmans, W. de Vrij, and W. N. Konings, *J. Bacteriol.* **171**, 3788 (1989).

[53] T. Shinbo, N. Kama, K. Kurihara, and Y. Kobataka, *Arch. Biochem. Biophys.* **187**, 414 (1978).

calculated, with the Nernst equation, from the distribution of TPP+ across the membrane. A correction for concentration-dependent binding of TPP+ to the membrane is applied according to the model of Lolkema *et al.*[54] ΔpH (inside alkaline) is measured with the membrane-impermeable fluorescent dye 8-hydroxy-1,3,6-pyrene trisulfonate (pyranine) (Molecular Probes).[32,46] Pyranine is a reliable indicator of the intravesicular pH because its fluorescent properties are similar when trapped in membranes or free in solution.[55] Pyranine (100 μ*M*) is trapped in the fused membranes during FTS. Membranes are passed through a Sephadex G-25 column (1-cm diameter × 5-cm height) to remove the untrapped pyranine. For fluorescence measurements, membranes (∼0.4 mg of protein) are diluted into a solution containing buffer A, 1 m*M* MgCl₂, and 20 μ*M* horse heart cytochrome *c* in a final volume of 2 ml. Energization is initiated by the addition of 200 μ*M* *N,N,N',N'*-tetramethyl-*p*-phenylenediamine and 10 m*M* ascorbate–KOH (pH 7.0). Excitation and emission of pyranine are at 460 and 510 nm, respectively. The fluorescence of pyranine is partially quenched by reduced cytochrome *c*. Calibration of the internal pyranine fluorescence intensity vs the external pH is achieved by adjusting the external pH in the presence of 100 n*M* nigericin under conditions in which cytochrome *c* remains reduced.

Reaction Centers. Δp-driven amino acid uptake and the generation of Δψ (inside negative) are measured simultaneously (see the previous section).[37] The incubation mixture contains buffer A, 2 μ*M* TPP+, 0.5 m*M* MgCl₂, 0.5 m*M* ascorbate–KOH (pH 7.0), 20 μ*M* horse heart cytochrome *c*, 400 μ*M* 2,3-dimethoxy-5-methyl-1,4-benzoquinone (ubiquinone-0), and fused membranes (0.5 mg of protein) in a total volume of 2 ml. The temperature of the reaction vessel is maintained at 20°. Actinic light is provided by a 150-W projector lamp via a glass fiber optic bundle, yielding a maximal light intensity of 1350 W m⁻². The light intensity can be varied by changing the voltage over the projector lamp. If required, anaerobic conditions are imposed by leading a constant flow of water-saturated oxygen-free N₂ over the surface of the incubation mixture. Energization is started by illuminating the suspension. Uptake of radiolabeled solutes is assayed as described in the previous section. The ΔpH formed on illumination is significant only at a pH above 8.[19] At this pH, pyranine is a poor indicator (i.e., p*K* ∼7.2). The magnitude of ΔpH is estimated from the increase in Δψ by the addition of 20 n*M* nigericin, assuming complete interconversion of ΔpH into Δψ.

[54] J. S. Lolkema, K. J. Hellingwerf, and W. N. Konings, *Biochim. Biophys. Acta* **681**, 85 (1982).

[55] N. R. Clement and M. J. Gould, *Biochemistry* **20**, 1534 (1981).

Bacteriorhodopsin. $\Delta\psi$ (inside positive) is measured with a TPP$^+$-selective electrode, using the lipophilic anion tetraphenylboron (TPB$^-$) at 20°.[56] The incubation mixture contains 50 mM HEPES–NaOH (pH 7.5), 50 mM NaCl, 2.5 mM MgSO$_4$, 1 μM TPB$^-$, 0.1 μM TPP$^+$, and fused membranes ($\sim$0.5 mg of protein) in a total volume of 1.5 ml. The stability of the electrode, as well as the permeability of the membranes for TPB$^-$, is enhanced by the presence of TPP$^+$.[26] Measurements are performed in a K$^+$-free solution to prevent the formation of an insoluble K$^+$/TPB$^-$ complex. Fused membranes are prepared in K$^+$-free buffers. Actinic light is supplied by a projector lamp (see previous section). $\Delta\psi$ is calculated according to the Nernst equation. Triphenyltin (40 nM) is used to dissipate ΔpH by electroneutral Cl$^-$/OH$^-$ exchange in K$^+$-free solutions. Nonactin (200 nM), an Na$^+$ ionophore, is used to collapse $\Delta\psi$. ΔpH (inside acid) is measured with pyranine (200 μM) trapped in the fused membranes during FTS. External pyranine is removed with a Sephadex G-25 column. Membranes ($\sim$0.25 mg of protein) are diluted into 50 mM HEPES–KOH (pH 7.5), 50 mM KCl, and 2.5 mM MgSO$_4$ in a total volume of 2 ml. The fluorescence quartz curvette is illuminated by a 1000-W xenon lamp, using fiber optics. Actinic light is passed through an orange cutoff filter ($>$550 nm), yielding a maximal light intensity in the cuvette of $\sim$50 W m^{-2}. Illumination of the fused membranes causes a decrease in pyranine fluorescence (internal acidification) that is reversed in the dark. Pyranine fluorescence is calibrated as described in the section, Cytochrome-*c* Oxidase. An analogous fused membrane system, that is, proteoliposomes containing bR and fused with *L. lactis* membrane vesicles at low pH, has been used to study the uptake of Ca^{2+} by a Ca^{2+}/H$^+$ antiport mechanism.[42,57]

With fused membranes, special care must be taken to use ionophores and protonophores to collapse Δp or its components. These compounds are effective at low concentrations but inhibit cytochrome-*c* oxidase and bR when used in excess. Standard assay conditions require only 20 nM nigericin to dissipate ΔpH with a compensatory increase in $\Delta\psi$. In the presence of 50 mM K$^+$, valinomycin collapses $\Delta\psi$ with an increase in ΔpH, at a concentration of 100 nM.

Miscellaneous Membrane Proteins. A membrane system in which the intravesicular ATP concentration can be controlled has been realized by fusion of membrane vesicles with liposomes containing the mitochondrial ATP/ADP antiporter.[58] Thus far, functional evidence for successful fusion has not been obtained.

[56] R. Casadio, G. Venturoli, and B. A. Melandri, *Photobiochem. Photobiophys.* **2,** 245 (1981).
[57] A. J. M. Driessen and W. N. Konings, *Eur. J. Biochem.* **159,** 149 (1986).
[58] D. Molenaar, unpublished results (1989).

Measurements of Transport Activity in Lipid-Enriched Membranes

Membrane fusion results in a significant incorporation of exogenous phospholipids into the bacterial membrane. Fusion therefore provides a straightforward technique to study the lipid requirement of intrinsic membrane proteins without the need for detergent extraction and protein purification. A disadvantage of this method is the low level of endogenous lipid remaining in the fused membranes. Lipid-enriched L. lactis membrane vesicles have been used in studies on the effect of the phospholipid (polar head groups[40] and fatty acyl chain[38]) composition and cholesterol content[43] of the membrane on the activity of the leucine–isoleucine–valine (LIV) carrier. Secondary effects of lipids on the generation of a Δp are excluded by measuring transport in the absence of Δp. With the LIV carrier of L. lactis, transport is assayed in the exchange mode with saturating concentrations of leucine present both on the inside and outside. Under these conditions, the carrier remains protonated and reorients in a Δp-independent fashion.[45] The initial rate of transport thus represents a direct measure for the activity of the transport system.

Counterflow Assay. Lactococcus lactis membrane vesicles are fused with liposomes of various lipid composition and incubated for 1 hr at 25° in 50 mM potassium phosphate, pH 7.0, containing 5 mM L-leucine. Valinomycin and nigericin are present, each at 1 nmol/mg of protein. Membranes are concentrated by centrifugation (45 min, 210,000 g_{max}, 4°), and aliquots of 2 μl (~10 mg of protein/ml) are diluted into 200 μl of 50 mM potassium phosphate, pH 7.0, containing 1.5 μM L-[^{14}C]leucine (12.4 TBq/mol; final concentration, ~50 μM). The initial rate of leucine uptake is assayed by rapid filtration.

Acknowledgments

This work was supported by grants from the Foundations for Biophysics (Biofysica), Fundamental Biological Research (BION), and Chemical Research (SON) with financial aid from the Netherlands Organization for Scientific Research (NWO).

[31] Liposome-Mediated Delivery of Nucleic Acids into Plant Protoplasts

By Paul F. Lurquin and Franco Rollo

Introduction

The molecular biology of plant cells and viruses has benefited enormously from the discovery that nucleic acids can be transferred from extracellular surroundings into the cytoplasm and nucleus of these cells. This chapter describes and discusses one of the techniques enabling such transfer, namely, the entrapment of nucleic acids into liposomes followed by their interaction with plant protoplasts. The technology described below has allowed transfection and transformation of plant cells.

Transformation can be defined as the phenomenon in which foreign genes are transferred, stably maintained, expressed, and transmitted to the progeny of the transformed cell or organism. In higher plants, this process can be achieved in a variety of ways and is now considered routine for a number of laboratory plants. Transformation techniques can be subdivided into two broad categories commonly referred to as the technique of cocultivation with *Agrobacterium tumefaciens* and the technique of direct gene transfer. The first approach relies on the natural ability of the soil bacterium *A. tumefaciens* to transfer genetic material to a large number of plant species and has been the subject of many review articles. Extensive experimental details of this procedure are given in Rogers *et al.*[1] and will not be considered in this chapter.

The direct gene transfer approach does not rely on a live gene donor, as does the first procedure. Rather, it consists of forcing the uptake of DNA (in general, genes cloned in plasmid vectors) into plant protoplasts, that is, plant cells from which the cell walls have been removed and that are therefore amenable to cell membrane manipulation. There are several ways in which the protoplast membrane can be made permeable to exogenous DNA or RNA, including electric discharges that disrupt the architecture of the bilayer, the use of the fusogen polyethylene glycol, and the entrapment of nucleic acids into liposomes. The latter approach will be described in this chapter. Aspects of DNA delivery into plant protoplasts have been reviewed.[2] Generally speaking, the conditions allowing the up-

---

[1] S. G. Rogers, R. B. Horsh, and R. T. Fraley, *in* "Methods for Plant Molecular Biology" (A. Weissbach and H. Weissbach, eds.), p. 423. Academic Press, San Diego, 1988.

[2] P. F. Lurquin, *in* "Biotechnology in Agriculture and Forestry" (Y. P. S. Bajaj, ed.), Vol. 9, p. 54. Springer-Verlag, Heidelberg, 1989.

take and replication of viral RNA in protoplasts have also been shown to allow uptake and expression of foreign DNA; this is particularly true in the case of liposome-mediated transfer. This chapter describes transfection of *Brassica rapa* protoplasts with liposome-encapsulated turnip rosette virus (TrosV) RNA and transformation of *Nicotiana tabacum* protoplasts with a liposome-encapsulated plasmid vector, with emphasis on the types of liposomes and conditions of incubation used in transfection and transformation experiments. Further details regarding subsequent protoplast handling and culture can be found in Roll and Hull[3] and Deshayes *et al.*[4]

Applicability

Liposome technology has been used more often to study expression of viral RNA genomes in protoplasts[3,5-10] than expression of foreign DNA. Liposomes have been used to transform plants stably through integration of the transformation vector[4] or to study the transient expression of a transgene in the absence of demonstrated integration.[11]

Methods

Protoplast Preparation

There is no universal protocol for the isolation of protoplasts from a variety of higher plants. The methods given here apply to the two particular examples described below. Other protoplast isolation techniques can be found in Potrykus and Shillito.[12]

Preparation of Brassica rapa cv. Just Right Protoplasts. A 20-cm long leaf from a 3-week-old greenhouse-grown plant is surface sterilized, washed with sterile distilled water, and cut with a razor into pieces of approxi-

[3] F. Rollo and R. Hull, *J. Gen. Virol.* **60**, 359 (1982).
[4] A. Deshayes, L. Herrera-Estrella, and M. Caboche, *EMBO J.* **4**, 2731 (1985).
[5] Z. Xu, C. S. Luciano, S. T. Ballard, R. E. Rhoads, and J. G. Shaw, *Plant Sci. Lett.* **36**, 137 (1984).
[6] T. Nagata, K. Okada, I. Takebe, and C. Matsui, *Mol. Gen. Genet.* **184**, 161 (1981).
[7] R. T. Fraley, S. L. Dellaporta, and D. Papahadjopoulos, *Proc. Natl. Acad. Sci. U.S.A.* **79**, 1859 (1982).
[8] T. Nagata, K. Okada, and I. Takebe, *Plant Cell Rep.* **1**, 250 (1982).
[9] A. A. Christen and P. F. Lurquin, *Plant Cell Rep.* **2**, 43 (1983).
[10] P. Rouze, A. Deshayes, and M. Caboche, *Plant Sci. Lett.* **31**, 55 (1983).
[11] N. Rosenberg, A. E. Gad, A. Altman, N. Navot, and H. Czosnek, *Plant Mol. Biol.* **10**, 185 (1988).
[12] I. Potrykus and R. D. Shillito, in "Methods for Plant Molecular Biology" (A. Weissbach and H. Weissbach, eds.), p. 355. Academic Press, San Diego, 1988.

mately 2 cm². Fragments are transferred into a flask containing 30 ml of 1% (w/v) cellulase Onozuka R10, 0.5% (w/v) Macerozyme, and 0.6% (w/v) mannitol, vacuum infiltrated, and incubated for 50–75 min at 30° on a rotary shaker set at 120 rpm. Protoplasts are then sieved through a nylon mesh to remove debris and washed three times by low-speed centrifugation in 0.6 M mannitol. An alternative protocol consists of cutting the leaf into two to three large pieces and rubbing them with carborundum to remove the outer epidermis. Fragments are then shaken at 60 rpm for 6 hr in digestion medium at 30°. Protoplasts are then harvested and purified as above and resuspended in 0.6 M mannitol at a concentration of 3×10^6 protoplasts/ml.

Preparation of Nicotiana tabacum cv. Xanthi Protoplasts. Leaves from greenhouse-grown plants are surface sterilized and washed with sterile distilled water; the lower epidermis is stripped with forceps. Fragments are then floated overnight on digestion medium containing half-strength Murashige–Skoog macronutrients, Heller's micronutrients, Morel–Wetmore vitamins, 0.45 M mannitol, 0.05 M sucrose, 5 μM benzyladenine, 16 μM naphthaleneacetic acid, 0.02% (w/v) Macerozyme, 0.1% (w/v) cellulase Onozuka R10, and 0.05% (w/v) Driselase (Fluka AG, Basel, Switzerland). The following day the debris is removed by filtration and low-speed centrifugation. The protoplasts are washed twice in 0.3 M KCl, 5 mM CaCl$_2$, 1 mM 2-(N-morpholino)ethanesulfonic acid (pH 5.7), and then resuspended in 0.5 M mannitol, 5 mM CaCl$_2$, 5 mM Tris-HCl (pH 7.6), at a density of 2×10^6 protoplasts/ml.

Liposome Preparation

There is general consensus in the literature that liposome-mediated nucleic acid transfer into plant protoplasts is most efficiently achieved through the use of negatively charged vesicles (REVs) produced by the reversed-phase evaporation technique.[13] In the examples below, REVs having different composition are used to transfer TrosV RNA and pLGV23*neo* to protoplasts.

Encapsulation of Turnip Rosette Virus RNA. L-α-Phosphatidylcholine (10 μmol), β-sitosterol (10 μmol), and dicetyl phosphate (0.2 μmol) dissolved in 3 ml of chloroform–diethyl ether–methanol (1:1:1) and 0.15 ml of sterile aqueous liposome buffer [2 mM Tris-HCl, 50 mM glucose, 25 mM KCl, 0.1 mM ethylenediaminetetraacetic acid (EDTA), pH 7.4], containing 5 to 20 μg purified TrosV RNA, are mixed in a 25-ml round-bottom flask. The resulting one-phase dispersion is placed in a rotary

[13] F. Szoka and D. Papahadjopoulos, *Proc. Natl. Acad. Sci. U.S.A.* **75**, 4194 (1978).

evaporator and the solvents are removed under reduced pressure. The liposome-encapsulated RNA is separated from unencapsulated material by flotation in a three-step Ficoll gradient as described by Fraley et al.[14] Liposomes (0.5 ml) are mixed with 1 ml of 30% (w/v) Ficoll dissolved in the above liposome buffer and transferred to a polyallomer ultracentrifuge tube. The suspension is then overlaid with 3 ml of 10% (w/v) Ficoll followed by 1 ml of liposome buffer and spun for 1 hr at 36,000 rpm at 4° in a swing-out rotor. At the end of the run the liposomes form a milky band in the top phase of the gradient.

Encapsulation of pLGV23neo DNA. Plasmid pLGV23neo contains a plant-expressible *neo* gene consisting of the coding sequence of the Tn5 aminoglycoside 3′ phosphotransferase II gene flanked by the *A. tumefaciens* T-DNA nopaline synthase promoter and terminator regions. This construct is inserted in pBR322. This vector confers kanamycin resistance to transgenic plants.[15]

Five micromoles of bovine brain phosphatidylserine and 2 μmol of cholesterol dried under vacuum are redissolved in 0.8 ml of diethyl ether. Two hundred micrograms of plasmid DNA, dissolved in 0.2 ml of 5 mM Tris-HCl (pH 7.6), 50 mM NaCl, 1 mM EDTA, and 0.44 M mannitol, is then added. The two phases are mixed by brief (no more than 30 sec) sonication in a bath-type sonicator and ether is removed by reduced pressure as described above. Liposomes are then diluted with 1 ml of the above buffer and separated from unencapsulated plasmid by flotation in Ficoll.

Using the above techniques, encapsulation of TrosV RNA was found not to exceed 12% of input whereas encapsulation of pLGV23neo was between 25 and 40%.

Liposome-Mediated Nucleic Acid Transfer to Protoplasts

Numerous studies[3-10] have demonstrated that the highest nucleic acid transfer efficiencies are observed when protoplasts and liposomes are incubated in the presence of polyethylene glycol (PEG) and calcium ions. These are used in the incubation medium described below.

Transfection with Encapsulated Turnip Rosette Virus RNA. About 3 × 10^6 *B. rapa* protoplasts resuspended in 1 ml of 0.6 M mannitol is mixed with 0.5 ml of a liposome suspension containing the equivalent of 3 μg TrosV RNA (based on a trapping efficiency of 12%). Then 1 ml of aqueous 20% PEG 1550 (w/v) containing 5 mM CaCl$_2$ is added to the protoplast–

[14] R. Fraley, S. Subramani, P. Berg, and D. Papahadjopoulos, *J. Biol. Chem.* **255**, 10431 (1980).
[15] L. Herrera-Estrella, M. de Block, E. Messens, J. P. Hernalsteens, M. Van Montagu, and J. Schell, *EMBO J.* **2**, 987 (1983).

liposome suspension. After 5 min at room temperature, protoplasts are washed twice (600 rpm for 4 min) with 0.6 M mannitol plus 1 mM CaCl$_2$ and resuspended in 5 ml B5 plus AB8 medium.[16] They are further cultivated at 22° under low light intensity for 72 hr and finally fixed and stained with anti-TrosV fluorescent antibodies. For this, protoplasts are pelleted at 600 rpm for 4 min at 4°, then spread on a glass microscope slide and air dried. The slide is then dipped in 90% ethanol for 15 min, rehydrated for 45 min in 0.8% (w/v) NaCl plus 10 mM sodium phosphate buffer, pH 7.4 (PBS), covered with a solution of rabbit anti-TrosV IgG, and incubated overnight at 3°. The next morning the slide is covered with goat anti-rabbit fluorescein-conjugated IgG and incubated for 6 hr at 3°. Finally the slide is washed for 1 hr in PBS and mounted in PBS–glycerol (1:1). Protoplasts are observed under a fluorescence microscope equipped with a camera.

Results indicate that under the above conditions, up to 12% of the treated protoplasts become infected. Even at low (0.025 μg) TrosV RNA concentrations, a small proportion (0.1%) of the protoplasts is found to be transfected. With unencapsulated RNA, it is necessary to use a 500-fold excess of nucleic acid to detect an equivalently small proportion of infected protoplasts.[3]

Transformation with Encapsulated pLGV23neo DNA. About 2×10^6 *N. tabacum* protoplasts suspended in 1 ml of buffer (see Preparation of *Nicotiana tabacum* cv. Xanthi Protoplasts, above) is mixed at room temperature with 50 μl of a liposome preparation containing the equivalent of 3 μg plasmid DNA. After 5 min, 5 ml of medium containing 22% (w/v) PEG 6000 is mixed with the protoplasts. Twenty minutes later, the mixture is diluted with 20 ml of 0.3 M KCl plus 5 mM CaCl$_2$ and protoplasts are sedimented by low-speed centrifugation. They are then plated in the medium described above in Preparation of *Nicotiana tabacum* cv. Xanthi Protoplasts (without enzymes) and cultivated for 1 week in the dark at 28°. Kanamycin-resistant transformants are then selected in the light in the presence of 70 μg of antibiotic/ml of C medium[17] and resistant plants are regenerated as in Muller and Caboche.[18]

Results show that transformation frequencies are on the order of 4×10^{-5} per viable protoplast.[4]

Detection of Gene Expression in Transfected and Transformed Cells

Obviously, the technique(s) used to detect transgene or viral genome activity in liposome-treated protoplasts or cells and plants derived from

---

[16] B. A. M. Morris-Krsinich, R. Hull, and M. Russo, *J. Gen. Virol.* **43**, 339 (1979).
[17] J. P. Bourgin, Y. Chupeau, and C. Missonier, *Physiol. Plant.* **45**, 288 (1979).
[18] J. F. Muller and M. Caboche, *Physiol. Plant.* **57**, 35 (1983).

them will vary with individual cases. In the case of transfection with viral RNA genomes, methods such as *in situ* immunofluorescence[3,9] or enzyme-linked immunosorbent assays (ELISAs)[10] have been used. In the case of stable transformation with plant-expressible vectors the presence and integration pattern of the transgene can be ascertained by Southern blot hybridization, and its expression can be verified by enzyme activity[19] or Northern blot hybridization. In one case of transient gene expression following the uptake of a plant-expressible *cat* gene (i.e., a short time after liposome-mediated plasmid transfer), protoplasts were processed for the convenient and sensitive chloramphenicol acetyltransferase assay.[11]

Discussion

In addition to constituting a tool to transfer nucleic acids to plant protoplasts efficiently, the liposome system also raises the question of what type of mechanism is responsible for this transfer. Several models have been put forth to explain liposome-mediated DNA and RNA delivery into plant protoplasts and are discussed briefly below.

Liposome–Protoplast Fusion

Several studies[3,6,7,9,10,20] have clearly demonstrated that the treatment of liposome–protoplast suspensions with the fusogens PEG or polyvinyl alcohol (PVA) in combination with calcium ions is a prerequisite to achieve efficient internalization of DNA or RNA molecules trapped in liposomes. As liposomes are known to display characteristics of model membranes, and as the above treatment is known to promote protoplast–protoplast fusion, the simplest way to explain liposome delivery of nucleic acids into protoplasts is to assume that the vesicles actually fuse with the protoplast plasmalemma. Although this model is attractive, there is so far no direct evidence to support it.

Endocytosis

Mixtures of PEG or PVA with calcium ions have been shown to stimulate the uptake of whole viral particles into plant protoplasts.[21] Because viral capsids are composed of proteins, it would seem unlikely that viral particles were taken up by fusion with the plasma membrane. Rather, internalization through the formation of endocytotic vesicles induced by

[19] M. Caboche and P. F. Lurquin, this series, Vol. 148, p. 39.
[20] Y. Fukunaga, T. Nagata, and I. Takebe, *Virology* **113**, 752 (1981).
[21] A. J. Maule, M. I. Boulton, C. Edmunds, and K. R. Wood, *J. Gen. Virol.* **47**, 199 (1980).

the chemical treatment or via transient permeabilization of the plasma-lemma seems to be the more reasonable explanation. Either mechanism could conceivably explain uptake of liposomes into plant protoplasts. Experiments performed by Fukunaga et al.[22] have shown the presence of intact liposomes in the cytoplasm of *Vinca rosea* protoplasts. The lipo-somes appeared to be surrounded by what was recognized by these authors to be membrane material originating from the plasmalemma. This type of evidence clearly speaks in favor of some kind of endocytotic mechanism. However, because endocytosis does not occur spontaneously in plant pro-toplasts, we propose the term *chemical endocytosis* to describe the phe-nomenon responsible for the uptake of viral particles and liposomes in the presence of fusogens. How nucleic acids are eventually released from liposomes sequestered in endocytotic vesicles is presently unknown.

Leakage of Liposome Contents after Contact with Plasmalemma

The possibility also exists that liposome-encapsulated nucleic acids may reach the interior of the protoplast without liposome fusion or actual uptake. According to this model, liposomes adsorb to the protoplast mem-brane and nucleic acids are transferred across both membranes through the formation of transient pores. Although studies on the interactions of posi-tively charged multilamellar vesicles with carrot protoplasts seemed to support the existence of such a mechanism,[23] attempts to observe leakage of a fluorescent dye (6-carboxyfluorescein) from positively charged uni- and multilamellar liposomes in the presence of PEG were unsuccessful (F. Rollo, unpublished observations, 1989).

Acknowledgments

This chapter was written while P.F.L. was on professional leave at the Department of Agricultural Genetics, University of Naples, Italy. Gratitude is extended to this department for institutional support.

[22] Y. Fukunaga, T. Nagata, I. Takebe, T. Kakehi, and C. Matsui, *Exp. Cell Res.* **144**, 181 (1983).
[23] P. F. Lurquin and F. Rollo, *Biol. Cell.* **47**, 117 (1983).

Author Index

Numbers in parentheses are footnote reference numbers and indicate that an author's work is referred to although the name is not cited in the text.

Subject Index

A

Golgi complex, transport to. *See* Endoplasmic reticulum-to-Golgi transport
GP4F cell line, 49. *See also* Erythrocyte–GP4F cell complexes
GP4f cells, expressing influenza virus hemagglutinin, delivery of liposome-encapsulated RNA to, 331–338

H

HA. *See* Influenza virus, hemagglutinin
Hamster, sperm–egg fusion in
detection of, 254
evidence of, 250
phase-contrast micrograph of, 251
H9 cells, chronically infected with HIV, 7–8
Hemagglutinating virus of Japan. *See also* Sendai virus
cell fusion activity of, 18
Hemolysis, virus–induced, as indicator of membrane fusion, 84–85
Hepatitis delta virus, liposome-encapsulated, HA-mediated delivery of, 339
Heterokaryons
binucleate, selection of, 386–391
formation, 19, 380–381
physical isolation of, in somatic hybrid selection, 390–391
somatic cell hybrid formation from, 19, 380–381
High mobility group-1
nuclear import, study using erythrocyte ghost–cell fusion for microinjection of proteins into cultured cells, 314–315
with plasmid DNA, cotransfer into mouse Ltk⁻ cells, by vesicle complex, 323–325
High-pressure freezing, 114–115
Hoechst 33342
detection of sperm–egg fusion using, 252–254
staining of living gametes, 250–251
HPTS
as indicator of intravesicular pH, 406
as probe of liposome uptake and intracellular processing, 365–370
Human immunodeficiency virus
biosafety with, 6

cell fusion induced by, CD4-dependent, 3–12
approaches not requiring infectious HIV, 11
assay, 6–12
inhibitors, 8–11
vaccinia virus vectors for study of, 11–18
cytopathogenicity, determinants of, 5
envelope glycoproteins, 4–5
expression, in absence of other genes, 11
fusion-mediating components of, 4–5, 94
transient expression of, using recombinant vaccinia viruses, 14–17
glycoprotein gp41, synthetic peptide corresponding to, effect on planar bilayer conductance, 86–87
internalization block, in murine and other cells, 17
HVJ. *See* Hemagglutinating virus of Japan
Hybrid cells. *See also* Plants, hybrid; Plants, somatic hybridization
formation, from heterokaryons, 19
human–mouse, disappearance of human chromosomes in, 19
interspecific, instability of chromosomal balance in, 19
selection, by use of tk⁻ and hprt⁻ mutants, 19
Hydrophobicity
Triton X-114 partitioning as indicator of, 78–79
of virus fusion proteins, changes in, in virus–cell fusion, 67–69, 72–73
Hydroxypyrene-1,3,6-trisulfonate. *See* HPTS

I

IgG-[¹²⁵I]DNP-BSA. *See* Anti-dinitrophenyl IgG, monoclonal, aggregated with DNP-derivatized radiolabeled BSA
I_m. *See* Membrane current
Immunoglobulin G. *See* Anti-dinitrophenyl IgG
Impedance
electrical, 290–291
membrane, 291
Inductance, electrical, 274–275

fluorescence, of erythrocyte–cell fusion, 52–53
of sperm–egg fusion, 249–252
Lipid mixing assay, of membrane adhesion and fusion, 90–92
probe dilution configuration of, 91–92
probe mixing configuration of, 90–92
Lipids, insertion into bacterial membranes, liposome-mediated, 394–408
Lipid vesicles, binding of ^{125}I-labeled annexins to, assay of, 195–196
Lipofectin, transfection mediated by, 303–304
Liposome–cell interaction, 362–370
Liposomes
aggregation and adhesion, lipid mixing assay for, 90–92
binding to cell surface
charge-mediated, 362–363
observation of, method for, 365–370
cationic, intracellular delivery of macromolecules by, 303–306
containing plasmid DNA, preparation of, 320–321
for delivery of nucleic acids into plant protoplasts, 409–415
preparation of, 411–412
fusion
with naked regions appearing on cells infected with SSPE, 38
with protoplast, in liposome-mediated DNA and RNA delivery to plant protoplasts, 414
internalization into endocytic pathway, 363–365
intracellular delivery of nucleic acids and transcription factors by, 303–306
intracellular processing of, 364, 375–376
observation of, method for, 365–370
leakage from, in liposome-mediated DNA and RNA delivery to plant protoplasts, 415
pH-sensitive
for delivery of macromolecules into cells, 361–376
efficiency of cellular delivery by, 373–376
effect of antibodies or other ligands on, 375–376
formulation of, 370–373

intracellular processing of, 375–376
preparation of, 372–373
stabilizing amphipath for, 371, 373
preparation of, 329–330
multilamellar vesicle method, 372–373
by reversed-phase evaporation, 320–321, 372–373
reconstitution of protonmotive force-generating systems into, 395–399
RNA encapsulated in
encapsulation method for, 330–331
HA-mediated delivery of, 327–339
for vesicle complexes, for cotransfer of DNA and nuclear proteins into cells, 318–319
troubleshooting, 325–326
N-(Lissamine) rhodamine B sulfonyl PE, in monitoring lipid mixing, by resonance energy transfer, 90–92
Lotus, protoplast fusion and somatic hybridization in, 389
Lymphocyte, mouse, membrane breakdown, with electrical pulse, 341–343
Lytechinus variegatus. See Sea urchin

M

Macromolecules. See also Nucleic acid; Transcription factors
delivery to cells
by electroinjection, 339–361
methods for, 317–318
by microinjection, erythrocyte ghost–cell fusion method for, 306–317
by pH-sensitive liposomes, 361–376
fluorescent, erythrocyte loading with, 48
uptake, in field-treated cells, mechanisms of, 354–356
Macrophage colony-stimulating factor, effect on phagosome–lysosome fusion in macrophages, 238
Macrophages
human peripheral blood monocyte-derived, fluorescence labeling of, 236
murine peritoneal, fluorescence labeling of, 236
phagosome–lysosome fusion in, fluorescence methods for monitoring, 234–238

ISBN 0-12-182122-6